PEDIATRIC RADIOTHERAPY

PLANNING AND TREATMENT

PEDIATRIC RADIOTHERAPY

PLANNING AND TREATMENT

Arthur J. Olch, Ph.D., FAAPM

CRC Press
Taylor & Francis Group
Boca Raton London New York

CRC Press is an imprint of the
Taylor & Francis Group, an **informa** business

CRC Press
Taylor & Francis Group
6000 Broken Sound Parkway NW, Suite 300
Boca Raton, FL 33487-2742

First issued in hardback 2019

CRC Press is an imprint of the Taylor & Francis Group, an Informa business

ISBN-13: 978-1-4200-8509-9 (hbk)

Library of Congress Cataloging-in-Publication Data

Olch, Arthur J.
Pediatric radiotherapy planning and treatment / Arthur J. Olch.
p. ; cm.
Includes bibliographical references and index.
ISBN 978-1-4200-8509-9 (hardcover : alk. paper)
I. Title.
[DNLM: 1. Child. 2. Neoplasms--radiotherapy. QZ 275]

RC281.C4
618.92'994--dc23 2013002172

Visit the Taylor & Francis Web site at
http://www.taylorandfrancis.com

and the CRC Press Web site at
http://www.crcpress.com

This book is dedicated to all the children with cancer who so bravely and optimistically endure their disease, and to the skill of those who carefully craft each radiotherapy treatment and ensure its safe delivery.

Contents

SECTION 2 Guide to Treatment Planning and Dose Delivery

Preface

Planning and treating children with radiotherapy requires an understanding of the differences between adult and pediatric cancers and the consequences of treatment, an awareness that is not generally taught in any great depth to the physics and treatment teams. I came to Children's Hospital Los Angeles about 14 years ago from a large adult radiotherapy center that treated a small number of children each year. At this prior facility, breast, prostate, and lung tumor treatments comprised the vast majority of treatments as in most radiotherapy centers. We in physics and dosimetry were quite accomplished in turning out treatment plans for these sites and needed little input from our physician colleagues regarding the planning objectives. However, on those rare occasions when a neuroblastoma, Wilms' tumor, or medulloblastoma presented, we felt uneasy about planning without considerable discussions with the radiation oncologist and therapists. We knew that the answers to the typical questions: what are the critical structures, what doses could they tolerate and to what volume, how should the patient be immobilized, what level of accuracy in daily setup is required, would be very different than for the last adult case we encountered. Pediatric cases regularly consumed large amounts of physics and dosimetry time compared only to the most difficult adult cases.

And then there are the children themselves. You see them sitting in the waiting room playing with their toys or being held by Mom, occasionally they are so ill that they are confined to a wheelchair or gurney. These children have barely begun life but already have an overwhelming obstacle to overcome. Their survival and quality of life are very much in our hands. Our ability to design the optimal treatment and to accurately deliver it can save lives. Critically important is designing a treatment that allows the surviving child to go to school with their age group, to hold a job later in life, and to be a fairly normal person. Also consider that a 5 year old cured by our treatment is given an average of seven or eight decades of life, a period far longer than that for the elderly patients whose breasts and prostates we treat with so much expertise. Although there is a sadness felt knowing that the child has a life-threatening disease, there is much gratification to see them coming back for follow-up visits with all the vigor and spirit of any other child.

Forty years ago, cancer was a death sentence for a child. Today, the average cure rate for childhood cancer is about 80%. Our responsibility as medical physicists, dosimetrists, radiation oncologists, and therapists is to be as knowledgeable about optimal treatment methods for childhood cancer as we are for adult cancer so that we can maintain and improve upon these survival statistics while also improving the quality of life.

Most radiation physicists and dosimetrists received their training in radiotherapy centers which treat exclusively adults. Here, the emphasis of the training was in competence in planning and delivery of the common adult cancers: prostate, breast, lung, head and neck, and metastases from these diseases. For many centers, pediatric patients are sent to a radiotherapy facility that is equipped to treat them; anesthesiologists are available in the department as needed and the radiation oncologists are experienced in prescribing radiation for pediatric cancers. The physics staff at these institutions often has received little formal training in pediatric cancer and its treatment; the complexities of these treatments are discovered on the job by experience provided by the handful of cases presenting each year.

The purpose of this work is to provide a systematic and comprehensive guide to the treatment planning and delivery of radiation therapy for these stricken children while providing brief clinical overviews as a backdrop and discussions of late effects as a reference frame for the planning process. Other texts focus on the clinical aspects of pediatric oncology, or even more specifically, pediatric radiation oncology, and their audience is primarily the physician. This book is written primarily for the treatment planning team, including physicists, dosimetrists, radiation oncologists, residents, therapists, and students, as a resource for understanding the treatment planning issues for the most commonly presenting childhood diseases and as a guide in the design of the treatment.

The stage is set in the first chapter, where the statistics of pediatric cancer incidence and survival is presented. A critical component to the planning of radiation treatment in children is the avoidance of late effects, including secondary cancers. An overview and thorough discussion of the current understanding of radiation-induced secondary malignancy is presented in the second chapter with the emphasis on the pediatric setting. The debate surrounding the use of intensity-modulated radiation therapy (IMRT) in children is explored and the dosimetric causality for secondary malignancy is presented. Site-specific information is provided for a variety of secondary malignancies. The following eight chapters are disease specific. Each chapter begins with a clinical overview of the disease, which is followed by treatment planning and delivery concepts and guidance, and ends with a survey of late effects and organ tolerance doses in the context of the particular disease. In most cases, the guide to treatment planning is provided with techniques that can readily be translated to any radiotherapy department. In many cases, the historical background underpinning current treatment paradigms is reviewed, revealing the tremendous creativity brought to bear by the field of radiation oncology to address some of the most difficult treatment dilemmas. The late effects sections inform the treatment planner of the consequences of our treatments. As of 2012, there is no pediatric version of QUANTEC (the ASTRO initiative on Quantitative Analyses of Normal Tissue Effects in the Clinic). The adult organ tolerance data in some cases translates to the pediatric patient but there are certainly many differences which the late effects sections attempt to illustrate.

A rich source of important and relevant historical and recent publications further elucidate the issues discussed in each chapter. The reader is encouraged to explore this literature where more detail is needed.

Many advances have occurred over the last 30 years in radiation oncology. One has to question the relevance of the late effects outcomes that are commonly reported based on treatments from decades in the past. With improvements in target definition, reductions in radiation dose and volume, and improvements in treatment precision, there is great promise of dramatic reductions in the most serious and debilitating late consequences of radiation treatments in children. Our job is to understand pediatric cancers and their optimal treatment and then to be able to accurately and safely implement these improvements to minimize late effects while maximizing the chance for cure or effective palliation.

I would like to gratefully acknowledge the contribution from Robert S. Lavey, MD, MPH for the clinical overviews and Kenneth K. T. Wong, MD for reviewing the clinically oriented portions of each chapter.

About the Author

Arthur J. Olch, PhD, is professor of clinical pediatrics and radiation oncology at the Keck School of Medicine at the University of Southern California in Los Angeles, California. He is also chief of physics for the Children's Hospital Los Angeles Radiation Oncology Program, one of only a few radiotherapy centers in the United States that treats exclusively children. Dr. Olch received his BS degree in engineering and his PhD in medical physics from the University of California Los Angeles. He is a Fellow of the American Association of Physicists in Medicine (AAPM). Dr. Olch has served as the chair of the Physics Subcommittee of the Children's Oncology Group Radiation Oncology Committee and the AAPM Subcommittee on Quality Assurance and Outcome Improvement as well as serving as chair or member of various AAPM Task Groups. He has been a radiation therapy physicist for over 30 years, and has authored or coauthored more than 40 journal articles and book chapters.

Section 1

Pediatric Cancers and Challenges for Radiotherapy

This section begins with a brief overview of pediatric cancer to provide the background necessary for the physics staff and other radiotherapy team members to understand the current incidence and survival statistics of childhood cancers. As medical physicists, dosimetrists, and others involved in the planning and delivery of these treatments, having this background will provide contrast to the differences between adult and pediatric cancer statistics, and focus our treatment planning efforts in the context of pediatric survival rates and late effects. The second part of the section provides an overview of the challenges of treating children with radiotherapy and some basic treatment planning concepts. It concludes with an extensive discussion of the issues that surround the induction of second malignancy in children treated with radiotherapy in general, and intensity-modulated radiation therapy (IMRT) and protons in particular.

Chapter 1

Overview of Childhood Cancer

Incidence, Survival, and Late Effects*

Although we have much to learn about how and why children get cancer, one thing we know for sure is that it is not because of excessive smoking and drinking. Environmental factors are suspected in some cancers but have not yet been proven as a cause. Childhood cancer is usually not inherited, although there are some well-studied syndromes such as neurofibromatosis and Li–Fraumeni, which predisposes to cancer, and some cancers, such as retinoblastoma, that are often heritable.

Although the incidence of childhood cancer is about one one-hundredth that of adults, cancer among children is nevertheless a substantial public health concern. Each year in the United States, about 150 children per million children younger than 20 years of age will be diagnosed with cancer, amounting to about 12,400 children per year (Figure 1.1). This leads to the statistic that about 1 in every 320 individuals in the United States will develop a cancer before the age of 20. For children between ages 1 and 19 years of age, cancer ranked fourth as a cause of death behind unintentional injuries, homicides, and suicides, but is the leading disease-related cause of death. About 2500 children with cancer die annually. The childhood cancers

* Statistics from the Surveillance Epidemiology and End Results (SEER) program of the U.S. National Cancer Institute (NCI) provides insights into the incidence of and mortality from childhood cancers for patients diagnosed between 1975 to 2000 (Ries et al. 1999; Bleyer et al. 2006). The statistics in this chapter are adapted in large part from this source.

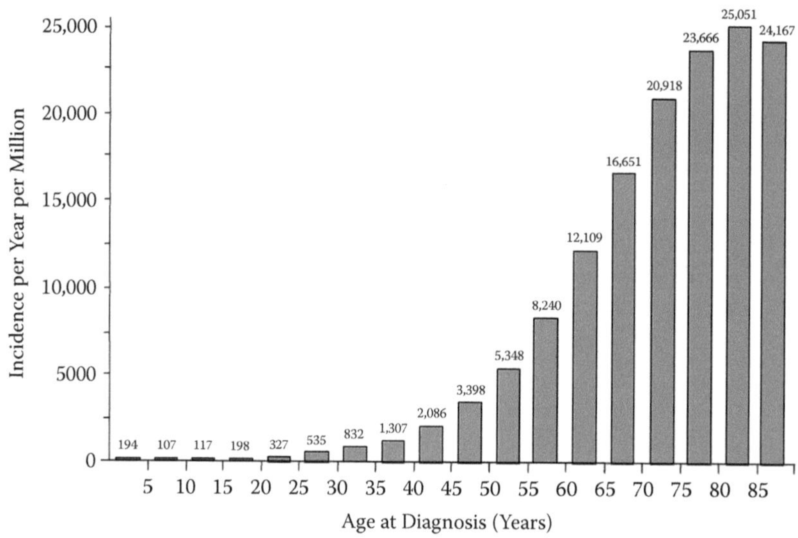

FIGURE 1.1 Incidence of all invasive pediatric and adult cancers per million of each age group per year, SEER 1975–2000. (After Bleyer, A., et al., eds., *Cancer Epidemiology in Older Adolescents and Young Adults 15 to 29 Years of Age, Including SEER Incidence and Survival: 1975–2000*, NIH Pub. No. 06-5767, Bethesda, MD: National Cancer Institute, 2006.)

demonstrate predilections for race, age, and gender depending on the type of cancer.

Childhood cancer incidence increased 19% between 1975 and 1990, mainly due to an increased incidence in acute leukemia. Since 1990, incidence rates have remained stable; however, mortality rates have continuously decreased (Figure 1.2). The age of peak incidence of cancer in children occurs between the ages of 1 and 5 and again between 15 and 20 years old. Figure 1.3 shows the incidence of each type of pediatric cancer by age. Many of these diseases are predominately found in the very young.

In the 1940s and 1950s, few children survived cancer. In the 1960s, however, researchers discovered new ways to use chemotherapeutic agents that could effectively treat leukemia and some other childhood cancers. Also, new combinations of treatment modalities (chemotherapy, radiation, and surgery) were designed to be more effective against childhood cancer. These new approaches resulted in increasing numbers of patients experiencing sustained remission and cures. The average annual percentage improvement in survival for patients under the age of 19 years old for the 25-year period between 1975 and 2000 has been about 2.7% compared to only 1.4% for adults (Bleyer et al. 2006).

Children get very different types of cancer than do adults (Table 1.1). In the typical radiotherapy department, breast, prostate, and lung cancer are the dominant diseases undergoing treatment. If a child presents with cancer in one of these sites, it is most likely a sarcoma rather than any of the

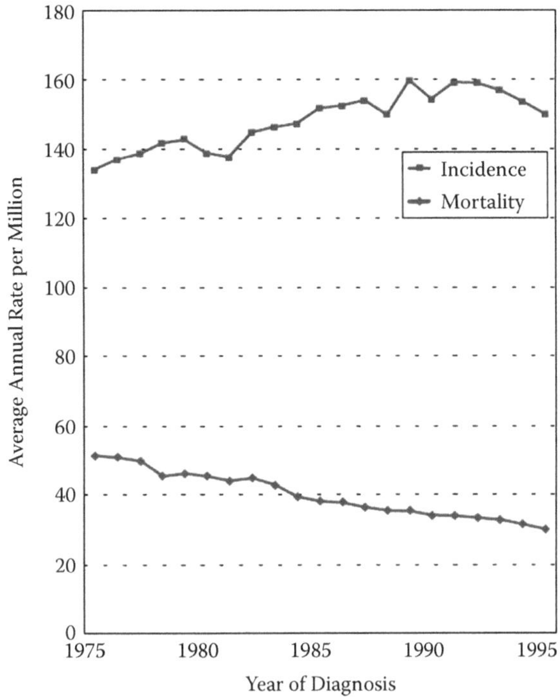

FIGURE 1.2 Five-year U.S. incidence and mortality for all childhood cancers, under 20 years of age, both sexes, all races. (After Ries, L. A. G., et al., *Cancer Incidence and Survival among Children and Adolescents: United States SEER Program 1975–1995*, NIH Pub. No. 99-4649, Bethesda, MD: National Cancer Institute, 1999.)

histologies for the adult cases. For children, leukemia is the most common cancer but infrequently treated by radiation therapy. The most commonly irradiated childhood tumors are those of the brain and central nervous system. Radiation therapy is given with palliative intent about 30% of the time in adults, compared to only about 5% of the time in children.

Smith et al. (2010) published a comprehensive overview and analysis of the SEER pediatric cancer data, providing incidence and survival rates by type and age of diagnosis for the years between 1975 and 2006. For all types of pediatric cancers combined, the chances of surviving one's cancer is now about 79% compared to just 63% 30 years ago. The decline in mortality was 64%, 85%, 75%, 35%, and 40% for leukemia, gonadal cancer, lymphomas, neuroblastoma, and bone cancer, respectively. Table 1.2 displays the improvements made in cancer-free survival over the 25-year interval from 1975 to 2000 for all the pediatric forms of cancer. Over 38,000 childhood cancer deaths were averted in the United States from 1975 to 2006. Although survival rates have continued to improve for leukemias and lymphomas, there has been no decline for certain cancers such as soft-tissue and bone sarcomas (Smith et al. 2010).

Today, about 1 in every 700 young adults is a childhood cancer survivor. It is estimated that there are over 300,000 survivors of childhood cancer

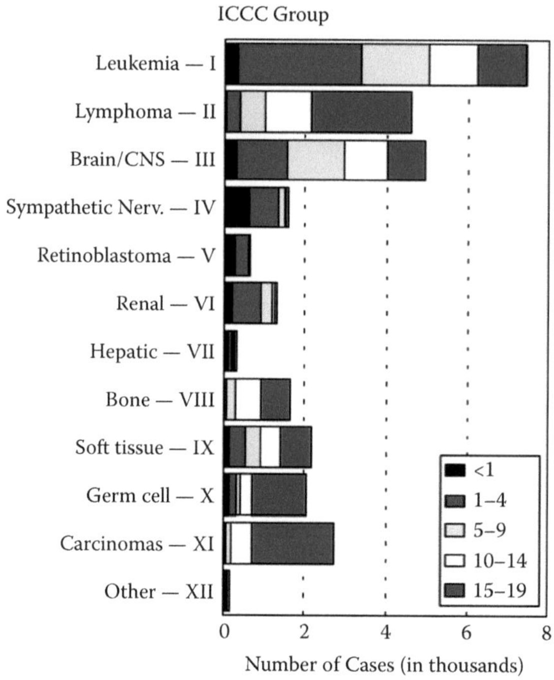

FIGURE 1.3 Number of cases of all childhood cancers by International Classification of Childhood Cancers (ICCC) and age group, all races, both sexes. (After Ries, L. A. G., et al., *Cancer Incidence and Survival among Children and Adolescents: United States SEER Program 1975–1995*, NIH Pub. No. 99-4649, Bethesda, MD: National Cancer Institute, 1999.)

in the United States with about one-fourth over age 40 (Dickerman 2007; Mariotto et al. 2009). Although it is fortunate that a large number of children with cancer survive, it is not without a cost in terms of loss of quality of life due to late effects of treatment. The Childhood Cancer Survivor Study (CCSS; http://ccss.stjude.org/published-research/publications) is a multi-institutional cohort of over 20,000 adults who have survived for at least 5 years after treatment for childhood cancer (Robison et al. 2002). In a comprehensive analysis of the CCSS data, Oeffinger et al. (2006) reported on the results of a survey of chronic conditions suffered by over 10,000 patients treated from 1970 to 1986 compared to their siblings. As a group, cancer survivors treated in the 1970s and 1980s were more than 8 times as likely as their siblings to have severe or life-threatening chronic conditions with a risk that is increasing with time. Thirty years after diagnosis, 73% of survivors have a chronic health condition (over 40% severe or life threatening) and one-third have multiple conditions. The relative risks of severe or life-threatening late effects for survivors compared to their siblings range from 4 to 54 and are shown in Table 1.3. Those who were treated with radiotherapy have a 7- to 10-fold risk of grade 3 or 4 chronic health problems when compared to their siblings and when combined with chemotherapy, the risk

TABLE 1.1 Incidence (per 100,000) of Various Cancers for Adults versus Children

Adult Cancer Incidence (Rate/100,000)		Childhood Cancer Incidence (Male and Female Ages 0–19)
Male (570.9)	Female (412.1)	All (16.7)
Prostate (180.1)	Breast (132.4)	Leukemias (4.4)
Lung and bronchus (82.7)	Lung and bronchus (49.2)	Brain and other nervous system (3.0)
Colon and rectum (64.0)	Colon and rectum (46.4)	Other lymphomas (1.6) [Hodgkin (1.2), Burkitt's and other (0.4)]
Urinary bladder (36.1)	Corpus and uterus (24.4)	Soft-tissue and other extraosseous sarcomas (1.2)
Non-Hodgkin lymphoma (23.7)	Non-Hodgkin lymphoma (15.5)	Germ cell, trophoblastic tumors, and neoplasms of gonads (1.1)
Melanoma of skin (20.4)	Ovary (14.2)	Malignant bone tumors (0.9)
Leukemia (16.4)	Pancreas (9.9)	Neuroblastoma and other peripheral nervous cell tumors (0.8)
Oral cavity and pharynx (16.4)	Thyroid (9.8)	Non-Hodgkin lymphoma (0.8)
Kidney and renal pelvis (15.5)	Cervix uteri (9.7)	Thyroid (0.7)
Stomach (13.1)	Leukemia (9.6)	Malignant melanoma (0.6)
Pancreas (12.7)	Urinary bladder (9.2)	Renal tumors (0.6)
Liver and intrahepatic bile duct (8.6)	Kidney and renal pelvis (7.6)	Retinoblastoma (0.3)
Brain and other nervous system (7.7)	Stomach (6.2)	Hepatic tumors (0.2)
Esophagus (7.6)	Brain and other nervous system (5.4)	Nasopharyngeal tumors (0.1)
Myeloma (7.1)	Myeloma (4.6)	
Thyroid (3.7)	Liver and intrahepatic bile duct (3.3)	

Sources: Adult data adapted from Edward, B. K., et al., Annual Report to the Nation on the Status of Cancer, 1975–2002, featuring population-based trends in cancer treatment, *Journal of the National Cancer Institute* 97:1407–1427, 2005. Childhood cancer data adapted from U.S. Cancer Statistics Working Group, United States Cancer Statistics: 1999–2004 Incidence and Mortality Web-Based Report, Atlanta: U.S. Department of Health and Human Services, Centers for Disease Control and Prevention and National Cancer Institute, 2007.

TABLE 1.2 Five-Year Relative Survival between 1975 and 2000 for the Major Pediatric Cancers

	5-Year Relative Survival (%)			
	1975–1979		1995–2000	
Type of Cancer	Male	Female	Male	Female
Bone and joint	43	57	71	64
Brain and other nervous system	57	60	72	75
Hodgkin lymphoma	86	88	96	96
Leukemia	44	53	75	78
Acute lymphocytic	52	64	82	84
Acute myeloid	23	21	46	54
Neuroblastoma	52	57	66	66
Non-Hodgkin lymphoma	42	58	79	82
Soft tissue	62	70	73	71
Wilms' tumor	73	76	92	92
All sites	58	68	77	81

Note: The relative survival rate represents the likelihood that a patient will not die from causes associated specifically with their cancer. It is always larger than the observed survival rate for the same group of patients.

TABLE 1.3 Relative Risk of Selected Severe or Life-Threatening or Disabling Health Conditions among Cancer Survivors, as Compared with Siblings

Late Effect	Percent Occurring in Survivors	Relative Risk
Major joint replacement	1.61	54.0
Congestive heart failure	1.24	15.1
Cognitive dysfunction, severe	0.65	10.5
Coronary artery disease	1.11	10.4
Cerebrovascular event	1.56	9.3
Renal failure or dialysis	0.52	8.9
Hearing loss not corrected by aid	1.96	6.3
Second malignant neoplasm[a]	5.5	6.0
Legally blind or loss of an eye	2.92	5.8
Ovarian failure	2.79	3.5

Sources: Oeffinger, K. C., et al., *New England Journal of Medicine* 355(15): 1572–1582, 2006; Friedman, D. L., et al., *Journal of the National Cancer Institute* 102(14):1083–1095, 2010.

[a] Second malignant neoplasm excludes nonmelanoma skin cancer.

rises to 10- to 13-fold. Chest, pelvic, or abdominal irradiation was associated with a greater than 10-fold increase in risk of severe or life-threatening conditions (Oeffinger et al. 2006). In a recent CCSS report, the cumulative incidence for all subsequent neoplasms 30 years after the diagnosis of childhood cancer was 21% and was 8% for malignant secondary neoplasms (Friedman et al. 2010).

Much has changed in the way we plan and deliver radiation therapy since the 1980s. Prescribed doses and target volumes have been reduced for some cancers and our ability to shape the dose distribution has dramatically improved. With a continuation of technological advances, there is the hope that the incidence of late effects will decrease. Part of our job is to understand this important fact and work hard to develop treatment plans and delivery methods that reduce the dose to normal structures as well as the volume receiving a given dose. Incidence, survival, and late effects considerations and statistics for each disease type will be discussed in the following chapters.

References

Bleyer, A., M. O'Leary, R. Barr, and L. A. G. Ries, eds. 2006. *Cancer Epidemiology in Older Adolescents and Young Adults 15 to 29 Years of Age, Including SEER Incidence and Survival: 1975–2000*. NIH Pub. No. 06-5767. Bethesda, MD: National Cancer Institute.

Dickerman, J. D. 2007. The late effects of childhood cancer therapy. *Pediatrics* 119 (3):554–68.

Edward, B. K., M. L. Brown, P. A. Wingo, et al. 2005. Annual Report to the Nation on the Status of Cancer, 1975–2002, featuring population-based trends in cancer treatment. *Journal of the National Cancer Institute* 97:1407–27.

Friedman, D. L., J. Whitton, W. Leisenring, et al. 2010. Subsequent neoplasms in 5-year survivors of childhood cancer: The Childhood Cancer Survivor Study. *Journal of the National Cancer Institute* 102 (14):1083–95.

Mariotto, A. B., J. H. Rowland, K. Robin Yabroff, et al. 2009. Long-term survivors of childhood cancers in the United States. *Cancer Epidemiology, Biomarkers & Prevention* 18 (4):1033–40.

Oeffinger, K. C., A. C. Mertens, C. A. Sklar, et al. 2006. Chronic health conditions in adult survivors of childhood cancer. *New England Journal of Medicine* 355 (15):1572–82.

Ries, L. A. G., M. A. Smith, J. G. Gurney, et al. 1999. *Cancer Incidence and Survival among Children and Adolescents: United States SEER Program 1975–1995*. NIH Pub. No. 99-4649. Bethesda, MD: National Cancer Institute.

Robison, L. L., A. C. Mertens, J. D. Boice, et al. 2002. Study design and cohort characteristics of the Childhood Cancer Survivor Study: A multi-institutional collaborative project. *Medical & Pediatric Oncology* 38 (4):229–39.

Smith, M. A., N. L. Seibel, S. F. Altekruse, et al. 2010. Outcomes for children and adolescents with cancer: Challenges for the twenty-first century. *Journal of Clinical Oncology* 28 (15):2625–34.

U.S. Cancer Statistics Working Group. 2007. United States Cancer Statistics: 1999–2004 Incidence and Mortality Web-Based Report. Atlanta: U.S. Department of Health and Human Services, Centers for Disease Control and Prevention and National Cancer Institute.

Chapter 2

Challenges of Treating Children with Radiation Therapy

2.1 Overview

Radiation therapy treatment planning for children with cancer is more challenging than for adults for a number of reasons. In adults, the vast majority of patients that are being treated curatively have breast, prostate, or lung cancer, and many of the rest, perhaps 30% to 50%, are being treated palliatively. Children being treated for cancer are overwhelmingly being treated curatively for a large number of cancers, many of which are rare to nonexistent in adults.

Many tissues in children have a lower tolerance to radiation than adults, but in many cases, these children have relatively large target volumes, which, depending on the diagnosis, require significant doses. For breast cancer, the most common female adult cancer, we typically can produce a good treatment plan with little effort to spare the normal tissues of the lung, heart, and contralateral breast. For the most common male adult cancer, prostate cancer, the most important organs at risk (OARs), the rectum and bladder, are readily kept below their tolerance doses with the use of modern planning and delivery techniques. In lung cancer, the third most common adult cancer behind breast and prostate, when the target is localized and treatable, our dosimetric challenge is to keep the normal lung tissue below its tolerance dose. Depending on the location of the lung tumor, we may also be challenged to keep the spinal cord below its tolerance dose. Head and neck cancer consistently presents the most challenging treatment planning exercise with several OARs to protect: parotid glands, spinal cord, mandible, airway, and brainstem, to name a few. In addition to these OARs, there may be multiple targets with different doses required. About 1 in 30 adult cancers are in the head and neck compared to 1 in 3 in the breast or prostate. Thus, although we do get our challenging cases when treating adults, they are generally infrequent and we are much more used to applying planning templates, which take little adjustment to produce a good plan.

In contrast, pediatric treatment planning is routinely as challenging as the adult head and neck cases. In many pediatric cases, there are unusual patient-specific features that make planning even more challenging. Contrary to a common belief, little kids do not usually present with little tumors. The primary reason for the high difficulty level for pediatric cases is that every tissue in the irradiated volume is potentially at risk for toxicity and the tolerance doses are frequently lower, sometimes much lower, than for the same tissues in adults, making it more frequently challenging to keep the normal tissue doses below tolerance. In addition, modern conformal treatments that avoid parallel opposed fields may cause more asymmetric growth, pitting the preservation of organ function against cosmesis. Some structures are critical to protect in children but can be largely ignored in adults. For example, in children, bone growth is arrested if more than about 18 Gy is given, whereas in adults, since full height has been reached, bone growth is not a concern. Also, in children, doses to the ovaries and testes of more than 5 to 10 Gy can render the child infertile, whereas adults who are past their reproductive age are not at risk from these doses. Another example is the growth effects on children when the pituitary is irradiated to more than about 20 Gy due to the resultant reduction in growth hormone production. In adults, again, growth is not a concern. Organs that can be damaged at lower doses in children than adults or are of increased concern due to the longer time that toxicity can manifest include the heart, carotid and other arteries, and brain. Conversely, when treating targets near the vertebral bodies in children, we have the unique requirement to increase our fields to uniformly treat the entire vertebral body so as not to cause differential bone growth. This requirement can sometimes hamper our ability to spare other nearby normal organs.

Brain tumors are the most frequent type of pediatric tumor that comes to radiotherapy. When treating adult localized brain tumors, we concentrate on giving the needed dose to the target volume while sparing the optic chiasm, optic nerves, lenses, and perhaps the pituitary gland, with little consideration for doses to surrounding normal brain because the adult brain is much less sensitive to induction of cognitive deficits. However, in the child, sparing normal brain tissue is of paramount importance and can drive the planning process. In the palliative setting, 30 to 36 Gy whole brain irradiation is given to adults for brain metastases without major cognitive ill effects; however, this dose given to a 3-year-old is devastating. Another difference is that, unlike breast and prostate cancer, children frequently get concurrent chemotherapy. This reduces tolerance, for example, increasing ototoxicity from cisplatin or cardiac toxicity from anthracyclines.

Target volume definition, target dose, normal tissue tolerance doses, and the treatment planning challenges for each of the disease types will be described in detail in their respective chapters.

A practical consideration is that children under about the age of 7 will need sedation. Daily sedation for several weeks is safely performed even with each period of sedation lasting up to an hour or more. Pentobarbital, chloral hydrate, or other drugs to achieve sedation have been used without intubation, and for some agents without the need for an anesthesiologist, but may require over an hour after treatment for the child to wake up. Propofol, a general anesthetic, is preferred because it can be titrated for the proper degree of sedation and it is short-acting once discontinued. However, an anesthesiologist must be present to administer and monitor the patient. Often, to achieve optimal ventilation, the child's chin needs to be lifted and kept in this position. Chin extension should be considered when making head immobilization devices. For an individual child, it is possible for the need for sedation to decrease over the course of treatment with distractions, parental support, and psychological counseling (Seiler et al. 2001; Stoh et al. 2001).

Another important consideration when planning a child's treatment is that the risk of producing a radiation-induced second malignancy (SM) is higher in children than in adults. This may be due to the increased susceptibility of normal tissue to the mutagenic effects of radiation at younger ages, the higher rate of cell proliferation during the early stages of development, and genetic susceptibility associated with some primary malignancies (Bhatia and Sklar 2002). The surviving child also has many more years to develop a second malignancy. Much of the rest of this chapter is dedicated to understanding the relationship between the induction of second malignancy and radiation therapy.

These challenges shape the way we think about planning and delivery of pediatric radiation treatments. Producing either a SM or chronic normal tissue injury or fatal complication in a child may lead to many more decades of harm and premature loss of life than for the 60-year-old adult cancer patient. We must always keep in mind that 60% to 70% of adults who survive childhood cancer will develop at least one medical disability caused by the therapy used to cure their primary cancer (Bhatia and Sklar 2002). Late toxicity from radiation therapy can be fatal although treatment failure is more commonly the cause of death (Mertens et al. 2008) (Figure 2.1).

Finally, and not at all a trivial consideration, is the psychological challenge to the physics and radiotherapy staff involved in the treatment of

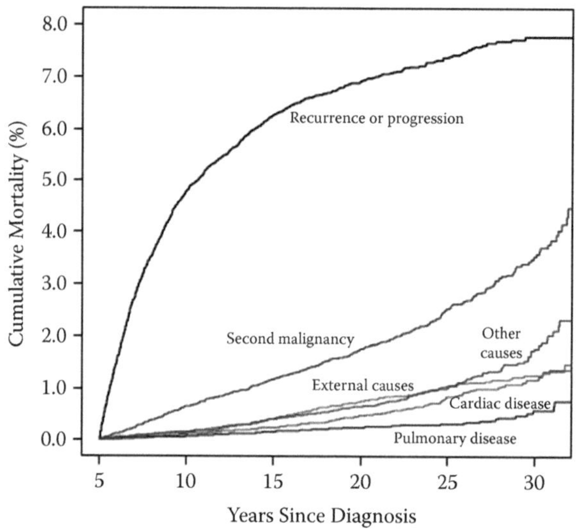

FIGURE 2.1 Cumulative mortality of pediatric cancer patients due to recurrence of cancer, second malignancy, cardiac disease, pulmonary disease, external causes, and all other cancers. (From Mertens, A. C., et al., *Journal of the National Cancer Institute* 100 (19):1368–79, 2008. With permission.)

children whose efforts in large part will determine the fate of the child. As the radiotherapy team members interact with the children, it is common to encounter feelings of sadness and despair that do not occur with the adult patients. These feelings are natural and should be dealt with as needed to maintain a professional but compassionate environment.

2.2 Treatment Planning Considerations

Before treatment planning starts, the planning team must make some basic decisions about how treatment will be delivered to achieve the goals of the treatment: adequately treat the target volume while keeping normal structures below their tolerance dose and ideally as low as possible. There are at least three decisions that have to be made, each of which determines to a great extent the design of the pediatric treatment, where applicable: (1) Will more than two beams be used? (2) Will noncoplanar beams be used? (3) Will intensity-modulated radiation therapy (IMRT) or protons be used? We will explore the considerations for each of these decisions.

All three of these planning approaches have a common real advantage as well as a common *hypothetical* disadvantage. The advantage in each case is that the high dose volume will be more conformal to the target volume. The potential disadvantage in each case, and here items 1 and 3 are interrelated, is that a greater volume of normal tissue will receive a small dose of radiation than if a small number of coplanar beams were used. The hypothetical problem with this redistribution of dose is that, as some believe, the risk of developing a second malignancy is attributable to the carcinogenic effects of low doses of radiation, on the order of a few gray. The treatment planning choices we make can increase the volume receiving these very low doses. This topic will be discussed thoroughly in a later section of this chapter.

The dosimetric benefit of using noncoplanar beams is that as one increases the number of beam directions in a three-dimensional (3D) space, the falloff of dose outside the target increases due to fewer paths of intersection between entrance and exit doses. However, as the number of beam paths in 3D space increases, the volume receiving low doses increases. Sometimes fewer noncoplanar beams are superior to more coplanar beams. These concepts of increasing the number of beam paths and avoiding overlap of entrance and exit doses applies for coplanar beams as well. Here, as one simply adds more beams, conformality increases at the same time that previously unirradiated tissues now become irradiated. Targets in the brain are the most obvious ones for consideration of noncoplanar beams because movement of the gantry and couch is unobstructed in the superior hemisphere. In the trunk, on the other hand, although noncoplanar beams may be helpful dosimetrically, collisions between the treatment delivery equipment and the patient significantly limit the degree of angulation one can achieve. Furthermore, superior-going beams for brain treatments exit into space rather than into other parts of the patient as would occur for treatment of targets in the trunk. Using noncoplanar beams is an extension of the principle of using multiple coplanar beams for improved conformality and is especially useful in more complex cases (Pugachev et al. 2001). IMRT is most commonly used in the context of multiple beams. This is necessary because each IMRT beam deposits an inhomogeneous dose. The more beam directions available to the dose-optimizing software, the more opportunities for constructing beams that compensate for hot or cold sleeves produced by other beams. The treatment of brain tumors is particularly interesting to study in terms of the consequences of treatment technique. In this site, we have the strongest motivation to use intensity modulation with noncoplanar beams.

Much has been written, from a theoretical perspective, about increasing the risk of producing SMs when using new technologies such as IMRT or even 3DCRT (three-dimensional conformal radiation therapy) compared to AP–PA (anteroposterior–posteroanterior) or other simpler treatments (Followill et al., 1997; Kry et al. 2005; Hall 2006). It is important for the radiation therapy planning and delivery team to understand the risk that radiation treatment adds to overall risk of producing a SM from cancer treatment and what features of the radiotherapy treatment plan may increase that risk. It is commonly assumed that cancer induction decreases significantly with increasing dose due to cell killing because cells that undergo mutation but then are killed by the next radiation dose cannot go on to produce a SM. Therefore, as the theory goes, the low doses produced by scatter, beam exits, and leakage are the major source of risk for SM. This theory has implicated IMRT, multibeam treatment plans, and neutron production during proton therapy in increasing the risk for SM and has shaped the treatment planning strategy for many centers. We face a dilemma with every child we treat: Do we do everything we can to cure the child of his or her cancer, and at the same time prevent or reduce morbidity that can cause lifelong suffering only to somewhat increase the risk of causing another cancer sometime later in life? The concern for SM, in some minds, overwhelms the known benefits of target conformality and normal tissue sparing, resulting in a hesitation to use these new technologies. Understanding the mechanisms, risks, and role of radiation therapy in the production of SM will enable better decision making when planning treatments. In the section that follows, the background for SM production will be presented as well as evidence that calls into question the assumptions made that cause trepidation when considering the use of IMRT for children.

2.3 Secondary Malignancies

2.3.1 History and Overview

Ionizing radiation is one of the most studied carcinogens, second only to tobacco. Clarence Dally, Thomas Edison's lab assistant, may have been the first reported instance of death due to radiation-induced cancer in 1904. He was frequently exposed to large doses of x-rays while demonstrating Edison's invention, the fluoroscope (Finch 2007). Marie Curie died in 1934 of aplastic anemia, thought to have been caused by her years of exposure to radium. The first report of leukemia due to radiation exposure was in atomic bomb survivors in 1952 (Folley et al. 1952). Alice Stewart reported increased leukemia in children exposed in utero to x-rays for diagnostic purposes, which was corroborated in a large epidemiological study (Giles et al. 1956). The late carcinogenic effects of ionizing radiation have been extensively documented for more than 50 years, based largely on studies of occupationally exposed workers, the atomic bomb survivors, and patients exposed to radiation either from diagnostic or therapeutic procedures. In Hodgkin lymphoma (HL), mortality due to second cancers is now the most common cause of death after HL itself. Many types of solid tumors can be caused by ionizing radiation, but the thyroid gland, female breast, and bone marrow are the most radiosensitive to the carcinogenic effects of radiation (Travis 2002).

There is a well-founded concern that children who are long-term cancer survivors can develop SMs and that the risk for them is greater than for cancer survivors treated as adults. The probability of cure from childhood cancer is now 80% overall, which means that large numbers of children will be living for 50 to 70 years or more beyond their treatment. The risk of producing a second malignancy in these children is greater than that for similarly treated adults partly due to this long period for the cancer to form and present itself and the consequences are more devastating due to the large number of years an affected individual will suffer or lose if the SM occurs or is fatal.

Changes in treatment strategies have been implemented or are now being tested to reduce toxicity and risk for SM in children. Examples are reduced dose or even elimination of cranial radiation for children with leukemia, the decreasing role of radiotherapy for Wilms' tumor and HL, the reduced craniospinal irradiation (CSI) dose and boost volume for medulloblastoma, and the reduced dose and volume for germinoma.

2.3.2 What Causes Radiation-Induced Second Malignancies?

Radiation-induced changes occur instantly when x-rays or other ionizing radiation ejects electrons from orbital paths around the atoms' nucleus as it passes through tissue. We are all continuously exposed to low levels of ionizing radiation from cosmic rays and terrestrial sources, such as the emissions from radioactive uranium and potassium in the soil, concrete, and our own bodies. The biological effects of ionizing radiation result largely from DNA damage, caused directly by ionizations within the DNA (about 30% of the time) or indirectly from the action of highly reactive hydroxyl radicals formed by ionization of intranuclear water molecules (70% of the time). DNA damage takes several forms; base damage, DNA protein cross-links, single- and double-strand breaks, and complex combinations of these. The most common type of DNA damage is single-strand breaks. They are usually repaired error free, rarely producing mutations. The ratio of single- to double-strand breaks for low linear energy transfer (LET) radiation is about 25:1 but approaches 1:1 with high LET radiation. Double-strand breaks are much more likely to be misrepaired, resulting in chromosomal aberrations. Even lesions that can be repaired may be repaired incorrectly if lesions are rapidly accumulating during high dose or dose rate irradiation, or if the cell undergoes DNA synthesis during the division cycle and while repair is in progress. DNA damage that is not repaired correctly can induce cell death, permanently prevent cell division, or create heritable chromosome aberrations including point mutations, deletions, and translocations of DNA sequences. Cancer is a result of the creation and subsequent reproduction of these mutations. Ordinarily, a high percentage of radiation-induced DNA damage is repaired by the cell. Repair depends on the correct operation of several pathways for which key genes have been identified (Finch 2007). Tubiana (2005) and Allan and Travis (2005) reviewed the diversity of mechanisms for cells to respond to low doses of radiation. Numerous genes are activated or inhibited in response to the radiation insult. Cells have defense mechanisms that protect them from damage from radiation including an adaptive response. Repair can be deficient in cells of individuals with mutated forms of critical genes, which may explain predisposition to cancer; retinoblastoma and Li–Fraumeni syndrome are most notable with the RB1 and the P53 tumor suppressor gene mutations, respectively.

2.3.3 Risk Estimates of Radiotherapy-Induced Second Malignancy Using Model-Based Calculations

The radiotherapy literature that calculates SM risk frequently relies on dose–response data from the Japanese atomic bomb (A-bomb) survivor cohort (Life Span Study, LSS), occupationally exposed worker studies, and others where the maximum absorbed doses are just a few gray and the mean dose is less than 1 Gy. Largely based on these data, the lifetime risk for developing a fatal radiation-induced malignancy is calculated to be about 5%/Sv (1 Sv is the radiation exposure equivalent of approximately 1 Gy absorbed dose) acute total body exposure, and the risk is linear with dose in the low dose range from about 10 to 50 mSv up to about 3 Sv (National Council on Radiation and Protection & Measurements 1993). The risk is often reduced by a factor of 2 for fractionated exposures. Complementing these data, studies with thousands of children treated with radiation for benign diseases such as tinea capitis and hemangioma showed that the risk for SM in the brain increased about 10-fold with doses less than 2.5 Gy. In the case of scalp treatment for tinea capitis, secondary brain tumors arose with a cumulative incidence of about 1% after 30 years from treatment (Ron et al. 1988; Karlsson et al. 1998). In these cohorts, when doses are less than about 3 Gy, the dose–response is linear. When high doses of radiation are used for cancer treatment, the assumption is frequently made that the volume receiving low doses distant from the target are responsible for the risk because higher doses would sterilize any transformed cells before they could develop into cancer (cell sterilization).

The cell sterilization hypothesis was developed by the radiobiologist Louis Gray in 1964, the man in whose honor the SI unit of radiation dose (Gy) was named. This theory says that cells that have been transformed into precancerous cells by low doses of radiation will be killed when their total radiation exposure is sufficiently high. The combination of increasing probability of mutation and cell sterilization with increasing dose is thought to lead to a theoretical semibell-shaped curve whose peak represents the dose that maximizes secondary cancer formation (Little 2001a).

Until recently, it has been assumed that the dose–response relations for carcinogenic mutations (at low doses) and for cell sterilization (at high doses) determine the overall dose–response relation for radiation-induced SM. Both effects are thought to increase linearly or in a linear-quadratic fashion as dose increases. Risk estimates for SM in patients treated with IMRT have been published. These are based on the assumption that most secondary cancers occur away from the primary treatment volume, low dose cancer incidence data, and the calculated or measured doses that occur distant from the treated volume (Followill et al. 1997; Kry et al. 2005; Hall 2006). For some, this risk outweighs the benefits for normal tissue sparing achievable with IMRT. But these risk estimates have been questioned by those who believe that the entire dose distribution needs to be included in the risk estimate (Sachs and Brenner 2005; Schneider et al. 2005).

In fact, large case-control studies of SM in childhood survivors of cancer show that risk continues to increase with dose at the site of the SM, making the assumptions used for risk calculations based only on the theory of mutation and cell sterilization questionable. To explain this deficiency in the theory, a third mechanism, repopulation or radiation-induced stem-cell accelerated repopulation during and after radiation treatment that compensates for losses due to cell killing, has been proposed to explain the clinical data showing risk increasing as dose increases. This three-component model applied to the entire dose distribution closely predicts the increase in SM risk in the breast and lung found in large epidemiological studies as dose increases well into the range where cell killing is expected (Sachs and Brenner 2005).

2.3.4 Is Intensity-Modulated Radiation Therapy (IMRT) Too Risky?

When deciding on a radiotherapy treatment strategy, should we be more concerned about limiting the volumes receiving the lowest doses or the high doses? Must we choose between possibly increasing the risk of SM by using IMRT or multibeam 3DCRT and increasing the risk of treatment toxicity by not using these technologies? To help answer these questions, we will review the dosimetric characteristics of IMRT and 3DCRT that cause concern about SM induction. The answers to these questions may shape our approach to treatment planning in children.

First, we will explore the basis for hypothesizing an increased risk of SMs when using multibeam 3DCRT or IMRT: increased integral dose (ID) or peripheral dose (PD). This hypothesis relies on the assumption that the majority of risk is due to low doses, especially those less than about 3 Gy. To the extent that IMRT and 3DCRT increase the volume receiving low doses, so the theory goes, risk increases. The ID is the product of the dose, the volume receiving that dose, and the density of that volume. It is expressed in Gy*kg and can be computed by binning the dose in 1 cGy bins, finding the volume receiving each dose level, then summing the products across all doses. It can also be expressed in joules because 1 Gy is 1 J/kg so 1 Gy*kg is 1 joule. Thus, the ID can be thought of as the total energy imparted into the patient. It can also be found by calculating the area under the differential absolute-dose, absolute-volume histogram.

There is a perception that increasing the number of beams (or beam energy) significantly increases the ID, which in turn increases the probability of developing a secondary malignancy. Risk of SM has been reported to increase with ID (Karlsson et al. 1998; Nguyen et al. 2008; Tukenova et al. 2011). Although the perception may seem intuitively obvious, several dosimetric studies have shown that increasing the number of beams does not increase the ID. D'Souza and Rosen (2003) compared normal tissue integral dose (NTID) for 2 through 36 equally spaced coplanar conformal beam plans all having the same tumor integral dose. With four or more beams, the variation in NTID was less than 1% and when two-beam plans were included, the variation was still less than 5%. They point out that decreasing target volume (by margin reduction, for example) provides the most significant NTID reduction and to a much lesser extent, increasing beam energy. Pirzkall et al. (2002) found that ID was within 2% for 4-, 6-, 9-, and 11-field 6 MV coplanar IMRT plans. Hermanto et al. (2007) found ID reduced by 7% to 10% with IMRT compared to 3DCRT for high-grade gliomas with equivalent target coverage and reduced normal tissue doses without significantly increasing the 0.5-5 Gy volume. Aoyama et al. (2006) compared seven equally spaced coplanar 6 MV conformal or IMRT beam plans with TomoTherapy (Accuray, Sunnyvale, California) for a prostate case and found the NTID was within 5% for all three. Treatment of the craniospinal axis by TomoTherapy produced the same NTID as when treated by conventional methods (Penagaricano 2006). When the analysis is extended to noncoplanar versus coplanar beams, depending on the situation and with the same number of beams in each plan, little difference is found in ID when the target coverage is comparable (Pirzkall et al. 2000). Whereas the low-dose volume does indeed increase with increasing number of beams, either coplanar or noncoplanar, the intermediate-dose volume decreases, while the high-dose volume remains constant or may also decrease due to increased conformality. Thus, the number of beams, whether they are intensity modulated or whether they are coplanar, does not significantly affect the ID (Reese et al. 2009). One exception is for very peripherally located targets; coplanar beams may give less normal brain integral dose (Marks et al. 1995).

Another attribute of IMRT that has been hypothesized to increase the risk of developing SMs is the increase in beam-on time compared to conventional treatment. Due to the relative inefficiency of fixed-field IMRT delivery, 3 to 6 times as many monitor units (MUs) are required as for nonmodulated plans. It should be noted though that 3DCRT plans that use 60 degree wedges can come close to the number of MUs from an IMRT plan. Because the head leakage (omnidirectional radiation that penetrates the lead shielding surrounding the x-ray target and beam-forming devices amounting to no more than about 0.1% of the dose from an open field at isocenter) is proportional to the total MUs delivered, not to the target dose; plans that require more MUs for the same target dose increase the total body dose to the patient.

Nomos Peacock TomoTherapy (NPT) in particular is known to require roughly 10 times the MUs compared to 3DCRT. Verellen and Vanhavere (1999), Followill et al. (1997), and Mutic and Low (1998) have computed the equivalent whole body dose from an NPT head and neck treatment to be from 406 to 1967 mSv, depending on the slice width mode used. This compares to 67 to 242 mSv with conventional parallel opposed and electron head and neck treatment.

In addition to head leakage, collimator and patient scatter dose both contribute to the dose outside the treated volume. Together, these three components make up the peripheral dose (PD), which is the very low dose component of the dose distribution distant from the target.

Peripheral dose within 60 cm of the field edges from fixed field IMRT has been compared to that produced with 3DCRT. Koshy et al. (2004) measured extra-target doses from multileaf collimator (MLC)-based IMRT or 3DCRT within 30 cm from the field edge and found no difference in dose to thyroid or breast from brain or head and neck treatment. Most of the PD was from internal scatter. Similarly, Mansur et al. (2007) measured PD in five brain or base of skull pediatric cases planned with mostly coplanar beams using both IMRT and 3DCRT with wedges. The MU ratio for IMRT to conventional planning was about 3.4 to 1. Using a pediatric-sized anthropomorphic phantom, ion chambers were placed at the location of the thyroid, breast, ovaries, and testes, and the PD was measured and scaled for a 5400 cGy treatment. For the thyroid, the closest peripheral normal tissue point, dose was lower for the IMRT plan, 25 cGy versus 37 cGy for 3DCRT over 30 fractions. For the testes measurement at a distance of 63 cm from the target, the measured IMRT dose was higher although the absolute doses were very small, 1.4 cGy for IMRT versus 0.4 cGy for 3DCRT. Averaging over all four interest points, the total body dose equivalent was 45 cGy for IMRT versus 44 cGy for 3DCRT. Sharma et al. (2006) also measured the peripheral dose up to 60 cm from field edge for IMRT versus 3DCRT, and found IMRT produced 2 to 15 times higher doses depending on field size and IMRT efficiency except in the first 15 cm from field edge where the dose was lower. Thus, there seems to be agreement that near the target volume, IMRT produces lower PD (where there may be a meaningful dose of about 1 Gy) but farther away, a higher PD (where there is a very small dose, less than 0.1 Gy). One explanation for the lower PD near the treated volume is that internal patient scatter and collimator scatter may be less with IMRT because of smaller field sizes and reduced average field intensities. These two factors dominate for tissues 15 to 30 cm away from field edge. For more distant tissue, head leakage predominates and is unavoidably higher due to the increased MUs for IMRT. If the MU ratio with IMRT gets too high, this could overwhelm the benefit of the first two factors. The reader is referred to the comprehensive review of dosimetric studies of peripheral dose and SM risk modeling related to external beam radiation therapy, including IMRT and proton therapy by Xu et al. (2008) and Palm and Johansson (2007).

Based on low dose mechanisms of cancer formation, several authors have calculated the increased cancer risk in adults from the leakage radiation from a 6 MV fixed field IMRT treatment. Hall (2006), Kry et al. (2005), and Followill et al. (1997) calculated that risk would double from about 1% to about 2% when converting conventional treatments to IMRT. When NPT was considered, the calculated risk rose to almost 10% due to the much higher number of MUs needed with that delivery system (Verellen and Vanhavere 1999). The risk estimates from these authors are based on the notion that SM largely occur outside the primary high-dose region and thus are due to low doses produced by head leakage and scatter. The increase in risk for children was assumed to be even higher commensurate with the known increased sensitivity to radiation-induced malignancy in children compared to adults. However, Ruben et al. (2008) and Schneider et al. (2005) point out that the risk of developing SM should be calculated considering the entire 3D dose distribution and not just the scatter and leakage low-dose volume, the latter overestimating the risk when the model discounts the effects of the high dose region. Applying their models to several clinical cases treated with either IMRT or 3DCRT, they concluded that the carcinogenic risk of IMRT was comparable to that of 3DCRT at all sites they studied. Thus, we have contradictory theories about the role of IMRT in the production of SM as well as dosimetric findings which contradict the theories.

We have seen that IMRT and multibeam treatment plans by themselves do not necessarily increase the ID and PD, although IMRT can increase the total body dose due to the increase in leakage radiation due to the need for increased MUs. Due to the large volume of irradiated bone marrow when the total body is irradiated, the risk of secondary leukemias becomes a concern. The LSS study of Japanese A-bomb survivors demonstrated linear dose–response for leukemia at up to the maximum dose included in the study of 4 Gy (Preston et al. 1994). However, the risk of leukemia or lymphoma was only about 1/20th that of solid tumors for an exposure of 1 Gy (Ron et al. 1994). There have been several studies comparing the risk of secondary leukemia in cancer patients treated with radiation therapy compared to the Japanese A-bomb data. In a case control study of breast cancer patients, Curtis et al. (1989) found no increased risk of leukemia for patients with a mean marrow dose of 3 Gy, and in a different study found that leukemia risk was lower than projected from risk estimates based on the A-bomb survivor data (Curtis et al. 1994), a conclusion that was also found by Little et al. (2001b) in children. A case controlled study of children treated for solid tumors in the 1980s and 1990s failed to find evidence of a role for the radiation dose to active bone marrow in the risk of secondary leukemia, myelodisplasia, or myeloproliferative syndrome when adjusted for the use of epipodophyllotoxin (i.e., etoposide) and anthracycline. The results of this study also contradict that of the A-bomb survivors study where leukemia risk in children was much higher than for adults (Allard et al. 2010).

Through a review of actual cases of SM in childhood cancer survivors, we will see if risk for SM increases or decreases with dose, and, correspondingly, whether SM most often occur in the treated volume or distant from the treated volume.

2.3.5 What Is the Dose–Response Relationship for Second Malignancy in Children?

One can debate the merits of various risk calculation models but the most compelling data are found in clinical studies of secondary cancer induction in children treated for cancer with radiation therapy. The present analysis includes studies exclusively on childhood cancer survivors who had radiotherapy as part of their treatment and developed a SM (most were solid). There are numerous other studies of SM incidence in adults not included in this analysis. A thorough literature search produced 37 papers that characterized either the dose at the location of the SM (20 papers) or the location of the SM with respect to the irradiated volume (21 papers) or both. The data in the Appendix are sorted by type of dose–response (10th column). It should be noted that SMs occurring after treatment for retinoblastoma were largely excluded from studies in this analysis but these cases have been addressed separately later in this chapter. Many of the reports are case-control studies that reported the risk of SM in large numbers of children (thousands to tens of thousands) from around the world treated by conventional methods, which generally included surgery, chemotherapy, and radiation. Some used statistical methods to calculate the risk of SM attributable to radiation alone. The dose to the location of each patient's SM was calculated from their radiotherapy records using a computer treatment planning system or by measurements. Twenty-eight of these papers stated that the type of dose–response had been formally calculated to be linear or less formally that the risk increased with dose up to the highest dose given, which was above 30 Gy in over 60% of the studies, and in many was over 50 Gy. Two studies specifically correlated ID to risk for SM and found risk to increase with ID, either linearly or by a linear-quadratic relation (Karlsson et al. 1998; Nguyen et al. 2008). One study found that patients had an increased risk of death from secondary sarcoma and carcinoma only if they received more than 150 J (Tukenova et al. 2011). Types of SM with a linear dose–response include the most common types: breast, osteosarcoma, soft-tissue sarcoma, thyroid, and brain.

In 21 studies that reported on the location of the SM relative to the treatment fields, 65% of the SMs were inside the radiation field. Notably, an even larger percentage of the secondary breast, thyroid, and sarcomas, some of the most common SM, arose in the field. About 20% of the SMs occurred at the field edge. When the SM is near the edge of the irradiated volume, it is sometimes unclear whether the SM originated inside or outside the high dose region because when it is found, it may already be several centimeters in size. Also, due to the high dose gradient at the edge of the field, there is a high degree of uncertainty in the dose adjacent to the field edge, for example, a 1-cm misidentification of the SM position can change the dose by a factor of 10. Overall, only a small percentage of SMs was noted to have occurred distant to the irradiated volume. Many of the patients in these studies were treated decades ago when kilovoltage treatments were common. These treatments gave much higher bone doses than the prescribed dose but this effect was taken into account in most studies where individual dosimetry reconstruction was performed.

Although data in the literature overwhelmingly supports a dose–response for induction of SM up to high doses, there are reports that contradict this that should be mentioned. Two studies were found in the literature in which dose to the site of a solid SM was not associated with risk. In a recent study of 4581 cancer survivors in the United Kingdom and France with 115 patients with SM, individual radiation doses were calculated at the site of SM as well as the distance from each SM to the irradiated volume. This analysis did not show a dose-response with doses up to 73 Gy and they found that the majority of SM occurred in the region near the field edge receiving intermediate doses (Diallo et al. 2009). An earlier large German study that identified 203 childhood cancer survivors reported findings similar to the study by Diallo (Dorr and Herrmann 2002). Also, in a meta-analysis, Little (2001b) compared the SM risk estimates for the Japanese atomic bomb survivor LSS cohort and a number of mostly adult patient cohorts who were irradiated for medical reasons. He found that for lung cancer, bone cancer, ovarian cancer, nonmelanoma skin cancer, and leukemia, the relative risks per unit dose of radiation in the comparable (age at exposure, time since exposure, sex matched) subsets of the Japanese data were significantly greater than those in the majority of second cancer studies. The interpretation of these findings was that low doses of radiation are more effective in producing SM than higher doses due to cell sterilization with higher doses.

In summary, clinical studies of tens of thousands of childhood cancer survivors show that there is a linear or near linear dose–response relationship for virtually every type of SM. For some types of SM, notably thyroid cancer, risk plateaus above 10 to 30 Gy. The most common SM, sarcomas, breast, and thyroid cancer, are found most commonly in the high dose volume within the radiation field. Few SM occur more than about 10 cm from the target where the dose is due to leakage, collimator, and patient scatter, which together amounts to about 3% of the tumor dose, generally less than about 2 Gy. About 20% occur under a block or adjacent to the radiation fields. There is increased risk (relative to the unexposed) at low doses, 2 to 4 Gy, but risk increases linearly as dose increases up to very high doses for many types of SM. Because the risk for SM appears to be highest in high-dose regions rather than low-dose regions, and correspondingly, most SMs occur in or near the treated volume, treatment techniques that decrease the volume receiving medium to high doses but increase the dose distant to the target where the dose is only a few grays are not likely to significantly increase the incidence of SM. If the dose–response is linear, then for the same ID, the shape of the dose-volume histogram (DVH) throughout the body does not affect the risk for SM. Decreasing the ID, the prescribed dose, or the target volume all would be predicted to decrease the risk for SM. IMRT or multibeam 3DCRT by themselves do not increase these key parameters but can significantly reduce normal tissue doses. The totality of these findings suggests that reducing the intermediate dose volumes (30% to 80% of tumor dose) that have a higher risk of producing toxicity and SM in exchange for increasing the lowest dose volumes that have a lower risk of toxicity and SM may be the safest for the patient.

2.3.6 Risk Estimates of Radiation-Induced Second Malignancies in Children

2.3.6.1 Methods of Assessing Risk

Several methods of expressing the risk of SM are used in the literature. Clinicians and particularly patients want to know the *absolute risk* of developing a SM. *Absolute risk* is the number of individuals who will get a SM per 10,000 exposed individuals per year or over their lifetime. Calculating this value is complicated because at long follow-up intervals the absolute background population risk of developing cancer is high, effectively reducing the relative risk associated with treatment during childhood. For example, the background risk of cancer by age 60 is almost 5% (Olsen et al. 1993; Horner et al. 2009). Epidemiologists often express risk in terms of *relative risk* (RR). The RR is the ratio of the risk for an exposed individual to the risk of the general (unexposed) population. Relative risk can be computed for specific contributory factors, for example, for radiation therapy alone or for chemotherapy alone. Thus, while the 1000-fold relative risk of developing a SM in children with a central nervous system (CNS) tumor and neurofibromatosis is highly statistically significant (Little et al. 1998), the 38% absolute risk for a child irradiated for hereditary retinoblastoma (Kleinerman 2005) is more meaningful to the patient and family. In another example of differences between RR and absolute risk, in a Hodgkin study the RR of secondary acute myeloid leukemia (AML) was greater than 20 (in part because of the rarity of AML), but the absolute risk was just 6 excess cases per 10,000 patients yearly (Dores et al. 2002). The terms *excess* relative or absolute risk (ERR or EAR) are found by subtracting 1 from the relative risk or subtracting the general population risk from the absolute risk.

Another common measure of SM induction is cumulative incidence. Cumulative incidence is the percentage chance that an exposed person will get a certain SM over a certain length of time. This figure usually includes an array of clinical factors that are not separated out as they can be when relative risk is stated but is useful when quantifying the overall risk of developing a SM.

In the discussion that follows, many large studies of SM rates found in the literature are discussed. The study populations range from hundreds to tens of thousands of children with cancer treated with radiotherapy and in most cases, surgery and chemotherapy. Many of these break out the overall risk for each type of SM found or the risk for a specific SM after treatment for a specific primary malignancy.

2.3.6.2 What Is Known about the Risk of Specific Radiation-Induced Second Malignancy in Children?

Several large epidemiological studies have reported on incidence and mortality rates for pediatric cancer survivors with specific SMs. Table 2.1 shows various first malignancies and the most commonly associated second malignancies. Table 2.2 gives the journal reference, group, and size for seven large cohorts that have been followed for decades. One such study is by the Surveillance, Epidemiology, and End Results (SEER) program of the National Cancer Institute (NCI). In the report covering the years 1973 to 2000, the cumulative incidence of a child developing any second cancer at 25 years after diagnosis of the primary cancer, adjusted for the competing risk of death due to other causes, was 3.5% (95% CI = 3.0%–4.1%) (Curtis et al. 2006). A similar value can be found in a number of large cohort studies (Hawkins et al. 1987; Olsen et al. 1993; Neglia et al. 2001) although there are also smaller single institution studies that estimate up to a 13% incidence at 30 years (Gold et al. 2003). In comparison, the 25-year cumulative risk of death, based on all deaths excluding deaths among patients with a second

TABLE 2.1 Most Common Second Malignancy after First Malignancy

Second Malignancy	First Malignancy
Bone tumors	RB, other bone tumors, Ewing's sarcoma, STS, ALL
Soft-tissue sarcoma	RB, STS, HD, Wilms' tumor, bone tumors, ALL
Breast cancer	HD, bone tumors, STS, ALL, brain tumors, Wilms' tumor, NHL
Thyroid cancer	ALL, HD, NB, STS, bone tumors, NHL
Brain tumors	ALL, brain tumors, HD
Carcinomas	ALL, HD, NB, STS
AML/ALL	ALL, HD, bone tumors

Source: Bhatia, S., and C. Sklar, Second cancers in survivors of childhood cancer, *Nature Reviews Cancer* 2 (2):124-32, 2002.

Notes: Retinoblastoma (RB) is the heritable type. STS, soft-tissue sarcoma; HD, Hodgkin disease; NB, neuroblastoma; NHL, non-Hodgkin lymphoma; ALL, acute lymphocytic leukemia; AML, acute myelogenous leukemia. Median latency is 9 to 20 years, except for ALL, ANLL, and AML, which is 3 to 5 years. References in Table 2.4.

TABLE 2.2 Recent Reports of SM in Large Studies of Childhood Cancer Survivors

Group	Number of Children	Number of SM Reported[a]	Reference
Nordic countries	25,120	196	Svahn-Tapper et al. 2006
Germany	24,203	238	Klein 2003
SEER (NCI)	23,819	352	SEER 1973–2000
CCSS	20,720	677	Bassal et al. 2006
Great Britain	16,541	278	Jenkinson et al. 2004
France, Great Britain	4401	124	Nguyen et al. 2008
LESG	1380 (HD)	212	Bhatia et al. 2003

[a] Not all SMs found in the entire cohort were included in each study. Except for Bhatia (Hodgkin disease), a variety of primary cancers were included in the studies.

cancer, was 37.6% (95% CI = 36.7%–38.5%). Most of these deaths were due to the first cancer (Curtis et al. 2006). Most common SMs were female breast, brain, bone, thyroid gland, and soft tissue. Solid cancers accounted for 81% of all SM. The average latency period was about 15 years, but this value increases with longer follow-up. These risk estimates were based on all diagnoses and treatments (Table 2.3). Figure 2.2 shows the cumulative incidence for SM by type of primary diagnosis (Curtis et al. 2006). The risk for secondary malignancy includes the effects of chemotherapy, radiotherapy, age, sex, endocrine and metabolic factors, environmental factors, and the underlying susceptibility to develop SM associated with each type of cancer. Table 2.4 gives risk and cumulative incidence data for the major types of SM. The primary diagnosis with the highest cumulative incidence for SM, at about 36% after 50 years from diagnosis, is hereditary bilateral

TABLE 2.3 SIR, EAR, and Cumulative Incidence of SM for Invasive Solid Second Cancers by Primary Diagnosis

Primary Diagnosis	SIR	EAR	% Cumulative Incidence
ALL	4.4	1.4	5.2
HL	8.7	6.9	18.4
NHL	4.1	1.8	5.8
Astrocytoma	4.3	1.6	4.7
PNET	7.3	2.7	7.8
Osteosarcoma	4.2	2.6	6.0
Ewing's sarcoma	8.5	5.1	10.1
Rhabdomyosarcoma	5.8	2.9	8.8
Retinoblastoma	59.2	60.7	36[a]
Neuroblastoma	6.9	1.6	5.9
Wilms' tumor	4.8	1.2	4.0

Source: From Friedman, D. L., J. Whitton, W. Leisenring, et al., Subsequent neoplasms in 5-year survivors of childhood cancer: The Childhood Cancer Survivor Study, *Journal of the National Cancer Institute* 102 (14):1083–95, 2010.

Notes: Within each primary diagnosis there are sites of SM that can be much higher risk than the overall risk. SIR is standardized incidence ratio. EAR is excess absolute risk per 1000 person years.

[a] Hereditary, 50 years from diagnosis. Kleinerman, R. A., M. A. Tucker, R. E. Tarone, et al., Risk of new cancers after radiotherapy in long-term survivors of retinoblastoma: An extended follow-up, *Journal of Clinical Oncology* 23 (10):2272–9, 2005.

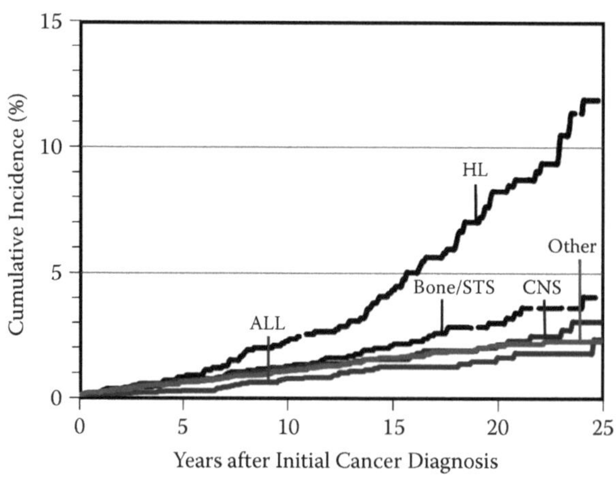

FIGURE 2.2 Cumulative incidence of developing a second cancer among children with selected primary cancers: Hodgkin lymphoma (HL), bone and soft-tissue sarcomas (Bone/STS), brain and other central nervous system cancers (CNS), acute lymphocytic leukemia (ALL), and other cancers (Other). (After Curtis, R. E., et al., eds., *New Malignancies Among Cancer Survivors: SEER Cancer Registries, 1973–2000,* NIH Pub. No. 05-5302, Bethesda, MD: National Cancer Institute, 2006.)

TABLE 2.4 Risk Estimates of Radiation-Induced Solid Second Malignancies by Type of Second Malignancy Excluding Retinoblastoma Patients

SM	PM	Cohort Size/ Number of SM	Cumulative Incidence	RR (All Diagnosis)	Reference
Bone	V	9170/64	2.8% (20 yrs)	133	Tucker et al. 1987
	V	13,175/55	0.9% (20 yrs[a])	–	Hawkins et al. 1996
	V	4400/32	1% (20 yrs)	–	Le Vu et al. 1998
	HD	1380/8	0.8% (30 yrs)	37	Bhatia et al. 2003
STS	V	13,581/32	NS	7	Neglia et al. 2001
	V	23,819/29	NS	15	SEER
	V	25,120/17	NS	30	Svahn-Tapper et al. 2006
	V	4400/25	0.6% (20 yrs[b])	54	Menu-Branthomme 2004
Breast	HD	480/29	20% (30 yrs)	57	Bhatia et al. 2003
	HD	670/16	12% (30 yrs)	17	Sankila et al. 1996
	HD	123/16	10.7% (40 yrs)	17	Guibout et al. 2005
	HD	3817/105	8.5% (30 yrs)	20	Travis et al. 2005[c]
	HD	770/48	NS	8	van Leeuwen et al. 2003
	V	7518/11	4.2% (25 yrs)	3	Jenkinson et al. 2004
	V	6304/60	NS	16	Neglia et al. 2001
Thyroid	V	9170/23	4% (26 yrs)	53	Tucker et al. 1991
	V	14,054/69	NS	16	Sigurdson et al. 2005
	HD	1380/19	4.4% (30 yrs)	36	Bhatia et al. 2003
	V	15,452/13	NS	17	Jenkinson et al. 2004
	V	4096/14	1.5% (30 yrs)	26	de Vathaire et al. 1999[d]
Brain	V	4400/12	NS	6	Little et al. 1998
	ALL	1612/21	1.4% (20 yrs)	NS	Walter et al. 1998
	V	14,361/116	NS	9	Neglia et al. 2006
	V	10,106/12	5.1% (25 yrs)	9	Hawkins et al. 1987
Carcinoma	V	14,372/71	0.5% (20 yrs)	5	Bassal et al. 2006

Notes: Not all patients in the cohorts had radiation therapy. RR is either overall or for highest radiation dose if data available, subsets can have higher values. V, various childhood malignancies; NS, not stated; STS, soft-tissue sarcoma.

[a] 5.4% after Ewing's sarcoma.

[b] 3% after Ewing's sarcoma.

[c] 20 to 40 Gy at age 15 extended field mantle, no alkylating agents.

[d] Thyroid cumulative incidence higher if neuroblastoma, 18% (30 yrs).

retinoblastoma, which mostly occurs in infants. Other examples are family history of early-onset cancers; Li–Fraumeni syndrome caused by genetic mutation of P53 tumor suppressor gene; and the ataxia telangiectasia gene mutation. The high relative risk of cancer in patients with retinoblastoma probably reflects the combination of genetic predisposition to multiple cancers, sensitivity to radiation-induced cancer, and the very low background cancer risk in the general population during the early years of life. Other large studies report a wide range of overall cumulative incidence rates and relative risks (Table 2.4), which indicate there are many variables responsible for these statistics. Depending on age, primary malignancy, combination of chemotherapy and radiotherapy (RT), relative risk can increase to over 100. For the solid secondary malignancies, risk continues to rise in most studies up to the longest follow-up, 30 years or more. We do not actually know the true lifetime incidence of radiation-induced SM because few patients have reached old age.

For most SM, risk decreases with increasing age at diagnosis. Tissues such as the brain, thyroid, bone, and breast appear to be more susceptible if exposed during periods of rapid growth and development. For 2 million cancer patients of all ages in the follow-up period 1973 to 2000, the overall RR for SM was 1.14, or 14% higher risk than the general population. The RR for ages 0 to 17 was 6.13 (excess absolute risk of 15 per 10,000 person-years), thus, children under age 17 were 5.4 times more likely to have a SM than older individuals. For the youngest patients, those less than 5 years old at diagnosis, the RR for all cancers was a factor of 2 to 3 higher than for those between 5 and 17 years old, 15 to 19 years after first cancer diagnosis (Curtis et al. 2006). Most other large epidemiological case-control studies also show that risk of SM is highest for the youngest children, especially for children less than 5 years old. In a multivariate analysis, Garwicz et al. (2000) showed a threefold increased risk with radiotherapy alone at ages less than 5 years old compared to older patients. From the A-bomb survivors LSS data with acute whole body exposure with a mean dose of less than 1 Gy, the ratio of ERR for death from SM for exposure at age 10 versus 50 years old was about 2 to 3 (UNSCEAR 2008). Bhatia and Sklar (2002) concluded that, overall, survivors of childhood cancer are at three- to sixfold higher risk of developing a SM than the general population. In children, genetic susceptibility and the child's growth rate at the time of exposure may contribute to these higher rates.

Certain chemotherapeutic agents, particularly alkylating agents and topoisomerase II inhibitors, are known to be mutagenic. Their mutagenicity is additive or supra-additive with radiation.

Radiation therapy generally has been associated with higher SM risks than chemotherapy. However, in a large Nordic study that analyzed the risk of SM by treatment era, risk was highest in the era of intensive multiagent chemotherapy after 1975 compared to the prechemotherapy era. They concluded that chemotherapy has an independent role as a risk factor for SM (Olsen et al. 1993). It is difficult to know the increased risk of SM in children attributable to radiation therapy alone compared to the risk for adults because chemotherapy and the other factors listed earlier are so different than in adults, and children sometimes have genetic syndromes that predispose them to getting SMs.

Although acute lymphoblastic leukemia (ALL) and CNS tumors are the most common childhood primary cancers, they are not the most common primary cancer for children who develop a SM. The most common primary cancers associated with SM are heritable retinoblastoma, soft-tissue sarcoma, and Hodgkin disease. Secondary bone tumors, sarcomas, breast cancer, thyroid cancer, and CNS cancer are all specifically associated with radiation therapy as a risk factor (Bhatia and Sklar 2002).

2.3.6.2.1 Bone and Soft-Tissue Sarcomas Osteosarcoma is the most frequent second cancer occurring within 20 years of treatment of a solid primary cancer in childhood (Olsen et al. 1993). Patients with retinoblastoma (RB), Ewing's sarcoma, and Hodgkin lymphoma (HL) suffer the highest risk for developing a SM, particularly secondary sarcomas (Strong et al. 1979; Kuttesch et al. 1996; Abramson and Frank 1998; Henderson et al. 2007). Cumulative risk of secondary bone cancer after Ewing's sarcoma was 22% at 20 years with a RR of 694 and a RR of 106 after HL in a study by the Late Effects Study Group (LESG), and in others, 5% to 7% at 20 years with relative risk also in the hundreds (Hawkins et al. 1996; Kuttesch et al. 1996). In the largest study of its kind, Henderson et al. (2007) reported on the incidence of sarcomas in the over 14,000 pediatric cancer survivors in the Childhood Cancer Survivor Study (CCSS) cohort. One hundred eight secondary sarcomas were diagnosed and in the 100 cases where histology was known, 31 were bone and 69 soft-tissue sarcomas. Fifty-six percent of the secondary sarcomas were in the field of radiation and 12% were distant. The LESG reported similar findings (Hawkins et al. 1996). Sixty-six percent of the deaths in the patients with SM were due to the SM. For patients who received radiation, a cumulative incidence of secondary sarcoma of 1.1% at 30 years after diagnosis was found, in agreement with other reports. For those who had a primary diagnosis of sarcoma, the cumulative incidence was higher, 2.2%. The relative risk was similar between radiation alone and higher dose alkylating or anthracycline chemotherapy alone (RR = 2 to 3) but about 4 to 34 when both were given (de Vathaire et al. 1989; Hawkins et al. 1996; Henderson et al. 2007). Other studies find the cumulative incidence at 30 years to be slightly higher, up to 3% (LESG). The LESG (Tucker et al. 1987), Childhood Cancer Research group (Hawkins et al. 1996), and the British and French cohort (Le Vu et al. 1998) found in their case-control studies that risk of second bone cancer increased substantially with increased exposure to radiation. Both Hawkins et al. (1996) and the LESG reported virtually no increased risk of secondary sarcoma with radiation doses less than about 10 Gy, and Diallo et al. (2009) found a higher median dose of 26 Gy for secondary sarcomas compared to 15 Gy for all other SM.

Retinoblastoma (RB) patients in particular are at a very high risk for developing a SM, most of which are bone or soft-tissue sarcomas. Abramson and Frank (1998) reported a cumulative incidence of any SM of 53% at 50 years after diagnosis in bilateral (inherited) RB patients. Three other important findings were reported. First, that no increased risk of SM was observed for tissues out of the field of radiation. Second, 71% of the SMs were in the field of radiation. Third, the risk for tumors in the field of radiation was heavily dependent on the age at which radiotherapy was given and may be acceptably small to the patient after the age of 12 months. In fact, the risk was no greater for patients irradiated after the age of 12 months than for those not irradiated. Abramson and Frank conclude that the increased risk for SM in irradiated RB patients is confined to those patients treated under the age of 12 months. For large secondary malignancy studies that have a number of children with RB, the risk for secondary sarcoma is so high that the risk statistics are often divided into those who have RB versus those who do not. Kleinerman et al. (2005) reported a very high cumulative incidence of SM in irradiated patients with heritable RB (38% in 50 years for heritable RB) but also showed that the risk of SM in nonheritable RB is no higher than for other types of cancer (5.7% in 50 years). They also reported that patients with heritable RB are at high risk of a SM even without radiation therapy, with a cumulative incidence of about 26% at 50 years. Wong et al. also showed that the risk of SM in RB patients increased with dose between 10 and 60 Gy. In another large study of SM in RB patients, the cumulative incidence of osteosarcoma as the SM after RB was about 10% at 20 years with a relative risk of nearly 400 (Hawkins et al. 1996).

2.3.6.2.2 Central Nervous System (CNS) Tumors The most common site for radiotherapy in children is the brain and other CNS sites (about 40% of cases). The majority of SM in the CNS occurs after treatment of a primary CNS tumor. Furthermore, in case-control studies of 13,000 to 25,000 pediatric cancer survivors, SMs of the CNS were the most common or one of the most common solid SMs after treatment to any site, representing 11% to 25% of all SMs (Neglia et al. 2001; Jenkinson et al. 2004; Svahn-Tapper et al. 2006). The cumulative incidence of a SM of the CNS was just 2% to 3% at 20 years from diagnosis of the primary cancer, and the incidence was associated with the radiation dose (Walter et al. 1998; Neglia et al. 2001). In leukemia patients, the most common childhood cancer, brain tumors were the most frequent SM that was apparently due to the cranial irradiation they received (Walter et al. 1998; Svahn-Tapper et al. 2006). Neglia et al. (2001) reported that 64% of the secondary CNS cancers arose in the patients with leukemia, who comprised only 34% of all patients in the cohort.

The excess relative risk for CNS SMs is only 0.14/Gy compared to the greatest value of 1.7/Gy for SM of the thyroid (Svahn-Tapper et al. 2006). A dose–response was seen in most studies (Walter et al. 1998; Bhatia and Sklar 2002; Neglia et al. 2006; Svahn-Tapper et al. 2006). One exception was Little et al. (1998), who found a dose–response only for benign SMs of the brain. Broniscer et al. (2004) reported on 1283 pediatric brain tumor patients treated at a single institution from which 24 SMs were found. The median dose to the site of the SM was 42 Gy. The 15-year cumulative incidence of SM was 4%; however, for patients with choroid plexus tumors, the rate was 20%. Seven of the 24 patients with SM were found to have genetic abnormalities such as neurofibromatosis or the TP53 germline mutation. Choroid plexus carcinoma in particular is known to carry the TP53 mutation that inhibits tumor suppression. Radiation oncologists are sometimes hesitant to prescribe radiotherapy for these patients. The results of the Childhood Cancer Survivor Study of nearly 3000 5-year survivors showed that patients who received cranial RT of 50 Gy or more had a cumulative incidence of SM within the CNS of 7.1% at 25 years from diagnosis compared to 1.0% for those who did not receive RT (Armstrong et al. 2009). The British version of the CCSS following 18,000 childhood cancer survivors found a twofold increased risk of secondary meningioma after less than 10 Gy increasing rapidly to 500-fold for doses above 30 Gy. The risk for secondary glioma or primitive neuroectodermal tumor (PNET) also increased linearly with dose with a fourfold increased risk for patients receiving more than 40 Gy to the brain (Taylor et al. 2010). In a study comparing the location of secondary brain tumors to the dose received by those tissues after whole brain or partial cranial irradiation, almost all the second tumors occurred in tissues within the target volume (Galloway et al. 2012). One can conclude from these many studies that the most effective way to reduce the risk of secondary brain tumors is to reduce the target dose and volume.

There are data that show that very small doses of radiation to the brain can cause neural tumors. Ron et al. (1988) investigated nearly 11,000 Israeli children given radiotherapy for tinea capitis (scalp ringworm) between 1948 and 1960. The mean dose to brain tissue was 1.5 Gy. In this matched case-control study, 60 neural tumors were found and the 30-year cumulative risk was 0.8%, with a relative risk of 8.4.

2.3.6.2.3 Breast Cancer Breast cancer as a SM is one of the three most frequent SMs in pediatric cancer survivors (Neglia et al. 2001; Svahn-Tapper et al. 2006). Radiation-induced breast cancer is most commonly attributed to treatment for HL but is reported after treatment for a variety of primary tumors. Besides the importance in the context of HL, it is important to understand secondary breast cancer statistics to be able to better plan treatments for diseases other than HL that arise in the mediastinum, chest, or even the spine such as in medulloblastoma. Bhatia et al. (2003) reported that in 212 SMs in the LESG cohort of 1380 childhood HL survivors, breast cancer was the most common with a cumulative incidence of 20.1% at 30 years and a relative risk of 57. Also, all of the 42 secondary breast cancers in that pool of HL patients received at least 26 Gy mantle irradiation. Other HL cohort studies report 10% to 19% cumulative incidence at 25 to 30 years from diagnosis (Sankila et al. 1996; Travis et al. 2005; Curtis et al. 2006; Constine et al. 2008); but in studies that include all primary cancers, a much lower cumulative incidence was found, just 2.9% at 30 years after diagnosis of the primary disease, only slightly higher than that for the general population (Guibout et al. 2005). This lower incidence is probably because the breast is rarely in the radiation field for primary cancers other than HL. In one study, 25% of the secondary breast cancers occurred in patients who did not receive chest irradiation (Kenney et al. 2004), consistent with out of field SM incidence. In another study, one-third of secondary breast cancers after HL were bilateral, with an association with dose and age at diagnosis greater than 12 years (Basu et al. 2008).

There does appear to be a dose–response relationship for radiation-induced breast cancer. Risk of second breast cancer increased substantially with increased exposure to radiation up to at least 40 Gy (Bhatia et al. 1996; Travis et al. 2003; van Leeuwen et al. 2003; Guibout et al. 2005; Constine et al. 2008). All or nearly all of the secondary breast cancers occurred in the field or near the field edge (Bhatia et al. 1996; Sankila et al. 1996; Travis et al. 2003; van Leeuwen et al. 2003).

It may be logical to assume that girls entering or just beyond puberty are most at risk for radiation-induced breast cancer due to the proliferation of breast tissue at that age, but a review of the literature does not consistently support that assumption. Neglia et al. (2001) found that risk of secondary breast cancer was actually higher in those diagnosed at ages 5 to 9 (RR = 1.8) than either 10 to 14 or 15 and older (RR = 1.5). Jenkinson et al. (2004) reported that girls less than 4 years old had slightly higher risk than those ages 10 to 14. Travis et al. (2003) found that the risk was not a function of age. After reanalyzing earlier data using different statistical methods, Bhatia et al. (2003) concluded that age at diagnosis of HL was not a risk factor for breast cancer. In contrast, Sankila et al. (1996) found no breast SM in patients less than 10 years old. Although we usually think of chemotherapy as being at least potentially additive with radiotherapy for toxicity as well as induction of SM, in the context of secondary breast cancers, alkylating agents stop ovarian estrogen production (which otherwise promotes tumorigenesis), creating a protective effect by a factor of 10 for breasts that received doses greater than 38 Gy (Travis et al. 2005). Pelvic irradiation has also been found to be protective for a similar reason (Kenney et al. 2004). The CCSS reported that the relative risk of secondary breast cancer was 11 for a breast dose of 40 Gy, increased linearly with dose, was not age related, and also found a decreased risk if the ovaries were also irradiated (Inskip et al. 2009).

2.3.6.2.4 Thyroid Cancer Thyroid neoplasms have a well-known association with exposure to radiation (Inskip 2001). Only about one-third of the secondary thyroid neoplasms are malignant (Acharya et al. 2003). In a study comprising 7 cohorts with 58,000 individuals exposed to low doses of radiation for treatment of tinea capitis (scalp ringworm), enlarged tonsil and thymus, childhood cancer, or exposure from the A-bomb, Ron et al. (1995) evaluated risk based on the 700 secondary thyroid cancers found. They reported an excess relative risk of 7.7/Gy, one of the highest values anyone has reported for any SM and convincing evidence for increased risk at doses as low as 0.1 Gy. However, they found a wide range of reported relative risks at 1 Gy, ranging from 1 to 30. In the Nordic study of over 25,000 childhood cancer survivors, the excess relative risk of 1.7 per gray of secondary thyroid cancer was the largest of any SM in their study and evidence of increased risk at doses as low as 1 Gy (Svahn-Tapper et al. 2006). Thyroid cancer was the second most common SM in the CCSS and other large childhood cancer survivor cohorts (Sankila et al. 1996; Neglia et al. 2001; Bhatia et al. 2003). Most thyroid SM occurred in leukemia or HD survivors, with patients treated for leukemia or CNS primaries at young age at highest risk (Socie et al. 2000; Neglia et al. 2001; Bhatia and Sklar 2002; Curtis et al. 2006). Other large cohort studies of childhood cancer survivors have found the risk of developing a secondary thyroid neoplasm relative to the general population to be greater than 20 (Tucker et al. 1991; Bhatia et al. 1996; Jenkinson et al. 2004; Svahn-Tapper et al. 2006; Chow et al. 2009) and over 100 for brain tumor patients treated with radiation alone (Hawkins et al. 1987). Several studies showed that relative risk rises with dose but then leveled off, after just 2 Gy in the LESG cohort (Tucker et al. 1991) and after 20 Gy followed by a decrease in risk in the CCSS cohort (Ronckers et al. 2006). This response was described as being consistent with the hypothesis of cell killing at the higher doses. However, de Vathaire et al. (1999) found no leveling off of risk up to doses greater than 30 Gy. Risk is reported to be higher in patients under 10 years old than in older patients (Rubino et al. 2003; Sigurdson et al. 2005). Nearly all thyroid SMs were in the radiation field (Hawkins et al. 1987; Bhatia et al. 2003) or near the field (Tucker et al. 1991). Cumulative risk has been reported to be about 4% at 25 years after diagnosis of the primary cancer (Tucker et al. 1991). In a report from the CCSS, survivors of ALL had a RR of 30 for developing thyroid cancer, most having received at least 10 Gy to the thyroid as exit dose from craniospinal irradiation (Chow et al. 2009).

2.3.6.2.5 Carcinomas In the over 13,000 CCSS patient cohort, 71 carcinomas were diagnosed. Sixty-seven percent of the secondary carcinomas arose in a previous radiation field. Thirty-five percent were in the genitourinary system, 32% in the head and neck, 24% in the gastrointestinal system, and 6% in the lungs. Ninety percent of the secondary carcinomas arising in the head and neck followed head and neck irradiation. Overall, the relative risk of a secondary carcinoma was 4.0 and was elevated for all primary childhood cancer diagnoses except CNS neoplasms. Survivors of neuroblastoma, soft-tissue sarcoma, and Wilms' tumor had the greatest relative risk of developing a subsequent carcinoma, with values of 24, 6, and 5, respectively. Associations were found for elevated risks of renal cell carcinoma among patients with neuroblastoma; gastrointestinal carcinomas among patients with Wilms' tumor and Hodgkin lymphoma; and head and neck carcinomas among patients with leukemia, neuroblastoma, and soft-tissue sarcoma. Overall cumulative incidence of developing a subsequent carcinoma was 0.45% at 20 years of follow-up (Bassal et al. 2006).

2.3.6.2.6 Hematologic Malignancies Radiation-induced leukemia was first reported over 100 years ago (Finch 2007). By 1952, the first report of increased leukemia in atomic bomb survivors was made (Folley et al. 1952). Since then, other reports have been published based on x-ray treatment of spondylitis and exposure from nuclear fallout from the atomic bomb, Chernobyl, and other nuclear incidents. Based on the A-bomb survivor's data, there is a fivefold increase in developing leukemia following 1 Sv of radiation (Finch 2007). Secondary leukemia is seen most frequently after HL, followed by non-Hodgkin lymphoma (NHL), Wilms' tumor, and neuroblastoma with an overall relative risk of about 7 (Neglia et al. 2001). In one cohort of HL patients, the relative risk of secondary leukemia was 17 (Sankila et al. 1996) but was 175 in another (Bhatia et al. 2003). The excess risk of secondary leukemia following childhood cancers seems to almost entirely be due to alkylating agents (Tucker et al. 1987; Bhatia et al. 2003) rather than radiotherapy. The alkylating agents induce a particular type of leukemia, acute nonlymphocytic leukemia (ANLL). The risk is not further increased by the addition of radiotherapy to chemotherapy in the treatment of HL or NHL (Lavey et al. 1990). The large differences in risk found across cohorts may be explained by differences in chemotherapeutic regimens used. The cumulative incidence of secondary leukemia ranged from 0.45%

to 2.1% with all cases seen within 8 and 14 years after treatment of the primary cancer (Breslow et al. 1995; Neglia et al. 2001; Bhatia et al. 2003). This pattern is distinctly different than for solid second malignancies for which risk continues with time. Another difference between secondary leukemia and secondary solid tumors is that the period between radiation therapy and the development of a secondary leukemia is typically 3 to 6 years compared to beyond 10 years for most secondary solid tumors (Curtis et al. 2006).

Secondary leukemia is also notably different from secondary solid cancers in its different dose–response characteristic. Risk for secondary leukemia has been found to rise with bone marrow dose linearly up to about 1 Gy (A-bomb data) and then plateau up to 16 Gy (Curtis et al. 1994). We have already seen that secondary solid tumor risk is linear with radiation dose up to doses well above 30 Gy. One hypothesis to explain the difference in risk patterns is that radiation leukemogenesis is primarily governed by initiation, inactivation, stem cell proliferation, and, uniquely for leukemia, *stem cell migration*. For leukemia, the repopulating stem cells are hematopoetic stem cells that have migrated through the bloodstream from minimally irradiated bone marrow compartments (having a smaller chance of being premalignant) to heavily irradiated compartments. This long range stem cell repopulation from unirradiated regions serves to reduce the repopulation of stem cells transformed to malignant ones in the high dose volume and thereby lower the risk of secondary leukemia compared to secondary solid tumors. Where bone marrow dose is low, more preleukemic hematopoietic stem cells are generated than are inactivated, explaining the incidence of leukemia in the A-bomb and other low dose cohorts. This model predicted the risk of leukemogenesis in a large cohort study of ovarian cancer patients treated with radiation therapy (Shuryak et al. 2006).

2.3.6.3 Risk of Second Malignancy from Acute Lymphoblastic Leukemia Therapy

The risk of solid cancers after therapy for childhood leukemia is related to both transplant therapy and treatment given before bone marrow transplant (BMT). In a cohort of 8831 children treated for ALL under Children's Oncology Group (COG) protocols using cranial or craniospinal irradiation and not BMT and total body irradiation (TBI), 63 patients developed a variety of SMs for a cumulative incidence of 2.1% at 15 years from diagnosis, consistent with estimates from some large multi-institutional cohorts (Neglia et al. 1991; Bhatia and Sklar 2002). Most of the solid SMs in the irradiated children were brain tumors. Other studies are reported in which the majority of patients were treated with BMT and TBI. In a cohort of over 19,000, about one-third of which were less than 20 years of age, only 50 excess cancers were found (Curtis et al. 1997). However, in a multinational cohort of 3182 children treated for ALL, 25 solid tumors were found and the cumulative incidence 15 years from diagnosis was reported to be 11% (Socie et al. 2000). In a single institution cohort of 2129 children, 29 SM were found with a 6% cumulative incidence 10 years after diagnosis (Bhatia et al. 2001). Leukemia, thyroid (Cohen et al. 2001; Faraci et al. 2005), and brain tumors were the most common second malignancies.

Radiation as well as radiation dose has been investigated as being a risk factor for SM after ALL therapy. In the COG patient cohort reported by Bhatia, radiation (mostly CSI) more than doubled the risk for SM and risk increased with dose (Bhatia and Sklar 2002). Many ALL patients receive TBI as part of their conditioning regimen prior to bone marrow transplant and some of these will have received prophylactic cranial irradiation prior to TBI. Higher doses of TBI have been associated with higher risk of solid cancers in multivariate analysis as well as age at diagnosis less than about 10 years old (Curtis et al. 1997; Socie et al. 2000; Cohen et al. 2001). However, Bhatia reported that TBI was only a risk factor for thyroid, liver, and oral cavity secondary malignancies. Competing risks from chemotherapy and graft versus host disease makes it difficult to determine the risk from TBI alone (Bhatia et al. 2001).

2.3.6.4 Risk of Second Malignancy from Proton Therapy

Proton beams are attractive for treatment in children because of their sharp characteristic Bragg peak that can better limit the dose to normal structures and significantly reduce the integral dose required to deliver the tumor dose compared to x-ray beam treatments (Lomax et al. 1999).

Despite the dosimetric advantages of proton beams, this modality has not escaped scrutiny with regard to risk for production of SM. The concern for proton therapy is that the passively scattered proton beams that most proton centers have today create stray neutrons as the protons interact with the beam modifiers and scattering foils in the treatment head. Neutrons produced within the patient by proton interactions contribute only about 20% of the total neutron dose (Taddei et al. 2009). For passive scattering proton beam systems (PSPT), there are many variables that influence the magnitude of stray neutrons: beam energy, field size, air gap, and type of scattering system. Spot scanning proton beam systems (SSPT) do not need the scattering foils and beam shaping devices of the passive scattering proton beam systems resulting in a 10-fold lower stray neutron dose (Newhauser et al. 2009; Schneider 2009). Another reason for concern is that the relative biological effectiveness (RBE), a multiplier for computing the photon-equivalent dose for neutrons, is not well established for the high-energy proton beams used today. Values in the literature range from 10 to 50 or more (Kellerer et al. 2006).

Several studies have shown, either by Monte Carlo calculation or measurements, that the neutron equivalent dose from passively scattered neutrons can be higher than the leakage dose from IMRT (Palm and Johansson 2007; Zacharatou Jarlskog et al. 2008; Zacharatou Jarlskog and Paganetti 2008). Fontenot et al. (2009) found that peripheral dose from proton therapy was about half that for IMRT near the treated volume but was twice as high in the tissues distant from the treated volume.

Shin compared neutron doses from PSPT and SSPT, and x-ray leakage from IMRT and 3DCRT. Depending on the details of the PSPT system, neutron dose can range from about the same as photon leakage to 10-fold lower, but SSPT was 10-fold lower than PSPT or IMRT (Shin et al. 2009). One study analyzed the risk of SM in a pediatric primary brain cancer patient based solely on the secondary neutron dose using a PSPT beam and a tumor dose of 70 Gy. The cancer risk over a lifetime in a number of organs was calculated and compared to the background risk. Although lifetime risks for most organs in adults were only a few tenths of a percent, a small fraction of the background risk, the risk for a 4-year-old to get a secondary thyroid cancer could be as high as about 1% and breast cancer about 3% (Zacharatou Jarlskog and Paganetti 2008).

SM risk estimates comparing proton therapy, IMRT, and conventional treatments for adults and children have been made by several investigators. Most of these analyses include both the PD and the primary tumor dose. Schneider et al. (2006) calculated the risk for SM for patients treated for prostate carcinoma, comparing IMRT with 6, 15, and 18 MV photons versus SSPT. The organ equivalent dose (OED) model was used, which utilizes the entire 3D dose distribution. He found that spot scanning reduced the risk by about 50% compared to the other modalities, which were not found to have significantly different risk from each other. They also compared 3DCRT, IMRT, and PSPT or SSPT in a study on a 14-month-old boy with rhabdomyosarcoma of the prostate. Risk was calculated for the whole body, bladder, and rectum. There was no significant difference in risk between any of the treatment modalities studied. Fontenot et al. (2009) found between a 26% and 39% reduction of relative risk for PSPT over IMRT for treatment of prostate cancer. These reductions are rather small in comparison to the known uncertainties in the assumptions made to calculate risk.

Miralbell et al. (2002) calculated the risk for SM from conventional photon, intensity-modulated photon, SSPT, and intensity-modulated spot-scanning proton (IMPT) treatments for pediatric whole spine treatment for medulloblastoma and for a rhabdomyosarcoma of the paranasal sinus. The risk relative to the conventional plan for the rhabdomyosarcoma case was 0.8, 0.7, and 0.4 for the 6MV photon IMRT, SSPT, and IMPT plans, respectively, and for the CSI case, 0.6 and 0.07 for the 15 MV photon IMRT and SSPT plans, respectively. Newhauser et al. (2009) also calculated the risk for SM in pediatric CSI cases, this time including both the whole brain and whole spine fields. The risk for secondary cancer from IMPT or PSPT were nearly the same (spot scanning only reduced the risk by 20% over passive scattering) with a six- to sevenfold lower risk than IMRT. They also found a 60% lower risk for IMRT than conventional x-ray therapy. Schneider et al. (2000) compared SSPT and IMPT to IMRT for HL. They report a 50% reduction of risk using scanned protons compared to photons. They also found that scanned protons could reduce the risk of secondary breast cancer by a factor of 10 using a single proton beam compared to conventional or IMRT photon treatments.

Using models that include the entire dose distribution and based on a linear or linear-quadratic dose–response relation, the risk of a SM was found to be predominately due to the in-field organ doses rather than dose to distant tissues. Only 10% of the risk was due to out-of-field dose, which renders somewhat moot the debate about the significance of stray neutrons during proton therapy (Fontenot et al. 2009). Thus, reductions in calculated risk for protons compared to photons are largely due to lower ID, not to differences in stray radiation doses. As can be seen from the clinical case reports, the risk for SM due to proton therapy compared to conventional or IMRT treatments is not always significantly lower and depends on the geometry of the patient and beams, the age, type of first malignancy, and type of second malignancy being considered.

In this chapter, we have seen that treating children with radiotherapy comes with greater challenges than for adult treatments. Treatment planning is more difficult in children due to the increase in number of critical structures present and in some instances their lower tolerance doses. Also, the technical parameters of treatment planning and delivery need careful consideration in light of the greater susceptibility for secondary malignancies in children. In this regard, we have seen that the changes in the dose distribution throughout the body created by treatment techniques that provide greater normal tissue sparing do not necessarily increase the risk of SM. In the chapters that follow, treatment planning and delivery techniques that maximally reduce toxicity will be emphasized.

Appendix

SECOND MALIGNANCY DATA BY TYPE OF DOSE–RESPONSE

Type of Study	Number of Patients/ Number of SM	Institutions	Type of FM	Type of SM	Dose (I, NS, P, E)[a]	Dose Range (Gy)	% SM in RT Field (IF, EF, Dist)[b]	Dose–Response (L, D, ND, NS)[c]	Comments	Reference
C-C	14,361/116	CCSS (U.S. and Canada)	Various	CNS	I	0–48	NS	L		Neglia et al. 2006
C-C	13,175/55	UK	Various	Bone	I	0 to >50	59% IF or EF	L		Hawkins et al. 1996
C-C	4401/124	France, UK	Various	Various	I	0–98	NS	L		Nguyen et al. 2008
C-C	4581/162	France, UK	Various	Various	E	0–103	NS	L		Guerin 2007
Cohort	5514/43	NWTS	Wilms'	Various	E	0–40	73% IF	L	Mostly abdominal RT	Breslow et al. 1995
C-C	4400/32	France, UK	Various	Osteosarcoma	I	0 to >50	NS	L		Le Vu et al. 1998
C-C	770/48	Netherlands	HD	Breast	I	0.26–56	NS	L		van Leeuwen et al. 2003
C-C	25,120/196	Nordic	Various	Various	I	>30	NS	L	Digestive tract, breast, bone, CNS, connective tissue, thyroid is <30 Gy	Svahn-Tapper et al. 2006
C-C	3817/105	International	HD	Breast	I	0–61.3	NS	L	Mantle field	Travis et al. 2003
Cohort	1814/16	France, UK	Various	Breast	I	0–88	NS	L	In 15/16, breast dose <20 Gy	Guibout et al. 2005
Cohort	28,008/88	Sweden	Skin hemangioma	Brain	I	0–11.5	NS	L	Ra^{226} or kV x-rays (mean dose = 0.07 Gy)	Karlsson et al. 1998

Various	58,000/700	Various	Various	Thyroid	P	0–47	NS	L	RR plateaus at 15 for doses >6 Gy	Ron et al. 1995
C-C	4400/12	France, UK	Various	Brain	I	0–89	NS	D	Benign brain tumors only	Little et al. 1998
C-C	4400/16	France, UK	Various	Soft-tissue sarcoma	I	0 to >61	88% IF/EF	D		Menu-Branthomme 2004
C-C	29,521/16	Nordic, France, UK	Various	Melanoma	I	0–51	NS	D		Guerin 2003
C-C	24,203/238	Germany	Various	Various	P	≤18 to ≥65	45% IF overall, 76% (13/17) thyroid IF	D	Dose–response only >65Gy	Klein 2003
Cohort	1380/88	LESG	HD	Various	I	0–47.5	16/17 breast IF or EF	D	Mantle field irradiation	Bhatia et al. 1996
C-C	9170/64	LESG	Various	Bone sarcoma	I	0 to >60	83% IF, 9% within 5 cm	D	No increased risk for dose <10 Gy	Tucker et al. 1987
C-C	9170/23	LESG	Various	Thyroid	I	0 to >60	68% IF, 27% 5–10 cm away, 5% > 10 cm	D	Most treatments with kV x-rays; RR = 14 > 2 Gy vs. <2 Gy, then plateau up to 60 Gy	Tucker et al. 1991
Cohort	1612/21	St. Jude, TN	ALL	Brain	E	0 to >30	100% IF	D	Dose to whole brain	Walter et al. 1998
C-C	634/32	France	Various	Various	I	0 to >50	64% IF	D		De Vathaire et al. 1989
C-C	14,054/69	CCSS	Various	Thyroid	I	0 to >40	NS	D	≤30 Gy then risk decreased; decrease risk after 20 Gy	Sigurdson et al. 2005, Ronckers et al. 2006

continued

Type of Study	Number of Patients/ Number of SM	Institutions	Type of FM	Type of SM	Dose (I, NS, P, E)[a]	Dose Range (Gy)	% SM in RT Field (IF, EF, Dist)[b]	Dose–Response (L, D, ND, NS)[c]	Comments	Reference
Cohort	266/16	Multi-institutional	Ewing's sarcoma	Various 10/16 sarcoma	P	0 to >60	100% (10/10) sarcoma IF, 67% (4/6) nonsarcomas IF	D	48–60 Gy	Kuttesch et al. 1996
Cohort	8831/63	COG (US and Canada)	Leukemia	Various	P	0–24	75% IF	D	Cranial or CSI	Bhatia and Sklar 2002
Cohort	3182/45	US	Leukemia (BMT)	Various	P	0 to >14 TBI	100% IF	D	87% had TBI	Socie et al. 2000
Cohort	19,229/80	International	Hematologic (BMT)	Various	P	<10 to >14	100% IF	D	TBI	Curtis et al. 1997
C-C	10,834/60	Israel	Tinea capitis	CNS	E	6	NS	D	5-field kV x-rays scalp treatment	Ron et al. 1988
Cohort	10,106/90	UK	Various CNS	Various thyroid	NS	NS	72% IF or EF; 62% IF (CNS FM); 86% IF (Wilms' FM)	NS	4/11 SM out of field were in NF patients, 4/8 SM out of field were in NF patients	Hawkins et al. 1987
Cohort	14,372/108	CCSS (U.S. and Canada)	Various	Sarcoma	NS	NS	56% IF, 12% Dist, 10% no RT, 22% unknown	NS		Henderson et al. 2007

Cohort	1641/62	Nordic	HD	Various	NS	NS	16/16 breast, 8/8 thyroid IF	NS	No breast SM in patients <10 years old	Sankila et al. 1996
Cohort	1380/212	LESG	HD	Various	NS	NS	100% breast IF or EF, 95% thyroid IF	NS	All breast SM had >26 Gy mantle RT	Bhatai et al. 2003
Cohort	14,372/677	CCSS	Various	Carcinomas	NS	NS	71% IF	NS	Most SM outside the field received alkylating agent	Bassal et al. 2006
C-C	2129/29	City of Hope	Hematologic (BMT)	Various	NS	NS	100% IF	NS	TBI was risk factor for some SM	Bhatia et al. 2001
Cohort	446/37	U.S.	Various	Various	NS	NS	59% IF, 24% EF, 14% Dist	NS		Gold et al. 2003
Cohort	4581/115	France, UK	Various	Various	I	0–73	12% IF, 66% EF, 22% Dist	ND	31% of SM <2.5 Gy	Diallo et al. 2009
Cohort	31,000/203	Germany	Various	Various	I (53)	1–65	10% IF, 50% EF, 12% Dist	RDD	kV or Co-60, rarely chemo, 43% of SM <6 Gy	Dorr and Herrmann 2002

Notes: FM, first malignancy; SM, secondary malignancy; C-C, case-control.

[a] Dose: I, individual dose to site of SM reconstructed from therapy records; NS, not stated; P, used prescribed dose; E, dose estimated to region.

[b] % SM in field: IF, in field; EF, at edge of field; Dist: >10 cm away from field edge.

[c] Dose–response: L, linear; D, risk increased with dose but not stated as linear; NS, not stated; ND, no dose response found; RDD, risk decreased with dose.

References

Abramson, D. H., and C. M. Frank. 1998. Second nonocular tumors in survivors of bilateral retinoblastoma: A possible age effect on radiation-related risk. *Ophthalmology* 105 (4):573–9; discussion 579–80.

Acharya, S., K. Sarafoglou, M. LaQuaglia, et al. 2003. Thyroid neoplasms after therapeutic radiation for malignancies during childhood or adolescence. *Cancer* 97 (10):2397–403.

Allan, J. M., and L. B. Travis. 2005. Mechanisms of therapy-related carcinogenesis. *Nature Reviews Cancer* 5 (12):943–55.

Allard, A., N. Haddy, M.-C. Le Deley, et al. 2010. Role of radiation dose in the risk of secondary leukemia after a solid tumor in childhood treated between 1980 and 1999. *International Journal of Radiation Oncology • Biology • Physics* 78 (5):1474–82.

Aoyama, H., D. C. Westerly, T. R. Mackie, et al. 2006. Integral radiation dose to normal structures with conformal external beam radiation. *International Journal of Radiation Oncology • Biology • Physics* 64 (3):962–7.

Armstrong, G. T., Q. Liu, Y. Yasui, et al. 2009. Long-term outcomes among adult survivors of childhood central nervous system malignancies in the Childhood Cancer Survivor Study. *Journal of the National Cancer Institute* 101 (13):946–58.

Bassal, M., A. C. Mertens, L. Taylor, et al. 2006. Risk of selected subsequent carcinomas in survivors of childhood cancer: A report from the Childhood Cancer Survivor Study. *Journal of Clinical Oncology* 24 (3):476–83.

Basu, S. K., C. Schwartz, S. G. Fisher, et al. 2008. Unilateral and bilateral breast cancer in women surviving pediatric Hodgkin's disease. *International Journal of Radiation Oncology • Biology • Physics* 72 (1):34-40.

Bhatia, S., A. D. Louie, R. Bhatia, et al. 2001. Solid cancers after bone marrow transplantation. *Journal of Clinical Oncology* 19 (2):464-471.

Bhatia, S., L. L. Robison, O. Oberlin, et al. 1996. Breast cancer and other second neoplasms after childhood Hodgkin's disease. *New England Journal of Medicine* 334 (12):745–51.

Bhatia, S., and C. Sklar. 2002. Second cancers in survivors of childhood cancer. *Nature Reviews Cancer* 2 (2):124–32.

Bhatia, S., Y. Yasui, L. L. Robison, et al. 2003. High risk of subsequent neoplasms continues with extended follow-up of childhood Hodgkin's disease: Report from the Late Effects Study Group. *Journal of Clinical Oncology* 21 (23):4386–94.

Breslow, N. E., J. R. Takashima, J. A. Whitton, J. Moksness, G. J. D'Angio, and D. M. Green. 1995. Second malignant neoplasms following treatment for Wilm's tumor: A report from the National Wilms' Tumor Study Group. *Journal of Clinical Oncology* 13 (8):1851–9.

Broniscer, A., W. Ke, C. E. Fuller, J. Wu, A. Gajjar, and L. E. Kun. 2004. Second neoplasms in pediatric patients with primary central nervous system tumors: The St. Jude Children's Research Hospital experience. *Cancer* 100 (10):2246–52.

Chow, E. J., D. L. Friedman, M. Stovall, et al. 2009. Risk of thyroid dysfunction and subsequent thyroid cancer among survivors of acute lymphoblastic leukemia: A report from the Childhood Cancer Survivor Study. *Pediatric Blood & Cancer* 53 (3):432–7.

Cohen, A., A. Rovelli, M. T. van Lint, et al. 2001. Secondary thyroid carcinoma after allogeneic bone marrow transplantation during childhood. *Bone Marrow Transplantation* 28 (12):1125–8.

Constine, L. S., N. Tarbell, M. M. Hudson, et al. 2008. Subsequent malignancies in children treated for Hodgkin's disease: Associations with gender and radiation dose. *International Journal of Radiation Oncology • Biology • Physics* 72 (1):24-33.

Curtis, R. E., J. D. Boice, Jr., M. Stovall, J. T. Flannery, and W. C. Moloney. 1989. Leukemia risk following radiotherapy for breast cancer. *Journal of Clinical Oncology* 7 (1):21–9.

Curtis, R. E., J. D. Boice, Jr., M. Stovall, et al. 1994. Relationship of leukemia risk to radiation dose following cancer of the uterine corpus. *Journal of the National Cancer Institute* 86 (17):1315–24.

Curtis, R. E., D. M. Freedman, E. Ron, et al., eds. 2006. *New Malignancies Among Cancer Survivors: SEER Cancer Registries, 1973–2000*. NIH Pub. No. 05-5302. Bethesda, MD: National Cancer Institute.

Curtis, R. E., P. A. Rowlings, H. Joachim Deeg, et al. 1997. Solid cancers after bone marrow transplantation. *New England Journal of Medicine* 336 (13):897–904.

D'Souza, W. D., and I. I. Rosen. 2003. Nontumor integral dose variation in conventional radiotherapy treatment planning. *Medical Physics* 30 (8):2065–71.

de Vathaire, F., C. Hardiman, A. Shamsaldin, et al. 1999. Thyroid carcinomas after irradiation for a first cancer during childhood. *Archives of Internal Medicine* 159 (22):2713–9.

de Vathaire, F., O. Schweisguth, C. Rodary, et al. 1989. Long-term risk of second malignant neoplasm after a cancer in childhood. *British Journal of Cancer* 59 (3):448–52.

Diallo, I., N. Haddy, E. Adjadj, et al. 2009. Frequency distribution of second solid cancer locations in relation to the irradiated volume among 115 patients treated for childhood cancer. *International Journal of Radiation Oncology • Biology • Physics* 74 (3):876–83.

Dores, G. M., C. Metayer, R. E. Curtis, et al. 2002. Second malignant neoplasms among long-term survivors of Hodgkin's disease: A population-based evaluation over 25 years. *Journal of Clinical Oncology* 20 (16):3484–94.

Dorr, W., and T. Herrmann. 2002. Second primary tumors after radiotherapy for malignancies: Treatment-related parameters. *Strahlentherapie und Onkologie* 178 (7):357–62.

Faraci, M., S. Barra, A. Cohen, et al. 2005. Very late nonfatal consequences of fractionated TBI in children undergoing bone marrow transplant. *International Journal of Radiation Oncology • Biology • Physics* 63 (5):1568–75.

Finch, S. C. 2007. Radiation-induced leukemia: Lessons from history. *Bailliere's Best Practice in Clinical Haematology* 20 (1):109–18.

Folley, J. H., W. Borges, and T. Yamawaki. 1952. Incidence of leukemia in survivors of the atomic bomb in Hiroshima and Nagasaki, Japan. *American Journal of Medicine* 13 (3):311–21.

Followill, D., P. Geis, and A. Boyer. 1997. Estimates of whole-body dose equivalent produced by beam intensity modulated conformal therapy. *International Journal of Radiation Oncology • Biology • Physics* 38 (3):667–72. Erratum appears in 39 (3):783, 1997.

Fontenot, J. D., A. K. Lee, and W. D. Newhauser. 2009. Risk of secondary malignant neoplasms from proton therapy and intensity-modulated x-ray therapy for early-stage prostate cancer. *International Journal of Radiation Oncology • Biology • Physics* 74 (2):616–22.

Friedman, D. L., J. Whitton, W. Leisenring, et al. 2010. Subsequent neoplasms in 5-year survivors of childhood cancer: The Childhood Cancer Survivor Study. *Journal of the National Cancer Institute* 102 (14):1083–95.

Galloway, T. J., D. J. Indelicato, R. J. Amdur, et al. 2012. Analysis of dose at the site of second tumor formation after radiotherapy to the central nervous system. *International Journal of Radiation Oncology • Biology • Physics* 82 (1):90–94.

Garwicz, S., H. Anderson, J. H. Olsen, et al. 2000. Second malignant neoplasms after cancer in childhood and adolescence: A population-based case-control study in the 5 Nordic countries. The Nordic Society for Pediatric Hematology and Oncology. The Association of the Nordic Cancer Registries. *International Journal of Cancer* 88 (4):672–8.

Giles, D., D. Hewitt, A. Stewart, and J. Webb. 1956. Malignant disease in childhood and diagnostic irradiation in utero. *Lancet* 271 (6940):447.

Gold, D. G., J. P. Neglia, and K. E. Dusenbery. 2003. Second neoplasms after megavoltage radiation for pediatric tumors. *Cancer* 97 (10):2588–96.

Gray, L. H. 1964. Radiation biology and cancer. In *Cellular Radiation Biology: A Collection of Works Presented at the 18th Annual Symposium on Experimental Cancer Research 1964,* pp. 7–25. Baltimore: Williams and Wilkins.

Guerin, S., A. Dupuy, H. Anderson, et al. 2003. Radiation dose as a risk factor for malignant melanoma following childhood cancer. *European Journal of Cancer* 39 (16):2379–86.

Guerin, S., C. Guibout, A. Shamsaldin, et al. 2007. Concomitant chemo-radiotherapy and local dose of radiation as risk factors for second malignant neoplasms after solid cancer in childhood: A case-control study. *International Journal of Cancer* 120 (1):96–102.

Guibout, C., E. Adjadj, C. Rubino, et al. 2005. Malignant breast tumors after radiotherapy for a first cancer during childhood. *Journal of Clinical Oncology* 23 (1):197–204.

Hall, E. J. 2006. Intensity-modulated radiation therapy, protons, and the risk of second cancers. *International Journal of Radiation Oncology • Biology • Physics* 65 (1):1–7.

Hawkins, M. M., G. J. Draper, and J. E. Kingston. 1987. Incidence of second primary tumours among childhood cancer survivors. *British Journal of Cancer* 56 (3):339–47.

Hawkins, M. M., L. M. Wilson, H. S. Burton, et al. 1996. Radiotherapy, alkylating agents, and risk of bone cancer after childhood cancer. *Journal of the National Cancer Institute* 88 (5):270–8.

Henderson, T. O., J. Whitton, M. Stovall, et al. 2007. Secondary sarcomas in childhood cancer survivors: A report from the Childhood Cancer Survivor Study. *Journal of the National Cancer Institute* 99 (4):300–8.

Hermanto, U., E. K. Frija, M. J. Lii, E. L. Chang, A. Mahajan, and S. Y. Woo. 2007. Intensity-modulated radiotherapy (IMRT) and conventional three-dimensional conformal radiotherapy for high-grade gliomas: Does IMRT increase the integral dose to normal brain? *International Journal of Radiation Oncology • Biology • Physics* 67 (4):1135–44.

Horner, M. J., L. A. G. Ries, M. Krapcho, et al. 2009. *SEER Cancer Statistics Review 1975–2006*. Bethesda, MD: National Cancer Institute.

Inskip, P. D. 2001. Thyroid cancer after radiotherapy for childhood cancer. *Medical & Pediatric Oncology* 36 (5):568–73.

Inskip, P. D., L. L. Robison, M. Stovall, et al. 2009. Radiation dose and breast cancer risk in the childhood cancer survivor study. *Journal of Clinical Oncology* 27 (24):3901–7.

Jenkinson, H. C., M. M. Hawkins, C. A. Stiller, et al. 2004. Long-term population-based risks of second malignant neoplasms after childhood cancer in Britain. *British Journal of Cancer* 91 (11):1905–10.

Karlsson, P., E. Holmberg, M. Lundell, et al. 1998. Intracranial tumors after exposure to ionizing radiation during infancy: A pooled analysis of two Swedish cohorts of 28,008 infants with skin hemangioma. *Radiation Research* 150 (3):357–64.

Kellerer, A. M., W. Ruhm, L. Walsh, A. M. Kellerer, W. Ruhm, and L. Walsh. 2006. Indications of the neutron effect contribution in the solid cancer data of the A-bomb survivors. *Health Physics* 90 (6):554–64.

Kenney, L. B., Y. Yasui, P. D. Inskip, et al. 2004. Breast cancer after childhood cancer: A report from the Childhood Cancer Survivor Study. *Annals of Internal Medicine* 141 (8):590–7. Summary for patients in 141 (8):I30, 2004; PMID: 15492332.

Klein, G., J. Michaelis, C. Spix, et al. 2003. Second malignant neoplasms after treatment of childhood cancer. *European Journal of Cancer* 39 (6):808–17.

Kleinerman, R. A., M. A. Tucker, R. E. Tarone, et al. 2005. Risk of new cancers after radiotherapy in long-term survivors of retinoblastoma: An extended follow-up. *Journal of Clinical Oncology* 23 (10):2272–9.

Koshy, M., A. C. Paulino, R. B. Marcus, Jr., et al. 2004. Extra-target doses in children receiving multileaf collimator (MLC) based intensity modulated radiation therapy (IMRT). *Pediatric Blood & Cancer* 42 (7):626–30.

Kry, S. F., M. Salehpour, D. S. Followill, et al. 2005. The calculated risk of fatal secondary malignancies from intensity-modulated radiation therapy. *International Journal of Radiation Oncology • Biology • Physics* 62 (4):1195–203.

Kuttesch, J. F., Jr., L. H. Wexler, R. B. Marcus, et al. 1996. Second malignancies after Ewing's sarcoma: Radiation dose-dependency of secondary sarcomas. *Journal of Clinical Oncology* 14 (10):2818–25.

Kyle, R. A., R. V. Pierre, and E. D. Bayrd. 1970. Multiple myeloma and acute myelomonocytic leukemia. *New England Journal of Medicine* 283 (21):1121–5.

Lavey, R. S., N. L. Eby, and L. R. Prosnitz. 1990. Impact on second malignancy risk of the combined use of radiation and chemotherapy for lymphomas. *Cancer* 66 (1):80–88.

Le Vu, B., F. de Vathaire, A. Shamsaldin, et al. 1998. Radiation dose, chemotherapy and risk of osteosarcoma after solid tumours during childhood. *International Journal of Cancer* 77 (3):370–7.

Little, M. P. 2001a. Cancer after exposure to radiation in the course of treatment for benign and malignant disease. *Lancet Oncology* 2 (4):212–20.

———. 2001b. Comparison of the risks of cancer incidence and mortality following radiation therapy for benign and malignant disease with the cancer risks observed in the Japanese A-bomb survivors. *International Journal of Radiation Biology* 77 (4):431–64. Erratum appears in 77 (6):745–60, 2001.

Little, M. P., F. de Vathaire, A. Shamsaldin, et al. 1998. Risks of brain tumour following treatment for cancer in childhood: Modification by genetic factors, radiotherapy and chemotherapy. *International Journal of Cancer* 78 (3):269–75.

Lomax, A. J., T. Bortfeld, G. Goitein, et al. 1999. A treatment planning inter-comparison of proton and intensity modulated photon radiotherapy. *Radiotherapy & Oncology* 51 (3):257–71.

Mansur, D. B., E. E. Klein, and B. P. Maserang. 2007. Measured peripheral dose in pediatric radiation therapy: A comparison of intensity-modulated and conformal techniques. *Radiotherapy & Oncology* 82 (2):179–84.

Marks, L. B., G. W. Sherouse, S. Das, et al. 1995. Conformal radiation therapy with fixed shaped coplanar or noncoplanar radiation beam bouquets: A possible alternative to radiosurgery. *International Journal of Radiation Oncology • Biology • Physics* 33 (5):1209–19.

Menu-Branthomme, A., C. Rubino, A. Shamsaldin, et al. 2004. Radiation dose, chemotherapy and risk of soft tissue sarcoma after solid tumours during childhood. *International Journal of Cancer* 110 (1):87–93.

Mertens, A. C., Q. Liu, J. P. Neglia, et al. 2008. Cause-specific late mortality among 5-year survivors of childhood cancer: The Childhood Cancer Survivor Study. *Journal of the National Cancer Institute* 100 (19):1368–79.

Miralbell, R., A. Lomax, L. Cella, and U. Schneider. 2002. Potential reduction of the incidence of radiation-induced second cancers by using proton beams in the treatment of pediatric tumors. *International Journal of Radiation Oncology • Biology • Physics* 54 (3):824–9.

Mutic, S., and D. A. Low. 1998. Whole-body dose from tomotherapy delivery. *International Journal of Radiation Oncology • Biology • Physics* 42 (1):229–32.

National Council on Radiation Protection & Measurments (NCRP). 1993. Report 116: Limitation of exposure to ionizing radiation. Bethesda, MD: NCRP.

Neglia, J. P., D. L. Friedman, Y. Yasui, et al. 2001. Second malignant neoplasms in five-year survivors of childhood cancer: Childhood Cancer Survivor Study. *Journal of the National Cancer Institute* 93 (8):618–29.

Neglia, J. P., A. T. Meadows, L. L. Robison, et al. 1991. Second neoplasms after acute lymphoblastic leukemia in childhood. *New England Journal of Medicine* 325 (19):1330–6.

Neglia, J. P., L. L. Robison, M. Stovall, et al. 2006. New primary neoplasms of the central nervous system in survivors of childhood cancer: A report from the Childhood Cancer Survivor Study. *Journal of the National Cancer Institute* 98 (21):1528–37.

Newhauser, W. D., J. D. Fontenot, A. Mahajan, et al. 2009. The risk of developing a second cancer after receiving craniospinal proton irradiation. *Physics in Medicine & Biology* 54:2277–91.

Nguyen, F., C. Rubino, S. Guerin, et al. 2008. Risk of a second malignant neoplasm after cancer in childhood treated with radiotherapy: Correlation with the integral dose restricted to the irradiated fields. *International Journal of Radiation Oncology • Biology • Physics* 70 (3):908–15.

Olsen, J. H., S. Garwicz, H. Hertz, et al. 1993. Second malignant neoplasms after cancer in childhood or adolescence. Nordic Society of Paediatric Haematology and Oncology Association of the Nordic Cancer Registries. *BMJ* 307 (6911):1030–6.

Palm, A., and K.-A. Johansson. 2007. A review of the impact of photon and proton external beam radiotherapy treatment modalities on the dose distribution in field and out-of-field; implications for the long-term morbidity of cancer survivors. *Acta Oncologica* 46 (4):462–73.

Penagaricano, J. A. 2006. Integral radiation dose to normal structures with conformal external beam radiation: In regards to Aoyama et al. (Int J Radiat Oncol Biol Phys 2006;64:962–967). *International Journal of Radiation Oncology • Biology • Physics* 65 (4):1274; author reply 1274–5.

Pirzkall, A., M. Carol, F. Lohr, et al. 2000. Comparison of intensity-modulated radiotherapy with conventional conformal radiotherapy for complex-shaped tumors. *International Journal of Radiation Oncology • Biology • Physics* 48 (5):1371–80.

Pirzkall, A., M. P. Carol, B. Pickett, P. Xia, M. Roach, 3rd, and L. J. Verhey. 2002. The effect of beam energy and number of fields on photon-based IMRT for deep-seated targets. *International Journal of Radiation Oncology • Biology • Physics* 53 (2):434–42.

Preston, D. L., S. Kusumi, M. Tomonaga, et al. 1994. Cancer incidence in atomic bomb survivors. Part III. Leukemia, lymphoma and multiple myeloma, 1950–1987. *Radiation Research* 137 (2 Suppl):S68–97.

Pugachev, A., J. G. Li, A. L. Boyer, et al. 2001. Role of beam orientation optimization in intensity-modulated radiation therapy. *International Journal of Radiation Oncology • Biology • Physics* 50 (2):551–60.

Reese, A. S., S. K. Das, C. Curie, and L. B. Marks. 2009. Integral dose conservation in radiotherapy. *Medical Physics* 36 (3):734–40.

Ron, E., J. H. Lubin, R. E. Shore, et al. 1995. Thyroid cancer after exposure to external radiation: A pooled analysis of seven studies. *Radiation Research* 141 (3):259–77.

Ron, E., B. Modan, J. D. Boice, Jr., et al. 1988. Tumors of the brain and nervous system after radiotherapy in childhood. *New England Journal of Medicine* 319 (16):1033–9.

Ron, E., D. L. Preston, K. Mabuchi, D. E. Thompson, and M. Soda. 1994. Cancer incidence in atomic bomb survivors. Part IV: Comparison of cancer incidence and mortality. *Radiation Research* 137 (2 Suppl):S98–112.

Ronckers, C. M., A. J. Sigurdson, M. Stovall, et al. 2006. Thyroid cancer in childhood cancer survivors: A detailed evaluation of radiation dose response and its modifiers. *Radiation Research* 166 (4):618–28.

Ruben, J. D., S. Davis, C. Evans, et al. 2008. The effect of intensity-modulated radiotherapy on radiation-induced second malignancies. *International Journal of Radiation Oncology • Biology • Physics* 70 (5):1530–6.

Rubino, C., E. Adjadj, S. Guerin, et al. 2003. Long-term risk of second malignant neoplasms after neuroblastoma in childhood: Role of treatment. *International Journal of Cancer* 107 (5):791–6.

Sachs, R. K., and D. J. Brenner. 2005. Solid tumor risks after high doses of ionizing radiation. *Proceedings of the National Academy of Sciences of the United States of America* 102 (37):13040–5.

Sankila, R., S. Garwicz, J. H. Olsen, et al. 1996. Risk of subsequent malignant neoplasms among 1,641 Hodgkin's disease patients diagnosed in childhood and adolescence: A population-based cohort study in the five Nordic countries. Association of the Nordic Cancer Registries and the Nordic Society of Pediatric Hematology and Oncology. *Journal of Clinical Oncology* 14 (5):1442–6.

Schneider, U. 2009. Mechanistic model of radiation-induced cancer after fractionated radiotherapy using the linear-quadratic formula. *Medical Physics* 36 (4):1138–43.

Schneider, U., A. Lomax, and N. Lombriser. 2000. Comparative risk assessment of secondary cancer incidence after treatment of Hodgkin's disease with photon and proton radiation. *Radiation Research* 154 (4):382–8.

Schneider, U., A. Lomax, P. Pemler, et al. 2006. The impact of IMRT and proton radio therapy on secondary cancer incidence. *Strahlentherapie und Onkologie* 182 (11):647–52.

Schneider, U., D. Zwahlen, D. Ross, and B. Kaser-Hotz. 2005. Estimation of radiation-induced cancer from three-dimensional dose distributions: Concept of organ equivalent dose. *International Journal of Radiation Oncology • Biology • Physics* 61 (5):1510–15.

Seiler, G., E. DeVol, K. Yasser, et al. 2001. Evaluation of the safety and efficacy of repeated sedations for the radiotherapy of young children with cancer: A prospective study of 1033 consecutive sedations. *International Journal of Radiation Oncology • Biology • Physics* 49 (3):771–83.

Sharma, D. S., Animesh, S. S. Deshpande, et al. 2006. Peripheral dose from uniform dynamic multileaf collimation fields: Implications for sliding window intensity-modulated radiotherapy. *British Journal of Radiology* 79 (940):331–5.

Shin, D., M. Yoon, J. Kwak, et al. 2009. Secondary neutron doses for several beam configurations for proton therapy. *International Journal of Radiation Oncology • Biology • Physics* 74 (1):260–5.

Shuryak, I., R. K. Sachs, L. Hlatky, M. P. Little, P. Hahnfeldt, and D. J. Brenner. 2006. Radiation-induced leukemia at doses relevant to radiation therapy: Modeling mechanisms and estimating risks. *Journal of the National Cancer Institute* 98 (24):1794–806.

Sigurdson, A. J., C. M. Ronckers, A. C. Mertens, et al. 2005. Primary thyroid cancer after a first tumour in childhood (the Childhood Cancer Survivor Study): A nested case-control study. *Lancet* 365 (9476):2014–23.

Socie, G., R. E. Curtis, H. Joachim Deeg, et al. 2000. New malignant diseases after allogeneic marrow transplantation for childhood acute leukemia. *Journal of Clinical Oncology* 18 (2):348–57.

Stoh, S. M., J. D. Derrickson, and R. S. Lavey. 2001. Sedation of young children receiving radiation therapy using pentobarbital sodium: Experience at Children's Hospital Los Angeles. *International Journal of Radiation Oncology • Biology • Physics* 51 (3):376.

Strong, L. C., J. Herson, B. M. Osborne, and W. W. Sutow. 1979. Risk of radiation-related subsequent malignant tumors in survivors of Ewing's sarcoma. *Journal of the National Cancer Institute* 62 (6):1401–6.

Svahn-Tapper, G., S. Garwicz, H. Anderson, et al. 2006. Radiation dose and relapse are predictors for development of second malignant solid tumors after cancer in childhood and adolescence: A population-based case-control study in the five Nordic countries. *Acta Oncologica* 45 (4):438–48.

Taddei, P. J., D. Mirkovic, J. D. Fontenot, et al. 2009. Stray radiation dose and second cancer risk for a pediatric patient receiving craniospinal irradiation with proton beams. *Physics in Medicine & Biology* 54:2259–75.

Taylor, A. J., M. P. Little, D. L. Winter, et al. 2010. Population-based risks of CNS tumors in survivors of childhood cancer: The British Childhood Cancer Survivor Study. *Journal of Clinical Oncology* 28 (36):5287–93.

Travis, L. B. 2002. Therapy-associated solid tumors. *Acta Oncologica* 41 (4):323–33.

Travis, L. B., D. A. Hill, G. M. Dores, et al. 2003. Breast cancer following radiotherapy and chemotherapy among young women with Hodgkin disease. *JAMA* 290 (4):465–75. Erratum appears in 290 (10):1318, 2003.

Travis, L. B., D. Hill, G. M. Dores, et al. 2005. Cumulative absolute breast cancer risk for young women treated for Hodgkin lymphoma. *Journal of the National Cancer Institute* 97 (19):1428–37.

Tubiana, M. 2005. Dose-effect relationship and estimation of the carcinogenic effects of low doses of ionizing radiation: The joint report of the Academie des Sciences (Paris) and of the Academie Nationale de Medecine. *International Journal of Radiation Oncology • Biology • Physics* 63 (2):317–9.

Tucker, M. A., G. J. D'Angio, J. D. Boice, Jr., et al. 1987. Bone sarcomas linked to radiotherapy and chemotherapy in children. *New England Journal of Medicine* 317 (10):588–93.

Tucker, M. A., P. H. Jones, J. D. Boice, Jr., et al. 1991. Therapeutic radiation at a young age is linked to secondary thyroid cancer. The Late Effects Study Group. *Cancer Research* 51 (11):2885–8.

Tukenova, M., C. Guibout, M. Hawkins, et al. 2011. Radiation therapy and late mortality from second sarcoma, carcinoma, and hematological malignancies after a solid cancer in childhood. *International Journal of Radiation Oncology • Biology • Physics* 80 (2):339–46.

UNSCEAR. 2008. *UNSCEAR 2006 Report. Annex A: Epidemiological Studies of Radiation and Cancer*. New York: United Nations.

van Leeuwen, F. E., W. J. Klokman, M. Stovall, et al. 2003. Roles of radiation dose, chemotherapy, and hormonal factors in breast cancer following Hodgkin's disease. *Journal of the National Cancer Institute* 95 (13):971–80.

Verellen, D., and F. Vanhavere. 1999. Risk assessment of radiation-induced malignancies based on whole-body equivalent dose estimates for IMRT treatment in the head and neck region. *Radiotherapy & Oncology* 53 (3):199–203.

Walter, A. W., M. L. Hancock, C. H. Pui, et al. 1998. Secondary brain tumors in children treated for acute lymphoblastic leukemia at St. Jude Children's Research Hospital. *Journal of Clinical Oncology* 16 (12):3761–7.

Wong, F. L., J. D. Boice, Jr., D. H. Abramson, et al. 1997. Cancer incidence after retinoblastoma. Radiation dose and sarcoma risk. *JAMA* 278 (15):1262–7.

Xu, X. G., B. Bednarz, and H. Paganetti. 2008. A review of dosimetry studies on external-beam radiation treatment with respect to second cancer induction. *Physics in Medicine & Biology* 53 (13):R193–241.

Zacharatou Jarlskog, C., C. Lee, W. E. Bolch, et al. 2008. Assessment of organ-specific neutron equivalent doses in proton therapy using computational whole-body age-dependent voxel phantoms. *Physics in Medicine & Biology* 53 (3):693–717.

Zacharatou Jarlskog, C., and H. Paganetti. 2008. Risk of developing second cancer from neutron dose in proton therapy as function of field characteristics, organ, and patient age. *International Journal of Radiation Oncology • Biology • Physics* 72 (1):228–35.

Section 2

Guide to Treatment Planning and Dose Delivery

This section discusses treatment planning and dose delivery by diagnosis. Each section begins with a clinical overview of the disease with anatomical drawings or magnetic resonance (MR) or computed tomographic (CT) images, incidence and survival statistics, and current treatment paradigm including surgery, chemotherapy, and radiotherapy. Following the clinical overview are treatment planning suggestions; case examples with beam arrangements; and planning parameters including intensity-modulated radiation therapy (IMRT) and three-dimensional (3D) conformal methods, special physics-related considerations, and practical points for accurate dose delivery.

Chapter 3

Leukemia

3.1 Clinical Overview

Leukemia is the most common childhood cancer, with approximately 4000 new cases per year, accounting for 30% of all cancer cases in children less than 15 years old. Eighty percent of leukemia cases are acute lymphoblastic leukemia (ALL), 15% to 20% are acute myeloid leukemia (AML), and about 5% are chronic leukemia. Children with Down syndrome have a 10- to 20-fold increased risk of developing leukemia, with a particularly high risk for AML.

Overall survival is about 85% for ALL and 60% for AML. The 5-year overall survival for high-risk or relapsed patients who are treated with an allogeneic bone marrow transplant with or without total body irradiation is approximately 50%. Leukemia is a hematological, or "liquid" malignancy of white blood cells (WBCs) from the bone marrow, rather than a solid tumor. The predominant blast cell in the bone marrow of ALL patients is the lymphocyte, one of the five types of WBCs. ALL is subcategorized as being comprised of malignant B-lymphocytes (85% of ALL cases) or malignant T-lymphocytes. Prognosis is generally better for those with B-cell ALL. The best prognosis category (with 80% 4-year event-free survival) is "standard-risk ALL," comprised of the two-thirds of B-cell ALL patients who present with a WBC count below 50,000/mL at age 1 to 10 years. The other one-third of B-cell ALL patients, classified as having high-risk ALL, have a 4-year event-free survival rate of 65%. The initial response to chemotherapy is also prognostic. Patients with a "rapid early response" (remission achieved after 1 to 2 weeks of the start of chemotherapy) have a better prognosis than those with a "slow early response" (remission not achieved until 1 month of chemotherapy has been given). Complete remission is obtained after 1 month of chemotherapy in over 95% of ALL and approximately 85% of AML patients. Patients younger than 20 years old generally do better than older patients (van Kempen-Harteveld et al. 2008).

Common presenting signs are fatigue, irritability, anorexia, pallor, bruising, bleeding, bone pain, enlarged lymph nodes, and low-grade fever. Evaluation of patients must include a biopsy of the bone marrow, evaluation of the cerebrospinal fluid (CSF), and, for boys, examination of the testes. The incidence of central nervous system (CNS) involvement at diagnosis is 10% to 15% among patients with T-cell and <5% for B-cell ALL.

Systemic chemotherapy is the keystone of treatment for all cases. As it is a systemic disease, surgical resection is not performed. The central nervous system and testes are "sanctuary sites" for leukemic cells in which the concentration of intravenously administered chemotherapy agents does not reach the levels achieved in other organs. The blood–brain barrier is partially overcome by the routine administration of chemotherapy intrathecally by lumbar puncture in addition to the intravenous route to all patients with acute leukemia.

Radiation therapy to the entire brain (cranial irradiation) was routinely given in the past to prevent CNS relapse. However, the negative impact of cranial irradiation on the cognitive development of young children has caused it to be supplanted by intrathecal chemotherapy. A cranial boost was shown to not be associated with a reduction in CNS recurrence, especially in patients with only hematologic disease at presentation (Alexander et al. 2005). Currently, only the 10% to 15% of ALL patients at highest risk for CNS relapse receives either prophylactic or therapeutic cranial irradiation. With chemotherapy alone, less than 5% of the better risk patients relapse in either the CNS or the testes. For patients with high-risk ALL, the small proportion of patients who present with leukemic blast cells and ≥5 WBCs/uL in their CSF receive cranial irradiation to a total dose of 18 Gy given in 1.8 Gy fractions. Patients who relapse in the CNS after chemotherapy only are treated with cranial or craniospinal irradiation (CSI) in addition to chemotherapy. A typical dose is 18 to 24 Gy to the cranium and 6 to 18 Gy to the entire spine, both in 1.8 to 2.0 Gy fractions. A full dose of 24 Gy to the spine is avoided because it may induce prolonged bone marrow suppression that prevents the delivery of necessary systemic chemotherapy. The reduced spinal radiation dose is usually effective when administered in conjunction with intrathecal chemotherapy. Prophylactic cranial irradiation is prescribed 12 Gy.

Those who present with gross testicular involvement that does not resolve after the first month of chemotherapy or who develop a testicular relapse undergo bilateral testicular irradiation to a total dose of 24 Gy in 2 Gy fractions.

Radiation therapy is rarely used as part of the initial treatment of children with AML. It may be considered in the unusual circumstance of a child presenting with a solid mass of AML cells (a chloroma) that is causing a neurological deficit, such as spinal cord compression or cranial nerve dysfunction, or is unresponsive to chemotherapy. Chloromas occur in about 5% of AML patients and 1% of chronic myelogenous leukemia (CML) patients. A retrospective evaluation of Children's Cancer Group data found that event-free survival and local recurrence rates were not significantly different between patients whose chloromas were electively given radiation therapy and those treated with chemotherapy alone. Current Children's Oncology Group (COG) protocols do not require that any chloromas receive radiation therapy, as they generally respond to chemotherapy. If radiation therapy is used, the standard dose is 20 to 24 Gy in 2 Gy fractions (Bakst et al. 2012).

3.2 Total Body Irradiation (TBI) as Preparation for Bone Marrow Transplant

Radiotherapy is mainly used in patients with involvement of the CSF or testes, or as total body irradiation (TBI) as part of the conditioning regimen prior to bone marrow transplant in patients with relatively poor prognosis leukemia. Chemotherapy is generally given prior to TBI for this type of conditioning. TBI destroys the recipient's bone marrow and tumor cells causing sufficient immunosuppression to avoid rejection of the donor bone marrow. There is evidence that grafted bone marrow can become better established if space is made in the marrow cavity by the cytotoxic effects of TBI. TBI works synergistically with chemotherapy by eradicating tumor cells in areas of the body that the drug cannot penetrate. Radiation therapy is not utilized for patients with standard risk ALL. Patients with relapsed or very high-risk leukemia may receive an allogeneic bone marrow transplant (BMT) or umbilical cord stem cell transplant. The BMT conditioning regimen consists of myeloablative chemotherapy alone or with TBI delivered in the days prior to intravenous infusion of the donor bone marrow or cord blood cells. The most common TBI regimens are 12 Gy in 2 Gy fractions or 13.5 Gy in 1.5 Gy fractions, each delivered twice daily with an interval of at least 6 hours between fractions. TBI is delivered twice daily to reduce the number of days between ablation of the bone marrow by radiation and chemotherapy and infusion of the salvage cells that will repopulate the marrow. Although the convention is to treat with a midplane dose rate of

≤0.1 Gy/min, controversy remains as to the impact of dose rate just above or below that figure once the treatment is fractionated.

3.2.1 Early TBI Dose and Fractionation

Until the early 1980s, 10 Gy single-dose TBI given at instantaneous dose rates of between 0.025 and 0.5 Gy/min was common. The relatively low dose rate was a result of the necessity to use long source–skin distances to broaden the radiation field to fully cover the patient. Use of a single-dose fraction was not motivated by radiobiology but by wanting to reduce the time to bone marrow transplant that reduced the chances for infection. The highest dose rate achievable for the treatment machines available was used to reduce the treatment time. Contemporary dose and fractionation will be discussed in Section 3.4.

3.3 Early Animal and *In Vitro* Radiobiological Data

By the late 1970s, it was possible to minimize and control infection; reducing pulmonary toxicity while decreasing the relapse rate became the focus of radiobiological studies in animals and man. Because of a high incidence of death due to normal tissue damage, radiobiologically motivated changes in TBI treatment schemes were of great interest. Based in part on Eric Hall's work on Chinese hamster V79 cells (1972), the dose rate effect was thought to be maximal at between 0.50 Gy/min and 0.01 Gy/min. Early experimental work showed that hemopoietic stem cells have limited to no capacity for repair of sublethal radiation damage and leukemic cells were thought to act similarly. In 1980, Peters made the case for fractionation during TBI and calculated that 10 Gy in a single fraction at 0.05 Gy/min (used in Seattle at the time) would produce a 3.6 log cell kill while 12 Gy in 2 Gy fractions delivered at 0.26 Gy/min would result in a 4.4 log cell kill and at the same time be less toxic to normal tissues due to their large repair capacity. Using an *in vitro* leukemia cell line irradiated with 15 Gy in 12 fractions (the Memorial Sloan-Kettering Cancer Center [MSKCC] regimen) but with a higher dose rate, repair but not regrowth between fractions was seen. They estimated that 15 Gy fractionated TBI at 0.1 to 0.2 Gy/min would result in 20 times more cell killing than a 10 Gy single fraction at 0.05 Gy/min (Shank 1993). It was also noted that fractionation might take advantage of the increased radiosensitivity of cells given the chance to move through the cell cycle. Based on these data, researchers and clinicians debated whether leukemic cells could in fact undergo repair and whether further lowering the dose rate in addition to fractionation would be advantageous or would increase the relapse rate. Consideration of both leukemic cell kill and normal tissue preservation were paramount. Perhaps once the treatment was sufficiently fractionated, the dose rate did not need to be aggressively lowered, or alternatively, with a low enough dose rate, perhaps a single or few fractions were safe and effective.

To answer these questions, several important animal studies were performed. In one, a single dose of 15.5 Gy given with dose rates between 0.02 and 1.0 Gy/min resulted in normal lung function for dose rates below 0.2 Gy/min. The relative biological effectiveness (RBE) for 0.04 Gy/min was about 50% less than for 0.4 Gy/min for this single fraction exposure (Depledge and Barrett 1982). In another study using dose rates between 0.01 and 0.25 Gy/min, the proportion of mice with histologic changes in the lung or kidney for single doses of 7 to 16 Gy was not statistically different for dose rates between 0.05 and 0.12 Gy/min (Travis et al. 1985). Peters (1985) showed that for single doses, there was no therapeutic benefit (nonhemopoietic lethality divided by the LD_{50} for bone marrow ablation) for a dose rate of 0.05 Gy/min compared to 0.25 Gy/min, but that ratio increased by more than a factor of 2 at dose rates less than 0.05 Gy/min. Studies in animals were also performed to determine if hyperfractionation along with low dose rate was advantageous. In one study in mice, 0.05 Gy/min was compared to 0.8 Gy/min for single dose, twice, or three times per day treatment for either the upper half body or the whole body. For the TBI group, no dose rate or fractionation sparing effect was seen on bone marrow survival while a significant increase in the lethal dose for 50% of the animals for days 30 to 180 after irradiation (representing normal organ effects) was seen for twice per day high- or low-dose rate treatments. This suggests that fractionation was a greater factor in normal organ sparing than dose rate. Further sparing was not seen going to three times daily treatments (Tarbell et al. 1987). In dogs, single dose versus 2 Gy three times per day doses of TBI of 10 to 18 Gy were delivered using 0.02, 0.05, 0.1, and 0.2 Gy/min followed by bone marrow transplant. The total dose for early death for single versus fractionated treatment was the same for each dose rate at or below 0.1 Gy/min showing that once the dose rate was low, fractionation was not beneficial. With fractionation, a dose rate of 0.2 Gy/min was no more acutely toxic than 0.1 Gy/min. But for long-term survival, fractionation but not dose rate was important (Deeg et al. 1988).

In contrast, when TBI prior to transplant was delivered using hyperfractionation and a 0.05 Gy/min dose rate in mice, higher total doses were required compared to single doses given at 1 Gy/min for equivalent engraftment indicating appreciable sublethal damage repair of the host marrow (Down et al. 1991). Others have also reported that in leukemic cell lines, the response of cells to irradiation decreased with decreasing dose rate, suggesting that fractionated treatments would require larger total doses to be equally effective (Weichselbaum et al. 1981; Rhee et al. 1985; Laver et al. 1987). Based on animal data showing that the repair capacity of malignant hemopoietic cells was not that different from normal or other malignant cells, Song et al. (1981) asserted that an isoeffective fractionated dose (for leukemic cells) for 10 Gy in a single low-dose rate fraction was 16 Gy in 2 Gy fractions, not 12 Gy, and that the theoretical therapeutic advantage of fractionation was overstated.

Thomas, Ashley, et al. (1959) showed that in dogs, 8 Gy permitted engraftment in some animals, but 10 Gy consistently permitted it. These data were largely the basis for the 10 Gy single fraction prescriptions used for the next couple decades.

3.4 Clinical Basis for TBI Prescription

In 1982, Thomas et al. published the results of their randomized trial of 53 adult acute nonlymphoblastic leukemia patients given 10 Gy in a single dose versus 12 Gy in 2 Gy/fraction once per day at a dose rate of 0.08 Gy/min for either regimen. There was a survival advantage for the fractionated group with each group having just one relapse and the same interstitial pneumonitis (IP) rate. The survival advantage was presumed to be due to the overall better sparing of normal tissues and the lack of any alteration in the anti-leukemic affect due to fractionation (Thomas et al. 1982). The idea of hyper-fractionation was then explored as a way to decrease the time to transplant to 3 or 4 days from 6 or more days. Shank reported on the results of MSKCC hyperfractionated TBI treatment for 48 patients with a mean age of 18 years old. They gave 13.2 Gy in 1.2 Gy fractions three times per day with a lung dose of 9 Gy. They compared this to prior cases treated with a single fraction of 10 Gy given at 0.05 to 0.07 Gy/min with no lung shielding. Survival rates were improved with fractionation due to significantly reduced incidence of deaths from IP (Shank et al. 1981, 1983). These studies were widely used as evidence for the benefits of fractionation. As a result of animal studies and some early human trials, by 1983, many TBI centers were using 4 to 11 fractions over 4 to 6 days (Shank et al. 1983).

However, other studies raised concerns about the changes being made to TBI treatments. In an adult patient series, relapses were significantly higher for doses below 9.9 Gy in three daily fractions along with dose rates below 0.04 Gy/min (Scarpati et al. 1989). Marmont et al. (1991) showed that there were fewer treatment failures in leukemia patients treated with dose rates greater than 0.14 Gy/min and more treatment failures with fractionation. In some studies, fractionated TBI resulted in lower IP rates but higher relapse rates (Socie et al. 1991). Increasing total dose to 15.75 Gy in 7 daily fractions from 12 Gy in 6 daily fractions in a randomized trial resulted in reduced chance of relapse but not improved survival due to increased mortality by causes other than relapse (Clift et al. 1990). Other studies have shown no dose rate effect (for dose rates between 0.08 and 0.19 Gy/min) for fractionated treatments (Oya et al. 2006). A meta-analysis by Kal et al. (2006) concluded that a biologically equivalent dose (BED) of at least 15 Gy was required for high relapse-free survival (8 × 1.65 Gy at 0.1 Gy/min, for example). Pediatric TBI treatment schedules follow those developed for the adult and mixed age studies.

3.5 TBI Delivery Method Evolution

Historically, the objective of TBI is to literally treat every cubic centimeter of the patient to the prescribed dose, within about ±10%. Even the skin is part of the clinical target volume (CTV), since it can harbor lymphatic vessels carrying tumor cells and can be theoretically invaded by them. Potentially leukemic cells circulating in the bloodstream are also targets. Radiation-induced IP was recognized very early on in the use of TBI and is the driver of the development of a myriad of treatment techniques reported since the 1930s. Since about 2005, reports have described a new paradigm for TBI, the use of TomoTherapy or volumetric modulated arc therapy (VMAT) methods of treating the total body or just the bone marrow, minimizing dose to all other tissues.

The TBI story goes back over 100 years. Total body irradiation was described by Dessauer in 1905 when he exposed the entire body of a patient to three x-ray tubes functioning at the same time, one from the anterior and one each from head to foot and foot to head. In 1925, Teschendorf described a method of treating the total body giving 250 Roentgen from a single 180 kV x-ray tube placed 1.8 m from the patient. In 1932, Heublein described one of the first dedicated TBI facilities in North America, working with Gioacchino Failla at the Memorial Hospital in New York. They described a method of treating leukemia with the x-ray tube placed near the ceiling in order to irradiate four patients simultaneously in adjoining shielded vaults with an open path between the patients and the x-ray source. The anterior surface would be exposed first followed by the posterior and the treatments spaced according to the rate of fall of the white blood cell count. The beds were about 5 m from the 185 kVp tube, which was operated continuously nearly all day. This was to be known as teleroentgentherapy. Exposure rates ranged from 0.7 R/h to 1.3 R/h and doses up to about 350 Roentgen were given over about 12 days (Heublein 1932). Up to about the 1950s, the motivation for total body irradiation was to eradicate leukemic cells with the hope that the patient's immune system would recover and be healed.

In 1957, E. Donnall Thomas, later to become a Nobel laureate, first reported the use of bone marrow infusion in humans following TBI or chemotherapy, and less than 1 year later he published his experience using TBI up to 600 Roentgen delivered with a patchwork of 250 kVp x-ray beams at 2.5 m with a dose rate of less than 2 R/hr, followed by bone marrow transplantation. By this time, it was known that radiation would kill the patient's leukemic cells; infusion of normal bone marrow would induce recovery in a patient irradiated to an otherwise lethal whole body dose. The infused cells would home in to the marrow spaces partially evacuated by the effects of the radiation and begin to produce the necessary blood products for survival (Thomas et al. 1957). In Thomas's words, "Thus, by a process of irradiation followed by marrow transplantation, the recipient becomes a chimera producing and tolerating cells of the blood type of the donor's and in general recognizing his tissues as friendly." A couple years later, Thomas noted that more uniform irradiation was needed, drawing on the experience of mouse studies where the total body could be irradiated uniformly. Thomas understood that prolonged low-dose rate uniform irradiation was needed to catch each cell at its most radiosensitive part of the cell cycle. He described a room with two opposing Co-60 sources for uniform total body irradiation that was being built at Columbia University in New York (Sahler 1959; Thomas, Lochte, et al. 1959). Management of infection was the chief problem faced at that time.

In 1960, Jacobs and Pape reported on the special TBI chamber being built at City of Hope, California. The chamber was 11 feet square with 12-foot deep wells at each corner. In each well, a steel rod tipped with a 300 Ci Cs-137 source could rise up to the unshielded treatment position to treat the patient lying between the four sources. Also in the early 1960s, Cunningham, at the Ontario Cancer Institute, reported on a Co-60 unit mounted on a track on the ceiling that could scan the length of the patient at about 120 cm source-to-surface distance (SSD) (Figure 3.1). Translating the table could provide constant velocity, variable velocity, and constant velocity with variable field size. It was noted that by scanning the field across the patient, the dependence of dose rate on distance was not by inverse-square but by an inverse linear function. Also, the percent depth dose (PDD) from a moving field was found to be greater than for a static field (Cunningham and Wright 1962). This work ushered in a number of other creative TBI facility designs, some of which used this moving source concept. Thomas, while at the Fred Hutchinson Cancer Research Center in Seattle, described their system of two track-mounted, mobile, parallel opposed Co-60 sources with specially designed collimators (Thomas et al. 1982). By the 1980s, dual source TBI systems were being used in several centers, most commonly with Co-60 sources but 4 MV linear accelerator sources were also employed (Lutz et al. 1983). These systems were convenient because the patient did not need to be turned from supine to prone. Other TBI-customized single source Co-60 units were being used as well (Leung et al. 1981).

For centers with limited space, systems employing a single source that could be scanned across the patient by translating the patient bed were developed. An early report by Quast in 1985 described a vertical Co-60 unit with a computer controlled constant velocity translating couch with a 1.8 m source to couch distance. Similar to the findings of Cunningham and Wright (1962), the dependence of the dose on distance was found to

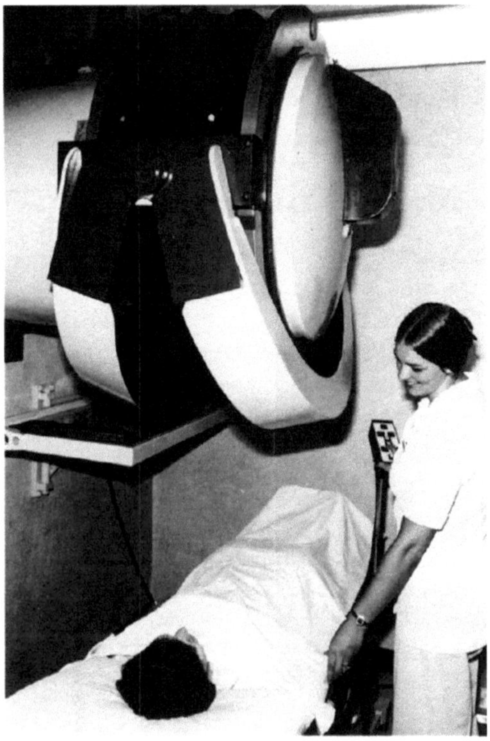

FIGURE 3.1 Co-60 unit designed for very large field treatments. (From Leung, P. M., et al., *International Journal of Radiation Oncology • Biology • Physics* 7 (6):705–12, 1981. With permission.)

vary by $1/(r^{1.4})$. Translation was shown to improve dose homogeneity for opposing beam TBI compared with stationary treatment. Lung blocks were mounted to a plate above the patient. Other implementations included placing lung blocks on the skin (Umek et al. 1996) or supporting lung blocks on a cradle that moves with the patient to lessen the smearing of the block shadow (Sarfaraz et al. 2001) (Figure 3.2). Variable speed versions of this system have also been described (Gerig et al. 1994; Chretien et al. 2000). The total dose delivered to the patient was a function of couch velocity, field size, and patient separation. Dynamic tissue maximum ratio (TMR) and PDD functions were needed in replacement for the conventional values. The variable speed allowed for the compensation of varying body thicknesses. Beam spoilers were used, and lung, liver, and kidney blocks were placed directly on the skin. The effect of these blocks was smeared over a several centimeter larger region than planned, both reducing shielding of the organ and introducing shielding of nearby tissues. This problem has been studied with solutions ranging from increasing the block dimensions (Gerig et al. 1994; Lavallee et al. 2008), moving the blocks as the patient moves (Papież et al. 1999), or using intensity-modulated radiation therapy (IMRT) techniques (Brown et al. 2010; Hussain et al. 2011). A comparison of the 3D

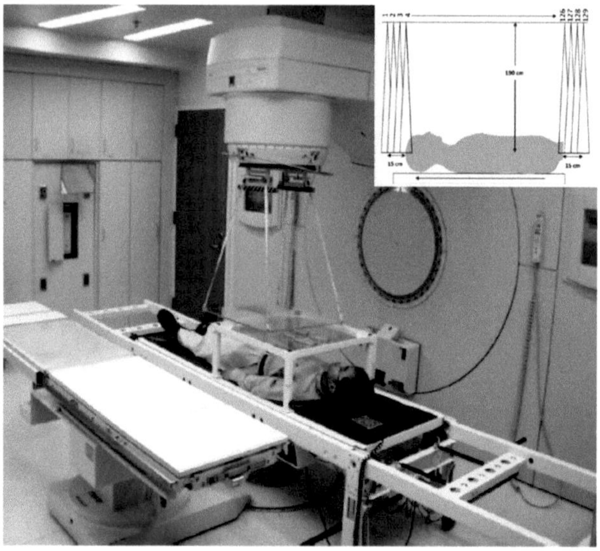

FIGURE 3.2 Translating couch technique for TBI with moving lung blocks. (After Sarfaraz, M., et al., *Journal of Applied Clinical Medical Physics* 2 (4):201–9, 2001. Inset from Hussain, A., J. E. Villarreal-Barajas, P. Dunscombe, and D. W. Brown, *Medical Physics* 38 (2):932–41, 2011. With permission.)

dose distributions from constant speed and variable speed couch translation showed that variable speed provided the best dose uniformity (Lavallee et al. 2011). For the translating bed IMRT technique, a vertical beam is shaped by the multileaf collimator (MLC) based on the missing tissue and inhomogeneities seen on the planning CT. In Hussain's technique, 129 AP–PA (anteroposterior–posteroanterior) aperture strips at 205 cm SSD are created each with their own MLC shape. The delivery of dose through each strip is synchronized with the longitudinal translating bed motion. This avoids the need for physical compensators and the associated penumbral issues. This method produced a more homogeneous dose than fixed open beam translating bed techniques, <5% versus 10% at midline, and lung dose reduction by 15% (instantaneous dose rate was 46 cGy/min). The method was also relatively insensitive to 2 cm setup errors (Hussain et al. 2011). Translating bed systems with dynamic MLC beam shaping have also been described (Brown et al. 2010).

A somewhat more practical method of treating at relatively short SSDs than translating the couch was sweeping the beam across the patient by rotating the treatment head (swivel) or the gantry. Pla (1983) described a column-mounted 4MV accelerator that could be programmed to continuously swivel the head back and forth. The patient lays supine and then prone for this treatment. Lead lung shields were placed directly on the patient's skin. With an SSD of 190 cm, the dose rate was 21 cGy/min. The variation of the midplane dose was less than ±5% for parallel-opposed beams. A similar approach with a Co-60 machine has been used, although a complex field flattener was necessary (Hussein and el-Khatib 1995). Using gantry rotation to sweep a beam across a patient lying on the floor, a gravity-oriented triangular shaped compensator has been used to make the beam more uniform. The 50 to 60 degree arc field covered up to a 180 cm long patient. TMR values were measured as lower and the penumbra in the superior-inferior direction was larger than for fixed fields. Lung blocks were used in addition to the compensator. Dose profiles were made uniform within 5% with this method (Chui et al. 1997). This group has subsequently replaced this method with intensity-modulated arcs, which both homogenize the dose profile and allow for lung dose compensation (Keane et al. 2000).

Total body compensation for opposed lateral fields was obtained by the use of compensator molds made of Styrofoam filled with tin granules. The compensator was designed based on a total body planning CT. The compensator was mounted on the couch and the patient was treated at an SSD of 390 cm in the supine position (Schneider et al. 2008).

Beginning in the mid-2000s, fractionated intensity modulated arc therapy methods were being developed and used in clinical trials for TBI and total marrow irradiation (TMI) treatments. Here, a new paradigm was being tested: Can normal organs be maximally spared in the case of TBI, and for TMI, can just the bone marrow be targeted, thereby allowing substantial sparing of all other tissues? The treatment dose rates of 4 to 8 Gy/min are 2 orders of magnitude greater than conventional TBI. Sparing of organs at risk (OARs) is achieved by dramatically lowering their total dose, which is hypothesized to compensate, along with fractionation, for the high dose rate. For TBI, the risk of underdosing leukemic cells in the circulation or in nonmarrow sites such as the skin, lymphatics, and other protected tissues is real and is being studied. However, patients with detectable malignant cells in the circulation are not candidates for this treatment (Hui et al. 2005). It has been estimated that the peripheral blood contains about 3% of the total lymphocyte pool (Field et al. 1972). Assuming there were leukemic cells in circulation, calculations of the risk of missing these cells when treating with a sequential treatment like helical TomoTherapy (HT) were performed showing that dose heterogeneity in circulating blood cells was clinically acceptable (Molloy 2010).

The initial studies utilizing HT focused on the dosimetric variations due to varying pitch, field width, and modulation factor. Protected OARs included the lungs, eyes, heart, liver, and kidneys. The CTV was either the whole body, excluding OARs (TBI) or the active bone marrow sites (outer edges of bones were contoured) throughout the body (TMI). The planning target volume (PTV) margins were 4 mm for TBI and 0 to 4 mm for TMI. A 1 cm margin outside the body was also used for TBI cases to account for setup error. The prescribed dose was a typical fractionated dose, 13.2 Gy in

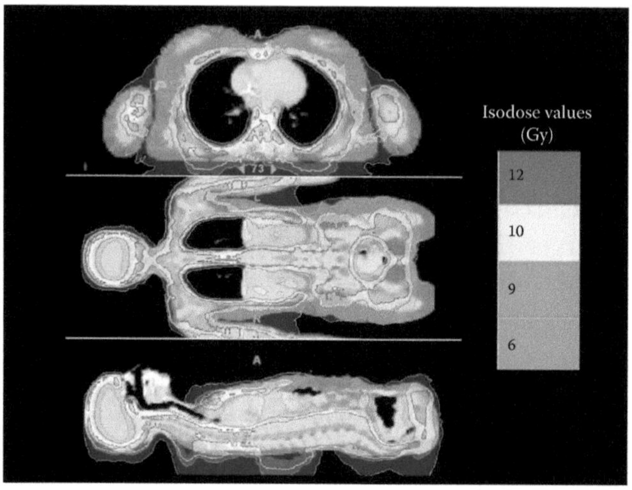

FIGURE 3.3 (See color insert.) Color wash of a TMI TomoTherapy plan to a prescribed dose of 12 Gy in a 20-year-old adult female. Relative sparing of dose to brain, oral cavity, thyroid, lungs, heart, soft tissue, and gastrointestinal tract is seen. (From Wong, J. Y. C., A. Liu, et al., *Biology of Blood and Marrow Transplantation* 12 (3):306–15, 2006. With permission.)

8 fractions, with planning goals of coverage of the PTV by the 95% isodose and ±10% dose homogeneity across the body. Vac-Lock total body and thermoplastic face mask immobilization was used, and kilovoltage CT (kVCT) to megavoltage (MVCT) image fusions were used to define volumes. HT treats from the head down to the mid femur and AP and PA parallel-opposed fields continue to the feet. Smaller field width (2.5 cm versus 5.0 cm), and greater modulation, improved the dose–volume histogram (DVH) modestly. Changing pitch weakly altered dose coverage. A field width of 2.5 to 5.0 cm, pitch of 0.3 to 0.45, and modulation factor of 2.0 to 3.0 were used for treatment, resulting in a beam-on time of approximately 16 to 50 minutes. About 85% of the CTV received the prescribed dose. It was found that doses to the sensitive organs were reduced by 35% to 85% of the target dose (Figure 3.3). This reduction is hypothesized to allow an increase in target dose up to 16 to 20 Gy, which should theoretically reduce the relapse rate without increasing toxicity above rates seen with conventional treatments (Hui et al. 2005; Wong et al. 2006; Schultheiss et al. 2007; Wong et al. 2009). HT has also been used as a boost after 12 Gy BID conventional treatments (Corvo et al. 2011). In a dosimetric comparison between conventional extended SSD TBI treatment methods and HT, 90% of the target volume received just 10 Gy for conventional, versus 12 Gy for HT. Lung doses were 9 Gy and 5 Gy, respectively (Figure 3.4). Planning time was about the same but HT delivery time was 40% longer (Zhuang et al. 2010). Because of the relatively long treatment times with HT, patient motion during treatment is a concern. Whole body optical motion tracking systems to monitor this motion have been described

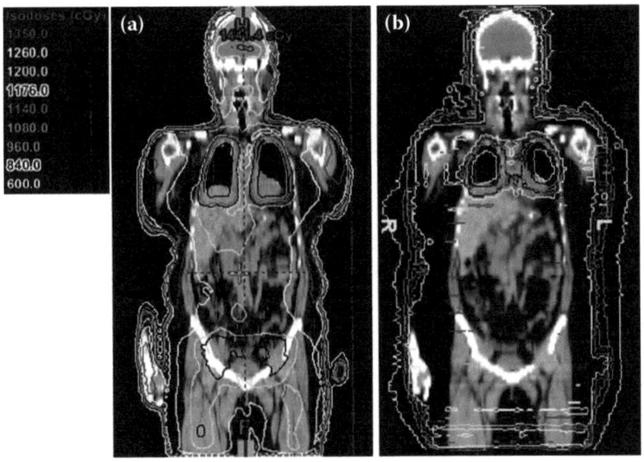

FIGURE 3.4 (See color insert.) TBI dose distribution for (a) a conventional extended distance plan and (b) an HT plan for the same patient. (From Zhuang, A. H., et al., *Medical Dosimetry* 35 (4):243–9, 2010. With permission.)

(Sharma et al. 2011). Adaptive planning and dose reconstruction methods have been applied to HT TBI to account for interfraction patient position changes using deformable image registration. In four cases, less than 3% differences in dose were found for six fraction treatments, but the technique could be used to adapt the plan if larger differences were seen before the end of treatment (Chao et al. 2011).

Conventional linear accelerator-based IMRT TMI treatments using nine coplanar 6 MV IMRT fields have been described. Plans used 2 to 3 isocenters distributed longitudinally with dose feathering between each. Their dosimetric results were comparable to that of HT (Wilkie 2008; Yeginer 2011) (Figure 3.5). Total marrow irradiation using RapidArc (Varian Medical Systems, Palo Alto, California) with eight overlapping coaxial arcs has been reported. OAR sparing was consistent with that found using HT, achieving a mean lung dose of less than 7 Gy for a 12 Gy prescribed dose. The total beam-on time was about 13 minutes compared to about 50 minutes for HT (Fogliata et al. 2011).

3.6 Conventional TBI Treatment Methods

By the 1960s, TBI treatments were being planned and delivered with opposing megavoltage fields at extended distance (>3 m), either lateral or AP–PA. There was considerable research and development of treatment techniques through the 1980s and 1990s, which formed the basis for today's treatment technique. Shielding of the lungs, brain, and kidney is often employed. The collimator was typically rotated 45 degrees so that the patient's long dimension was aligned with the diagonal of the field, further increasing the useful

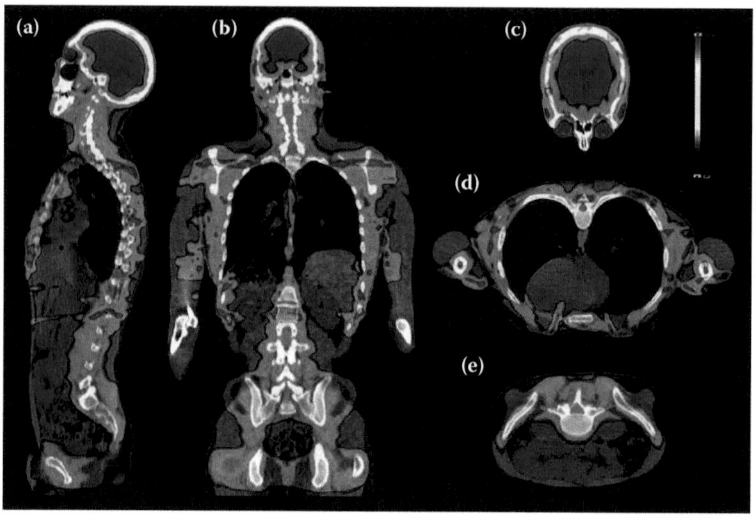

FIGURE 3.5 (See color insert.) Intensity modulated total marrow isodose distribution using a conventional linear accelerator, (a) sagittal, (b) coronal, (c–e) transverse planes. Doses range from 9.6 Gy (blue) to 14.4 Gy (red); the prescribed dose is 12 Gy. (From Yeginer, M., et al., *International Journal of Radiation Oncology • Biology • Physics* 79 (4):1256–65, 2011. With permission.)

dimension. A large plastic sheet placed in front of the patient (beam spoiler) was typically used for eliminating skin sparing, increasing the skin dose to match the tumor dose. The prescribed dose is at midplane at the umbilicus level or at the thickest part of the patient with a ±10% dose uniformity goal. The dose rate is generally ≤0.1 Gy/min at the prescription point. Beam energy ranges from Co-60 to 25 MV x-rays. Six MV x-rays are commonly used, having an acceptable PDD at an SSD of ≥3 m even for opposed lateral beams in children. This describes the vast majority of TBI treatments given in the past and continues to be the most common treatment method. Memorial Sloan Kettering popularized their approach, which used AP–PA beams with the patient standing in a cage, with lung blocks followed by an electron chestwall boost (Glasgow et al. 1989; Shank et al. 1990; Miralbell et al. 1994). They gave 13.2 Gy in 11 fractions, most at three times per day. The main driver in the decision to use opposing lateral or AP–PA beams is consideration of the lung dose and also patient comfort during a treatment that could take well over an hour. Opposed laterals are simpler and more comfortable, permitting the patient to sit in a chair or lay on a gurney with arms at the sides to provide lung compensation. The brain dose will be larger than the midpelvic dose due to the smaller thickness, so brain compensation is often used. AP–PA treatments are generally performed with the patient standing inside a booth with mounting plates in front for lung blocks and a beam spoiler and in back for holding a film cassette for verifying lung block positioning. The brain dose may be lower than prescribed, especially for children, because the maximum AP separation of the head is generally larger than at the umbilicus level. Lung dose is a key consideration in the design of any treatment method for TBI. It was well known by the 1960s that IP was a large threat to the patient's survival, should the radiation treatment and BMT render the patient in remission. Because the lung is only about one-third as dense as muscle, the dose needed at the pelvic midline would result in an approximately 25% higher dose in the lungs. Strategies to compensate this dose increase are based on whether the objective is to reduce the lung dose to equal the prescribed dose or to lower it further.

Opposed lateral beam techniques are generally used to maximize patient comfort and to compensate the lung dose to a value equal to the prescribed dose. The patient, seated in a chair with their arms at their sides, was seen as a built-in method of achieving this. However, studies have shown the pitfalls in this method. First, the upper arms do not cover the entire thickness of the chest. Second, the thickness of the arm may not be a very good match for the missing tissue due to the lung. Third, the arm inferior to the lung base will decrease the dose to the spleen and other tissues that could be harboring leukemic cells. A CT study with thermoluminescent dosimetry (TLD) dose verification showed that some of the bone marrow at the level of the manubrium was underdosed by more than 20% due to the presence of the shoulder, the thicker inferior sections of lung were undercompensated by the thinner portions of the arm, and the anterior areas outside the arm were overdosed by up to 30% (Hui et al. 2004) (Figure 3.6). Although most will not have this high an energy beam, it is interesting to note that a 24 MV x-ray beam not only improves homogeneity relative to 6 MV, but also increases the dose to bone marrow by 6% to 11% due to pair production and backscatter from the surrounding bone (Bradley et al. 1998). Lung blocks are not usually used for opposed lateral treatments because of the presence of the arm and shoulder already in the beam path through the lungs. Glasgow and Mill (1980) described a method for opposed laterals designed for small

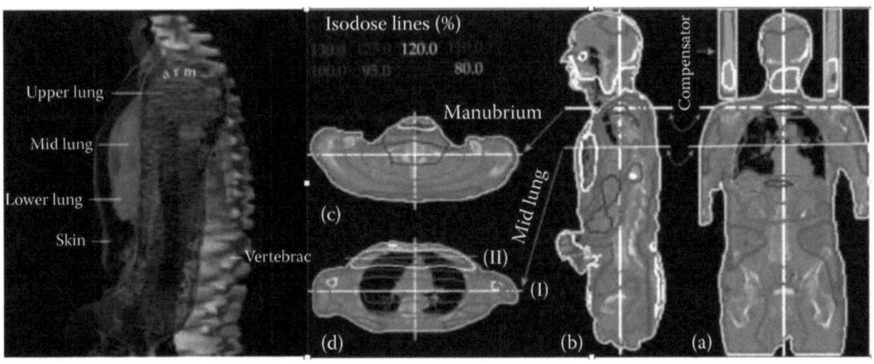

FIGURE 3.6 (See color insert.) Beam's eye view of one lateral field. (Left) Lung is in pink and arm is in blue. (Right) Isodoses at shoulder and chest level. (After Hui, S. K., et al., *Journal of Applied Clinical Medical Physics* 5 (4):71–79, 2004.)

treatment rooms where the patient sits in a chairlike structure angled along the edge of the floor, with the gantry angled downward toward it. With this method, the SSD is greater than can be achieved with a true lateral.

For AP–PA treatments, a special TBI stand has been used as shown in Figure 3.7. The patient is asked to partially sit on a small seat and hold on to rails but patient stability is still a concern. This method is problematic for children and untenable for sedated patients. The ability to mount partial transmission lung blocks on a tray in front of the patient allows for reduction of lung dose below the prescription. But this beam attenuation also decreases the dose to the ribs, which of course contains bone marrow, the main target of the treatment. In some cases, this underdosage was ignored, but more commonly, an electron beam boost was given to the chest wall in the standard geometry with the patient lying on the treatment couch. A typical dose was 600 cGy in two fractions. These AP and PA fields were shaped by the lung block shapes and a dose to the ribs was delivered that was calculated to replace the missing dose due to the lung blocks. About 1 Gy of additional lung dose was delivered with these electron beams as they penetrated beyond the ribs. For AP–PA, an alternative patient position is decubitus, with the patient lying on a gurney on their side. A thermoplastic shell can be used to reproduce the patient position and partial transmission lung blocks

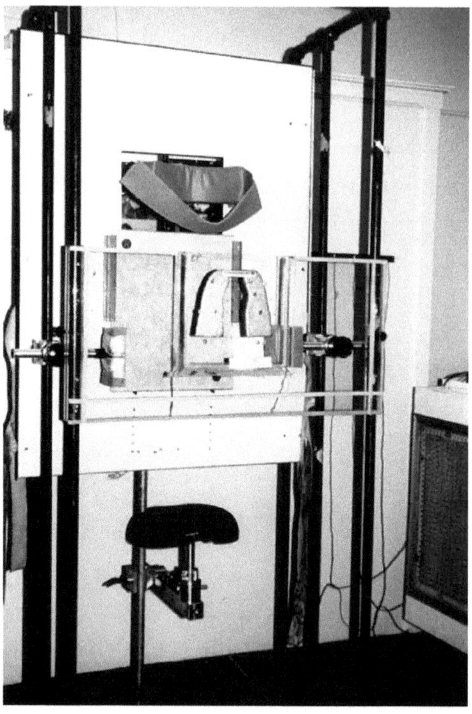

FIGURE 3.7 TBI stand for AP–PA treatments. Bicycle seat, lung blocks, and film holder can be seen. (From Ho, A., S. Kishel, and G. Proulx, *Medical Dosimetry* 23 (4):299–301, 1998. With permission.)

can be attached to the shell. Dosimetric studies of AP–PA versus opposed lateral beams using energies ranging from Co-60 to 25 MV x-rays have been made. Measurements and calculations were made with each technique in a Rando phantom at the levels of the head, chest, and pelvis. It was concluded that AP–PA was more homogeneous than opposed laterals (Svensson et al. 1980; Van Dyk et al. 1986). Differences in the dose homogeneity of the techniques diminish as the beam energy increases.

3.7 Conventional TBI Commissioning and Dosimetry Requirements

There are several good references for starting up a TBI program (Glasgow et al. 1980; Van Dyk et al. 1986; Briot et al. 1990; Niroomand-Rad 1991; Smith et al. 1996; Fiorino et al. 2000). The methods used by the Radiologic Physics Center (RPC) to verify TBI dose at institutions participating in clinical trials were described and the RPC found similar dose agreement for TBI setups as for standard treatment setups, with a maximum difference of 4% in 16 institutions (Kirby et al. 1988). Although many are more than 20 years old, these references describe physics principles and dosimetric properties that have not changed. These and others should be used to supplement the guidance given below. It is assumed that a 40 cm × 40 cm field size, collimator angle of 45 degrees, and SSD of greater than 3 m will be used for all patients.

Measurements needed (in the TBI geometry) to commission a TBI technique include:

1. Dose/monitor unit (MU) at dmax (depth of maximum dose) or dose specification point—A 0.6 cc ionization chamber in a 30 cm × 30 cm × 30 cm solid water phantom can be used to measure this value. Most patients have an equivalent square area less than 30^2 cm^2. The chamber can be placed at dmax or at a depth of 10 cm for this calibration. The phantom should be set up relative to the beam exactly as a patient would be treated. At a distance of about 3 m, the output will be reduced by about a factor of 10, so 1000 MU provides a reasonable chamber response. The chamber reading can be converted to dose using TG-51 methods or by intercomparison to readings at the standard distance. Because a low dose rate will be used, perhaps 100 MU/minute instead of 600 MU/min for other treatments, the dose rate measured by the chamber will be low as well, about 10 to 15 cGy/min. Although ion recombination is not usually a problem at these rates, currents produced in the long length of irradiated chamber cable could be large enough to cause a 1% to 2% error (Van Dyk et al. 1986; Fiorino et al. 2000). The induced cable current is due to a net removal of charge due to dose buildup within

the cable. The effect increases with increasing energy and has been found to be proportional to the length of irradiated cable. The effect can be reduced by a factor of 20 by wrapping the cable in a material that fully builds up the dose to the cable for the beam energy used. Minimizing the length of irradiated cable will also reduce the effect. One should also avoid using very small volume chambers where a small chamber measurement signal will be more greatly affected by charge induced by the cable. The inverse square law should not be used to calculate the output at the TBI distance based on the output from the standard distance. Scatter from the walls and floor in the TBI geometry can cause large percentage differences from the calculated value. For patient MU calculations, the phantom scatter factor for the equivalent square area of the patient will need to be ratioed to that of the phantom to correct the calibrated dose rate.

2. Dose profile at dmax and 10 cm depth across the diagonal of the field—Because flattening filters were not designed to flatten the beam along the diagonals, the dose profile at extended distance over the length of the diagonal is highly nonuniform and changes with depth as beam softening increases peripherally (Glasgow et al. 1980; Svensson et al. 1980; Briot et al. 1990). Measurements of the beam profile can be made at points with an ion chamber positioned at relevant locations inside the 30 cm × 30 cm × 30 cm phantom, varying depth and longitudinal position, as it is moved across the field. As an alternative, several sheets of radiochromic film can be used in the movable phantom or, if available, inside a much larger phantom that spans the entire diagonal of the field. The distance from the light field edge to the 90% isodose at dmax is 8 to 15 cm for linear accelerator beams and over 30 cm for Co-60 beams (Svensson et al. 1980; Briot et al. 1990; Niroomand-Rad 1991).
3. PDD or TMR—PDD or TMR should be measured in the TBI treatment geometry. When these data measured at standard distance are transformed by calculation to the extended distance, the calculated values can be 1% to 3% higher than measured (Van Dyk et al. 1986; Niroomand-Rad 1991; Smith et al. 1996). Measurements can be made in a water phantom or solid water. Off-axis PDD should be measured at intervals (at least every 20 cm) from the central axis toward the edge of the field to quantify the effects of beam softening. The PDD for a 6 MV x-ray beam at 3 m is nearly that of a 25 MV beam at 1 m due to reduced dose fall off by inverse-square for the TBI treatment. For a 36 cm thick patient, opposed 6 MV beams would result in dmax doses within 10% of the midline dose, equivalent to treatment with a 25 MV beam at 1 m. Once the PDD at the extended SSD and phantom geometry is measured, to obtain the PDD for the patient, a

correction needs to be made to account for the different equivalent square area and thus the different scattering conditions. The central axis PDD is not very sensitive to the range of equivalent square areas encountered with TBI and monitor unit calculators can generally do a good job of calculating the PDD for various irradiated areas at extended SSD. A verification of the absolute dose at dmax, 10 cm, 20 cm, and 30 cm depths in a 30 cm × 30 cm × 30 cm phantom in the TBI geometry with the beam spoiler should be performed to assess the total deviation of dose on the central axis when compared to calculations.

4. Design of blocks used for shielding critical structures—Shielding blocks can be placed either on a blocking tray in the accessory slot, on a tray at a short distance from the patient, or directly on the skin surface. Blocks placed in the blocking tray will be smaller but will create greater geometric error with patient movement and have a larger penumbra than those placed close to the patient. Typically, full thickness blocks are not used; instead, partial transmission blocks are used with a transmission factor of 50% to 80%, requiring 0.5 to 2.0 cm thick lead or cerrobend. Attenuation measurements must be made for the block thicknesses to be used, again, in the TBI geometry (Ho et al. 1998). Attenuation measurements are usually made in narrow beam geometry, but data suggests that broad beam (TBI) geometry can result in 20% greater transmission, requiring an additional 2 mm lead thickness. Broad beam attenuation coefficients have been found to vary linearly with field size (Galvin et al. 1980; Van Dyk et al. 1986; el-Khatib and Valcourt 1989). Also, due to beam softening with distance from central axis, attenuation per centimeter of lead will be greater for shielding of the brain and other peripheral locations. Another consideration is that the dose under the block will increase with depth in the patient and will vary with the tissue densities under the block. To fully characterize the attenuation of lead shielding blocks, measurements of dose under varying size blocks for on and off axis locations and relevant depths should be made. For compensation only of differences in body thicknesses for bilateral treatments, rice, water bags, or other similar material has been laid around the supine patient, often framed by a box for consistent and equalized compensator placement (Lin et al. 2001).
5. Beam spoiler dosimetric effects, surface and exit dose—The surface dose for a 40 cm × 40 cm 6 MV x-ray beam at extended distance is much higher than at standard distance, about 70% compared to just 45%, but not high enough to satisfy the ±10% dose uniformity requirement. For TBI, the surface dose for 18 MV is about 60%. At least an additional 0.5 cm of buildup for 6 MV and 1 cm for 18 MV

x-rays increases the surface dose to greater than 95% (Briot et al. 1990; Niroomand-Rad 1991). This is generally achieved by placing a 3 × 6 foot plastic sheet (beam spoiler) of the required thickness no more than about 30 cm in front of the patient (Planskoy, Bedford, et al. 1996). The surface dose increases by about 5% for every 10 cm closer the beam spoiler is to the patient (Briot et al. 1990). With opposing fields and use of a beam spoiler, the surface dose can be greater than the midline dose, so 100% surface dose from each beam is not necessary to achieve the dose uniformity requirement. The beam spoiler should be present during the output calibration measurement so that its attenuation effect can be included. The exit skin dose from each beam is less than the exit dmax dose due to lack of backscatter. The exit dose depends on the distance from the wall to the patient and the relative size of the beam and patient, and is typically 5% to 10% higher for TBI than treatment at the standard distance. The exit skin dose for one TBI field is 90% to 97% of the exit dmax dose for 6 to 25 MV beams and is 100% if 1 cm of backscatter is added (Glasgow et al. 1980; Briot et al. 1990; Planskoy, Bedford, et al. 1996). With summation of opposing field doses with the beam spoiler but without backscatter added, the surface dose on each side will be about 95% of the midline dose.

3.8 Treatment Planning and Verification

3.8.1 TBI

Monitor unit calculations and shielding block design are performed based on a current CT or caliper measurements of relevant body sections. Most commonly, the dose is prescribed at the midplane at the level of the umbilicus so a separation at that level is necessary. For shielding design calculations, other separations that may be needed are at the level of the chest, brain, and kidney. An equivalent square area of the patient can be used to correct the dose/MU for the phantom used for the calibration measurement. For AP–PA treatments, measure the patient width, for opposed laterals measure the AP separation, and in both cases, measure the length of the patient from the neck to the pelvis to calculate the irradiated equivalent square area. The ratio of phantom scatter factor (Sp; based on the equivalent square area) for the patient and for the calibration geometry corrects the calibration dose/MU.

One formulation of the calculation for monitor units for an SSD treatment is

$$\text{MU} = \text{Rx dose}/[(\text{PDD}_{d,s}) \times (\text{dose/MU}_c) \times (\text{Sp}_p/\text{Sp}_c) \times \text{SF}]$$

where Rx dose is the prescribed dose, $\text{PDD}_{d,s}$ is the TBI PDD for the depth and side of the patient's equivalent square area, dose/MU_c is the TBI calibration output, (Sp_p/Sp_c) is the ratio of the Sp for the patient and the calibration phantom, and SF is the beam spoiler transmission factor. By always using the same beam energy, collimator size and angle, SSD, and position of the patient relative to the room, the calculation is greatly simplified (Curran et al. 1989). The dose to specific regions of the body can be more accurately calculated by considering the equivalent square area of the region (Kirby et al. 1988).

Calculations for the shielding of the lung and other critical structures can be performed in a number of ways. For lung, these include manual calculations based on dose correction factors related to caliper measurements of the separation across the chest, or CT-based methods (Galvin et al. 1980; Glasgow et al. 1980; Van Dyk et al. 1986; el-Khatib and Valcourt 1989; Hussein and Kennelly 1996; Ho et al. 1998; Mangili et al. 1999). Without a CT scan, one can approximate the lung density for a given patient based on their age. It has been shown that the average density of lung decreases linearly with age, with the average density at 5 and 80 years old being 0.35 and 0.19 g/cm^3, respectively (Van Dyk et al. 1982). Taking the lung thickness as a percentage of the chest separation or using lung correction factors as a function of patient chest thickness is less accurate and should only be used if the patient CT data is not available (Van Dyk et al. 1986; Ho et al. 1998). To be most accurate, the density and thickness of the patient's lungs should be determined from a recent CT scan. The density of lung can be determined from the CT pixel information converted to relative electron density (by use of the calibration curve in the planning system) or can be assumed based on the patient's age. The average physical thickness of lung can be found most accurately from measurements using the patient's recent CT scan. An alternate patient-specific determination of missing tissue due to the lung is to use the exit dose information from a radiograph taken in the TBI geometry (Hussein and Kennelly 1996).

Dose correction factors for lung were found to vary by up to 20% depending on lung thickness and whether linear attenuation or equivalent TAR calculation methods were used (Van Dyk et al. 1986). For the lateral field technique, a procedure for compensation for the thinner body parts, head, neck, legs, and feet has been described. In the method of Galvin et al. (1980), with the patient in the treatment position, the outline of the patient is marked on the clear plastic tray in the accessory mount of the linear accelerator by viewing the shadow on the patient as one draws. Lead strips, whose thickness has been calculated for each section to be shielded, are attached to the blocking tray and are positioned based on the outline (Galvin et al. 1980; Liu et al. 1983). Other common methods used with AP–PA treatments employ cerrobend blocks mounted to a blocking tray placed within about 30 cm of the patient. Although block placement for both of these methods can be verified radiographically, these methods are susceptible to errors due to patient motion during treatment. In consideration of this, fairly large margins between the lung and block edges have been used; 2.5 cm margin from lateral rib cage and vertebral bodies and 3.5 cm from diaphragm and lung apex (Miralbell et al. 1994).

Kidney shielding is also performed for patients treated in the semistanding position using AP–PA fields. However, the blocks are designed using supine position CT scans. In one study, intravenous contrast was administered before either lying supine or placed in a seated position on a conventional simulator. The image intensifier was used to record radiographs in both positions and the changes in kidney position and size relative to the vertebral bodies were determined. From supine to upright patient position, kidneys were found to shift inferiorly by an average of 3.6 cm (0.5 to 7.5 cm range), and laterally by 0.9 cm to 4.9 cm. The range of width changes of the kidney was a broadening by 1.2 cm to a decrease in width by 1.8 cm (Reiff et al. 1999).

One concern with shielding any organ during TBI is that neoplastic stem cells will also be shielded, potentially causing treatment failure. In one study, a significantly higher rate of relapse was found when both the lungs and right lobe of the liver were shielded. The conclusion was that liver shielding was to blame since the liver was more likely to harbor neoplastic cells and increased relapse had not been reported by those centers shielding the lungs and not the liver (Anderson et al. 2001).

3.8.2 Cranial Irradiation

The field design for whole cranial irradiation, either prophylactically or for known CNS leukemia is the same as for medulloblastoma or other CNS diseases with dissemination, except that the posterior half of the eyes are also included in the fields. See Chapter 4 for a description of these fields.

3.8.3 Dosimetric Verification

Because of the large differences between physical factors in the TBI and standard geometry and their dependence on patient-specific features, uncertainty in the actual attenuation from shielding blocks and compensators, and need for detection of setup errors, *in vivo* measurements are typically made at a number of beam entrance and exit locations (head, chest, abdomen, pelvis, and legs) on the patient during the first fraction. These measurements are used to fine-tune the MUs or compensator and shield thicknesses. In addition, if compensation or shielding is used, films are taken to verify the correct placement. The same film that is used to verify lung shielding can be used to verify correct placement of detectors at the entrance and exit surfaces of the chest. The entrance and exit doses recorded by the *in vivo* dosimeters are used to infer the midline dose. It is the midline dose that is required to be within 10% of the prescribed dose. For some locations, the detector will be on the skin with just full buildup, whereas for locations with tissue-equivalent compensation, such as the lung and brain, the detector may be on the skin but with the compensation material overlying it or may be placed on the surface of compensation. If detectors are always placed on the skin surface, both entrance and exit doses under the compensation can be evaluated. One method of converting the entrance and exit doses to midline dose is (Ribas et al. 1998)

$$D_{mid} = [(D_{exit} + D_{entrance})/2] \times CF,$$

where CF is a correction factor given by

$$CF = PDD_{midline}/[(PDD_{exit} + PDD_{entrance})/2]$$

The PDD values are for the same depth as the corresponding measurement. Typically, CF is <1 because for the large separations encountered for TBI, especially for lateral opposed treatments, the midline dose accumulated from both fields will be smaller than the dmax dose. Behind lung blocks, the PDD_{exit} will need to be based on the water equivalent depth with an inverse square correction. This formulation can predict midline doses from entrance and exit doses to within 3% of the measured midline dose.

TLDs have been used by many centers for TBI dose verification (Galvin et al. 1980; Liu et al. 1983; Kirby et al. 1988; Sánchez-Doblado et al. 1995; Planskoy, Tapper, et al. 1996; Hui et al. 2004). Because TLDs are time-consuming to use, difficult to use accurately, and require a time delay between irradiation and readout, real-time readout detectors such as diodes and MOSFETs (metal-oxide-semiconductor field-effect transistors) have been used (Bloemen-van Gurp et al. 2007), both with acceptable accuracy. Alanine-electron paramagnetic resonance detectors have also been used (Schaeken et al. 2010). More recently, optically stimulated luminescent detectors (OSLDs) have begun to be used, taking advantage of their relative ease of use, nearly tissue equivalent composition, accuracy, and cost effectiveness (Jursinic 2007; Viamonte et al. 2008; Yukihara et al. 2008; Jursinic and Yahnke 2011; Kerns et al. 2011). Each of these detectors has pros and cons that must be considered before their use.

Diode detectors are the most commonly used alternative to TLD. Diodes have been used successfully for many years but have to be understood to be used accurately. Diodes offer real-time readout so that treatment corrections can be made sooner than with TLD. However, unlike TLD, they are not nearly tissue equivalent and, depending on the type, may overrespond by several percent to low energy scattered radiation, an effect that worsens with field size and depth. They also tend to have a temperature dependence that requires about a 1% dose correction, an angular dependence that can amount to about 3% for greater than 60 degree angle of incidence, and an instantaneous dose rate dependence of about 1% across the range of pulse repetition rates found on most linear accelerators (Planskoy, Bedford, et al. 1996). With proper calibration in the TBI geometry, diodes have generally been found to provide dose accuracy comparable to TLD (Torrisi et al. 1990; Sánchez-Doblado et al. 1995; Ribas et al. 1998; Mangili et al. 1999). TLD and diode measurements and their uncertainty have been comprehensively described in a study of 84 patients with over 1000 dose measurements (Planskoy, Tapper, et al. 1996). For diodes used on 60 patients with 360 measurements, the standard deviation of the ratio between calculated and measured exit doses have been found to be larger than for entrance doses, around 4% versus 1%, respectively, leading to setting an action level of 5% for entrance and midplane, and 10% for exit doses (Ribas et al. 1998).

Those who have studied the degree of agreement between *in vivo* measurements and doses predicted from calculations generally find good agreement but have reported 10% deviations for lung dose in a significant proportion of patients (Briot et al. 1990). Mangili et al. (1999) reported that about 45% of TBI patients needed at least a minor compensator or MU correction based on *in vivo* dosimetry on the first fraction. It should be noted that a potential large source for differences between *in vivo* measurements and predicted doses is positioning of the detectors, especially when measuring exit dose through lung. Based on CT, for a patient in the lateral position, the lung closest to the couch was found to be about 30% more dense than regions toward the ceiling and speculated for the standing patient, that the inferior lung would be denser than the superior lung for the same reason. There also was age dependence to this effect (Briot et al. 1990).

CT-based treatment planning for TBI has been investigated and found to offer increased accuracy over caliper-based manual calculations. Algorithms used in Eclipse (Varian Medical Systems, Palo Alto, California) (Mangili et al. 1999; Hussain et al. 2010) and Pinnacle (Elekta, Sunnyvale, California) (Hui et al. 2004; Lavallee et al. 2009; Schaeken et al. 2010; Lavallee et al. 2011) planning systems commissioned for standard treatment geometry have been tested using TBI geometry and have been sufficiently accurate to use clinically. Once verified, the treatment planning system (TPS) can be used for MU calculations, perhaps with correction factors, and for relative dose calculations that guide organ-shielding design. DVHs of the shielded organs have demonstrated that large volumes of lung, up to 40%, are not shielded when using standard lung block margin recommendations (Briot et al. 1990; Hui et al. 2004).

3.9 Organ-at-Risk Doses and Late Effects

As mentioned earlier, the most important factor driving the decision on TBI treatment method is the lung dose. In most cases, a secondary albeit important consideration is the dose to the brain and kidneys, followed by liver and lenses. Acute nausea and vomiting, skin rash, and erythema are expected among patients given TBI. Premedication with an antiemetic is mandatory. About half of patients experience transient fever/chills and oral mucositis (Chou et al. 1996).

3.9.1 Pulmonary Toxicity

The dose limiting toxicity for TBI in children as in adults is lung interstitial pneumonitis (IP), which can be fatal. A distinction is generally made between IP caused by virus or pathogens and cases where the cause is not known, called idiopathic pneumonitis, which is generally attributed to lung damage due to radiation. Reports indicate that idiopathic IP represents less than half all IP (Keane et al. 1981). The incidence of pneumonitis is strongly associated with dose per fraction and total dose, a finding that motivated the early research and patient trials on the optimal treatment schedule. Other factors are also important, such as type of chemotherapy, older age, presence of graft versus host disease (GVHD), allogenic versus autologous (lower rate of IP) transplant, and performance status prior to transplant (Pino y Torres et al. 1982; Weiner et al. 1986; Ozsahin et al. 1992; Kelsey et al. 2011). Based on CT-aided dosimetry with lung density corrections, Van Dyk et al. (1981) demonstrated that the onset of IP occurred at about 8 Gy given in a single high-dose rate fraction without chemotherapy and reaches 100% incidence at about 11 Gy. The treatment paradigm of the 1970s to 1980s was a single 10 Gy fraction with a low dose rate of about 0.05 Gy/min to limit lung toxicity. With chemotherapy, probit curves of lung density corrected doses versus incidence of idiopathic IP after TBI and allogeneic transplants from several TBI centers show that for low dose rates, 0.03 to 0.15 Gy/min, the total single fraction lung dose for a 5% incidence was about 10 Gy, an improvement over the 8 Gy high dose rate limit. However, due to the steepness of the dose–response curve and the lack of accounting for corrected lung dose, many centers were reporting unacceptably high rates of IP. Lower lung doses (4.5–9.0 Gy at 0.03–0.50 Gy/min) resulted in a dramatic reduction in fatal idiopathic IP (Keane et al. 1981). However, 10 Gy at 0.125 Gy/min with lung shielding to limit the lung dose to 6 Gy resulted in a 25% recurrence rate compared to 0% for an 8 Gy lung dose. Presumably, this difference was due to shielding of bone marrow in the thoracic wall where 2% to 3% of the bone marrow resides (Labar et al.

1992; Girinsky et al. 1994). These observations led to the fractionation of TBI treatments with a compensatory increase in prescribed dose and for many, the goal of limiting the lung dose to between 8 and 12 Gy.

The benefits of fractionation over single-fraction TBI were reported in numerous studies, where a reduction of the rate of IP was 50% or more (Pino y Torres et al. 1982; Vitale et al. 1983; Blume et al. 1987; Shank et al. 1990). Results vary as to whether partial shielding of the lungs to produce a lung dose of about 9 Gy during fractionated TBI lowers the incidence of IP in comparison to an unshielded dose of 12 Gy (Weshler et al. 1990; Oya et al. 2006; Soule et al. 2007; Schneider et al. 2008). In a randomized study in adults, 10 Gy single-fraction TBI given at 0.125 Gy/min with a lung dose of 8 Gy was compared to 14.85 Gy hyperfractionated TBI given at 0.25 Gy/min with a lung dose of 9 Gy. Survival and IP rates were the same, indicating that as long as the prescribed dose was high enough and the lung dose remained low, fractionation and dose rate do not impact on IP rate (Girinsky et al. 2000). Omission of electron boosts of the chest wall after lung shielding has been shown to still result in acceptable engraftment and IP rates (Weshler et al. 1990; Miralbell et al. 1994).

Although most of the aforementioned data are for adults, some pediatric data exists. In children treated to a total TBI dose of 12 Gy in 2 Gy fractions without lung shielding and a dose rate of 0.15 Gy/min, the incidence of IP was 22% to 25% (Chou et al. 1996; Schneider et al. 2008). With the same prescribed dose, dose rate, and fractionation, the IP rate in children was reduced to 4% for an 11 Gy lung dose (Schneider et al. 2008). The literature is mixed as to whether children have a lower (Shank et al. 1990; Granena et al. 1993; Morgan et al. 1996) versus the same (Schneider et al. 2008) IP rate as adults for the same TBI regimen. Shank et al. (1990) reported a 4% fatal IP rate in children for their 15 Gy in 12 fraction 9 Gy lung dose TBI regimen using up to 0.19 Gy/min.

There are conflicting data as to whether dose rate has an influence on the development of IP for fractionated treatments. For 12 Gy in 6 fractions twice per day and lung doses between 9 and 12 Gy, studies have reported a reduced IP rate for dose rates <0.04 Gy/min (Beyzadeoglu et al. 2004), or for 0.075 Gy/min vs. 0.15 Gy/min (Carruthers and Wallington 2004), but others have reported no difference for dose rates between 0.03 and 0.09 Gy/min (Gogna et al. 1992) or 0.12 Gy/min versus 0.17 Gy min (Izawa et al. 2011). In a study by Oya et al. (2006) using 12 Gy in 6 fractions over 3 days, they reported no difference in IP rate, survival, or relapse rate for a lung dose of 8 Gy by shielding or lung dose of 12 Gy without shielding at either 0.08 Gy/min or 0.19 Gy/min. The overall IP rate was 20%. A retrospective review of 26 patient cohorts and 1090 total patients concluded that 12 Gy in 6 daily fractions resulted in about an 11% IP rate without lung shielding and no dose rate effect was seen (Sampath et al. 2005).

Clearly, the interplay between fractionation, lung dose, and dose rate are complex as they relate to the incidence of IP. Fractionation clearly decreases pulmonary toxicity but so does reducing lung dose. The additional benefit of reducing dose rate, especially below 0.2 Gy/min, is not clear.

3.9.2 Cognitive Impairment

The late effects of 18 to 24 Gy cranial irradiation combined with intravenous and intrathecal chemotherapy in children with ALL are well documented. Particular aspects of cognitive function most affected are attention, working memory, processing speed, reasoning, and visual spatial skills although 18 Gy cranial irradiation per se was not an independent toxic agent for cognitive outcome (Waber et al. 1995; Janzen and Spiegler 2008). Compromised function in these domains is associated with learning difficulties at school, particularly in mathematics. IQ scores tend to decrease serially over decades posttreatment (Schatz et al. 2000; Edelstein et al. 2011). These same deficits occur in ALL patients treated with chemotherapy alone but tend to be more pronounced with the addition of cranial irradiation (Moleski 2000; Nathan et al. 2006; Janzen and Spiegler 2008; Kadan-Lottick et al. 2010). The degree of cognitive impairment is positively correlated with the radiation dose and negatively correlated with the patient's age at the time of treatment, with significantly greater treatment effect in children under 5 years of age (Silber et al. 1992; Edelstein et al. 2011). The dose to the brain due to TBI alone can be either higher or lower than the TBI prescribed dose, higher for opposed lateral (lateral separation of the brain is smaller than abdomen) and lower for AP–PA (AP separation of the head is often larger than for the abdomen) treatments. For opposed lateral treatment, compensation is generally applied in the form of plastic blocks or slabs of bolus to reduce the dose to the brain so as not to exceed the prescribed dose.

3.9.3 Growth

Spinal irradiation is associated with acute nausea, vomiting, and anorexia due to exit dose to the gastrointestinal tract anterior to the spine. There is a dose-related acute and chronic suppression of bone marrow function from sacral and vertebral body irradiation. Because this may impact the patient's ability to receive chemotherapy that is essential to cure, irradiation of the spine is generally reserved for special situations of CNS relapse. Patient growth is unlikely to be significantly impacted by a total dose of 6 Gy but may be by doses of 12 Gy and above. TBI doses above 12 Gy can similarly affect growth and also impair tooth development and craniofacial growth in most young patients (Vesterbacka et al. 2012).

3.9.4 Endocrine Dysfunction

Growth hormone production by the pituitary gland is highly radiosensitive. Growth hormone deficiency was found following BMT in 34% of patients treated with TBI compared with none of the patients treated with chemotherapy alone (Felicetti et al. 2011). Patients given TBI or 24 Gy cranial irradiation have been demonstrated to achieve a shorter adult height. The impact of TBI on height is greater after single-dose than fractionated TBI (Tarbell et al. 1990; Schriock et al. 1991; Cohen et al. 1999). Hypothyroidism, either as a result of pituitary or thyroid dysfunction, occurs in 10% to 20% of leukemic children following cranial irradiation or fractionated TBI and in 45% following single-dose TBI (Barrett et al. 1987; Leung et al. 2000; Faraci et al. 2005). Primary hypothyroidism was found in 34% of children in whom TBI was included in the bone marrow transplant regimen, compared to in 6% of children transplanted without TBI (Felicetti et al. 2011).

3.9.5 Secondary Malignancy

TBI increases the risk of secondary solid tumors above that seen in patients given bone marrow transplant conditioning with chemotherapy alone (Curtis et al. 1997; Cohen et al. 2001). Secondary malignancy risk of 11% to 14% at 15 years following transplant have been reported (Socie et al. 2000; Pommier et al. 2009). The risk was 4 times higher in patients under 10 years old at the time of TBI than in patients between 10 and 17 years old. The excess incidence of secondary malignancies in young patients was mainly accounted for by thyroid and CNS malignancies. The majority of patients with secondary brain tumors had received cranial irradiation prior to TBI (Socie et al. 2000; Mody et al. 2008).

3.9.6 Fertility

Testicular irradiation to a dose of 24 Gy usually causes permanent sterility and often results in permanently decreased testosterone levels and delayed sexual maturation (Brauner et al. 1988). Among boys given TBI, all had azoospermia and 30% had delayed development of secondary sexual characteristics (Sanders 1990). The level of follicle stimulating hormone was found to be elevated in 100% of boys treated with TBI compared to in 36% of boys given BMT without irradiation (Felicetti et al. 2011). Chemotherapy for BMT, with or without TBI, may cause ovarian failure, although a number of successful pregnancies have occurred in women who underwent TBI either prior to or following puberty. Following TBI, 38% of prepubertal girls achieved menstruation and 29% of postpubertal girls recovered ovarian function (Sanders et al. 1996).

3.9.7 Cataract Formation

Henk et al. (1993) demonstrated that the dose that causes a 50% probability of visual impairment in adults is approximately 15 Gy when fractionated and delivered at a dose rate of about 1 Gy/min. Cataract induction is dependent on dose rate as well as fractionation. Cataracts developed in all 27 patients receiving 8 Gy single-fraction TBI delivered at 0.2 Gy/min (van Weel-Sipman et al. 1990), but when 10 Gy single-fraction TBI is delivered more slowly, at 0.04-0.08 Gy/min, the rate was 24% to 85%. When 12 Gy in six fractions was delivered at these same dose rates, the cataract rate was just 0% to 34% (Deeg et al. 1984; Ozsahin et al. 1994; Benyunes et al. 1995; Belkacemi et al. 1998; Zierhut et al. 2000; Thomas et al. 2001; Beyzadeoglu et al. 2002). Somewhat higher dose rates, up to about 0.15 Gy/min, seemed to be well tolerated in other studies (Chou et al. 1996; Schneider et al. 2008). The dependence of cataract formation on age is not clear (Fife et al. 1994; Belkacemi et al. 1998). Steroid usage has been associated with a higher cataract rate (Deeg et al. 1984; Benyunes et al. 1995; van Kempen-Harteveld et al. 2002). Shielding the eyes during TBI has been associated with local relapse (van Kempen-Harteveld et al. 2008) but also with decreased cataract incidence (van Kempen-Harteveld et al. 2003).

3.9.8 Kidney Dysfunction

Another potentially serious toxicity of TBI is chronic renal dysfunction, which has been reported in between 2% and 25% of children with ALL following fractionated TBI (Tarbell et al. 1990; Chou et al. 1996; Borg et al. 2002; Cheng et al. 2008; Esiashvili et al. 2009; Gerstein et al. 2009; Kal et al. 2009). Acute renal dysfunction can occur at much higher rates (Esiashvili et al. 2009). The renal dysfunction usually occurs a few months after TBI and frequently resolves within 1 year (Gerstein et al. 2009). In some cases, renal dysfunction can be chronic and require dialysis (Lawton et al. 1997; Cheng et al. 2008). Antifungal agents, antibiotics, GVHD, infection, and chemotherapy all play some role in nephrotoxicity (Miralbell et al. 1996; Lawton et al. 1997). Acute renal dysfunction after TBI was more common in children under 5 years of age than in older children in one study (Esiashvili et al. 2009), but no age relationship was found in a meta-analysis (Cheng et al. 2008). Whether there is a dose–response is unclear. There is data that indicates that risk increases as dose increases from 10 to 14 Gy and the recommendation was to limit the kidney dose to 10 Gy (Miralbell et al. 1996; Lawton et al. 1997; Kal et al. 2009). Other studies show no dose–response (Cheng et al. 2008; Esiashvili et al. 2009). Blocking the kidneys to 10 Gy did not reduce overall survival despite necessarily blocking other tissues (Igaki et al. 2005; Kal et al. 2009). Dose rate below 0.1 Gy/min was found protective, but no dose rate effect was seen above that rate (Cheng et al. 2008).

3.9.9 Liver Dysfunction

Veno-occlusive disease of the liver, in which fibrosis around collapsed veins leads to hepatic dysfunction, is reported in 13% to 23% of children given TBI for ALL. It also occurs in patients undergoing BMT without radiation, although less commonly. It is fatal in about one quarter of cases (Deeg et al. 1988; Chou et al. 1996; Schneider et al. 2008). Most TBI treatment techniques do not include liver blocking.

References

Alexander, B. M., D. Wechsler, T. M. Braun, et al. 2005. Utility of cranial boost in addition to total body irradiation in the treatment of high risk acute lymphoblastic leukemia. *International Journal of Radiation Oncology • Biology • Physics* 63 (4):1191–6.

Anderson, J. E., F. R. Appelbaum, G. Schoch, et al. 2001. Relapse after allogeneic bone marrow transplantation for refractory anemia is increased by shielding lungs and liver during total body irradiation. *Biology of Blood and Marrow Transplantation* 7 (3):163–70.

Bakst, R., S. Wolden, and J. Yahalom. 2012. Radiation therapy for chloroma (granulocytic sarcoma). *International Journal of Radiation Oncology • Biology • Physics* 82 (5):1816–22.

Barrett, A., J. Nicholls, and B. Gibson. 1987. Late effects of total body irradiation. *Radiotherapy and Oncology* 9 (2):131–5.

Belkacemi, Y., M. Labopin, J. P. Vernant, et al. 1998. Cataracts after total body irradiation and bone marrow transplantation in patients with acute leukemia in complete remission: A study of the European Group for Blood and Marrow Transplantation. *International Journal of Radiation Oncology • Biology • Physics* 41 (3):659–68.

Benyunes, M. C., K. M. Sullivan, H. J. Deeg, et al. 1995. Cataracts after bone marrow transplantation: Long-term follow-up of adults treated with fractionated total body irradiation. *International Journal of Radiation Oncology • Biology • Physics* 32 (3):661–70.

Beyzadeoglu, M., B. Dirican, K. Oysul, F. Arpaci, and Y. Pak. 2002. Evaluation of fractionated total body irradiation and dose rate on cataractogenesis in bone marrow transplantation. *Haematologia (Budap)* 32 (1):25–30.

Beyzadeoglu, M., K. Oysul, B. Dirican, et al. 2004. Effect of dose-rate and lung dose in total body irradiation on interstitial pneumonitis after bone marrow transplantation. *Tohoku Journal of Experimental Medicine* 202 (4):255–63.

Bloemen-van Gurp, E. J., B. J. Mijnheer, T. A. M. Verschueren, and P. Lambin. 2007. Total body irradiation, toward optimal individual delivery: Dose evaluation with metal oxide field effect transistors, thermoluminescence detectors, and a treatment planning system. *International Journal of Radiation Oncology • Biology • Physics* 69 (4):1297–1304.

Blume, K. G., S. J. Forman, D. S. Snyder, et al. 1987. Allogeneic bone marrow transplantation for acute lymphoblastic leukemia during first complete remission. *Transplantation* 43 (3):389–92.

Borg, M., T. Hughes, N. Horvath, M. Rice, and A. C. Thomas. 2002. Renal toxicity after total body irradiation. *International Journal of Radiation Oncology • Biology • Physics* 54 (4):1165–73.

Bradley, J., C. Reft, S. Goldman, et al. 1998. High-energy total body irradiation as preparation for bone marrow transplantation in leukemia patients: Treatment technique and related complications. *International Journal of Radiation Oncology • Biology • Physics* 40 (2):391–6.

Brauner, R., P. Caltabiano, R. Rappaport, G. Leverger, and G. Schaison. 1988. Leydig cell insufficiency after testicular irradiation for acute lymphoblastic leukemia. *Hormone Research* 30 (2–3):111–4.

Briot, E., A. Dutreix, and A. Bridier. 1990. Dosimetry for total body irradiation. *Radiotherapy and Oncology* 18 (Suppl. 1):16–29.

Brown, D. W., K. Knibutat, N. Edmonds, et al. 2010. A novel translational total body irradiation technique. *Journal of Medical Devices* 4 (3):031003–6.

Carruthers, S. A., and M. M. Wallington. 2004. Total body irradiation and pneumonitis risk: A review of outcomes. *British Journal of Cancer* 90 (11):2080–4.

Chao, M., J. Penagaricano, Y. Yan, E. G. Moros, P. Corry, and V. Ratanatharathorn. 2011. Voxel-based dose reconstruction for total body irradiation with helical TomoTherapy. *International Journal of Radiation Oncology • Biology • Physics* 82 (5):1575–83

Cheng, J. C., T. E. Schultheiss, and J. Y. C. Wong. 2008. Impact of drug therapy, radiation dose, and dose rate on renal toxicity following bone marrow transplantation. *International Journal of Radiation Oncology • Biology • Physics* 71 (5):1436–43.

Chou, R. H., G. B. Wong, J. H. Kramer, et al. 1996. Toxicities of total-body irradiation for pediatric bone marrow transplantation. *International Journal of Radiation Oncology • Biology • Physics* 34 (4):843–51.

Chretien, M., C. Cote, R. Blais, et al. 2000. A variable speed translating couch technique for total body irradiation. *Medical Physics* 27 (5):1127–30.

Chui, C. S., D. P. Fontenla, E. Mullokandov, A. Kapulsky, Y. C. Lo, and C. J. Lo. 1997. Total body irradiation with an arc and a gravity-oriented compensator. *International Journal of Radiation Oncology • Biology • Physics* 39 (5):1191–5.

Clift, R. A., C. D. Buckner, F. R. Appelbaum, et al. 1990. Allogeneic marrow transplantation in patients with acute myeloid leukemia in first remission: A randomized trial of two irradiation regimens. *Blood* 76 (9):1867–71.

Cohen, A., A. Rovelli, B. Bakker, et al. 1999. Final height of patients who underwent bone marrow transplantation for hematological disorders during childhood: A study by the Working Party for Late Effects-EBMT. *Blood* 93 (12):4109–15.

Cohen, A., A. Rovelli, M. T. van Lint, et al. 2001. Secondary thyroid carcinoma after allogeneic bone marrow transplantation during childhood. *Bone Marrow Transplantation* 28 (12):1125–8.

Corvo, R., M. Zeverino, S. Vagge, et al. 2011. Helical TomoTherapy targeting total bone marrow after total body irradiation for patients with relapsed acute leukemia undergoing an allogeneic stem cell transplant. *Radiotherapy and Oncology* 98 (3):382–6.

Cunningham, J. R., and D. J. Wright. 1962. A simple facility for whole-body irradiation. *Radiology* 78:941–9.

Curran, W. J., Jr., J. M. Galvin, and G. J. D'Angio. 1989. A simple dose calculation method for total body photon irradiation. *International Journal of Radiation Oncology • Biology • Physics* 17 (1):219–24.

Curtis, R. E., P. A. Rowlings, H. J. Deeg, et al. 1997. Solid cancers after bone marrow transplantation. *New England Journal of Medicine* 336 (13):897–904.

Deeg, H. J., N. Flournoy, K. M. Sullivan, et al. 1984. Cataracts after total body irradiation and marrow transplantation: A sparing effect of dose fractionation. *International Journal of Radiation Oncology • Biology • Physics* 10 (7):957–64.

Deeg, H. J., R. Storb, G. Longton, et al. 1988. Single dose or fractionated total body irradiation and autologous marrow transplantation in dogs: Effects of exposure rate, fraction size, and fractionation interval on acute and delayed toxicity. *International Journal of Radiation Oncology • Biology • Physics* 15 (3):647–53.

Depledge, M. H., and A. Barrett. 1982. Dose-rate dependence of lung damage after total body irradiation in mice. *International Journal of Radiation Biology and Related Studies in Physics, Chemistry, and Medicine* 41 (3):325–34.

Dessauer, F. 1905. Eine neue Anordnung zur Roentgenbestrahlung. *Archiv für physikalische Medizin und medizinische Technik* 2:218–23.

Down, J. D., N. J. Tarbell, H. D. Thames, and P. M. Mauch. 1991. Syngeneic and allogeneic bone marrow engraftment after total body irradiation: Dependence on dose, dose rate, and fractionation. *Blood* 77 (3):661–9.

Edelstein, K., N. D'Agostino, L. J. Bernstein, et al. 2011. Long-term neurocognitive outcomes in young adult survivors of childhood acute lymphoblastic leukemia. *Journal of Pediatric Hematology/Oncology* 33 (6):450–8.

el-Khatib, E. E., and S. Valcourt. 1989. Calculation of lung shielding for total body irradiation. *International Journal of Radiation Oncology • Biology • Physics* 17 (5):1099–102.

Esiashvili, N., K. Y. Chiang, M. D. Hasselle, C. Bryant, R. H. Riffenburgh, and A. C. Paulino. 2009. Renal toxicity in children undergoing total body irradiation for bone marrow transplant. *Radiotherapy and Oncology* 90 (2):242–6.

Faraci, Maura, S. Barra, A. Cohen, et al. 2005. Very late nonfatal consequences of fractionated TBI in children undergoing bone marrow transplant. *International Journal of Radiation Oncology • Biology • Physics* 63 (5):1568–75.

Felicetti, F., R. Manicone, A. Corrias, et al. 2011. Endocrine late effects after total body irradiation in patients who received hematopoietic cell transplantation during childhood: A retrospective study from a single institution. *Journal of Cancer Research and Clinical Oncology* 137 (9):1343–8.

Field, E. O., H. B. Sharpe, K. B. Dawson, V. Andersen, S. A. Killmann, and E. Weeke. 1972. Turnover rate of normal blood lymphocytes and exchangeable pool size in man, calculated from analysis of chromosomal aberrations sustained during extracorporeal irradiation of the blood. *Blood* 39 (1):39–56.

Fife, K., S. Milan, K. Westbrook, R. Powles, and D. Tait. 1994. Risk factors for requiring cataract surgery following total body irradiation. *Radiotherapy and Oncology* 33 (2):93–8.

Fiorino, C., P. Mangili, G. M. Cattaneo, and R. Calandrino. 2000. Polarity effects of ionization chambers used in TBI dosimetry due to cable irradiation. *Medical Dosimetry* 25 (3):121–6.

Fogliata, A., L. Cozzi, A. Clivio, et al. 2011. Preclinical assessment of volumetric modulated arc therapy for total marrow irradiation. *International Journal of Radiation Oncology • Biology • Physics* 80 (2):628–36.

Galvin, J. M., G. J. D'Angio, and G. Walsh. 1980. Use of tissue compensators to improve the dose uniformity for total body irradiation. *International Journal of Radiation Oncology • Biology • Physics* 6 (6):767–71.

Gerig, L. H., J. Szanto, T. Bichay, and P. Genest. 1994. A translating-bed technique for total-body irradiation. *Physics in Medicine and Biology* 39 (1):19–35.

Gerstein, J., A. Meyer, K. W. Sykora, J. Fruhauf, J. H. Karstens, and M. Bremer. 2009. Long-term renal toxicity in children following fractionated total-body irradiation (TBI) before allogeneic stem cell transplantation (SCT). *Strahlentherapie und Onkologie* 185 (11):751–5.

Girinsky, T., E. Benhamou, J. H. Bourhis, et al. 2000. Prospective randomized comparison of single-dose versus hyperfractionated total-body irradiation in patients with hematologic malignancies. *Journal of Clinical Oncology* 18 (5):981–6.

Girinsky, T., G. Socie, H. Ammarguellat, et al. 1994. Consequences of two different doses to the lungs during a single dose of total body irradiation: Results of a randomized study on 85 patients. *International Journal of Radiation Oncology • Biology • Physics* 30 (4):821–4.

Glasgow, G. P., and W. B. Mill. 1980. Cobalt-60 total body irradiation dosimetry at 220 CM source-axis distance. *International Journal of Radiation Oncology • Biology • Physics* 6 (6):773–7.

Glasgow, G. P., W. B. Mill, G. L. Phillips, 2nd, and G. P. Herzig. 1980. Comparative 60Co total body irradiation (220 cm SAD) and 25 MV total body irradiation (370 cm SAD) dosimetry. *International Journal of Radiation Oncology • Biology • Physics* 6 (9):1243–50.

Glasgow, G. P., S. Wang, and J. Stanton. 1989. A total body irradiation stand for bone marrow transplant patients. *International Journal of Radiation Oncology • Biology • Physics* 16 (3):875–7.

Gogna, N. K., G. Morgan, K. Downs, K. Atkinson, and J. Biggs. 1992. Lung dose rate and interstitial pneumonitis in total body irradiation for bone marrow transplantation. *Australasian Radiology* 36 (4):317–20.

Granena, A., E. Carreras, C. Rozman, et al. 1993. Interstitial pneumonitis after BMT: 15 years experience in a single institution. *Bone Marrow Transplantation* 11 (6):453–8.

Hall, E. J. 1972. Radiation dose-rate: A factor of importance in radiobiology and radiotherapy. *British Journal of Radiology* 45 (530):81–97.

Henk, J. M., R. A. Whitelocke, A. P. Warrington, and E. M. Bessell. 1993. Radiation dose to the lens and cataract formation. *International Journal of Radiation Oncology • Biology • Physics* 25 (5):815–20.

Heublein, A. C. 1932. A preliminary report on continuous irradiation of the entire body. *Radiology* 18 (6):1051–62.

Ho, A., S. Kishel, and G. Proulx. 1998. Partial lung shield for TBI. *Medical Dosimetry* 23 (4):299–301.

Hui, S. K., D. Henderson, R. K. Das, and B. Thomadsen. 2004. CT-based analysis of dose homogeneity in total-body irradiation using lateral beam. *Journal of Applied Clinical Medical Physics* 5 (4):71–9.

Hui, S. K., J. Kapatoes, J. Fowler, et al. 2005. Feasibility study of helical TomoTherapy for total body or total marrow irradiation. *Medical Physics* 32 (10):3214–24.

Hussain, A., E. Villarreal-Barajas, D. Brown, and P. Dunscombe. 2010. Validation of the Eclipse AAA algorithm at extended SSD. *Journal of Applied Clinical Medical Physics* 11 (3):3213.

Hussain, A., J. E. Villarreal-Barajas, P. Dunscombe, and D. W. Brown. 2011. Aperture modulated, translating bed total body irradiation. *Medical Physics* 38 (2):932–41.

Hussein, S., and E. el-Khatib. 1995. Total body irradiation with a sweeping 60Cobalt beam. *International Journal of Radiation Oncology • Biology • Physics* 33 (2):493–7.

Hussein, S., and G. M. Kennelly. 1996. Lung compensation in total body irradiation: A radiographic method. *Medical Physics* 23 (3):357–60.

Igaki, H., K. Karasawa, H. Sakamaki, et al. 2005. Renal dysfunction after total-body irradiation: Significance of selective renal shielding blocks. *Strahlentherapie und Onkologie* 181 (11):704–8.

Izawa, H., H. Hirowatari, Y. Yahata, et al. 2011. Effect of dose fractionation on pulmonary complications during total body irradiation. *Journal of Radiation Research (Tokyo)* 52 (4):502–8.

Jacobs, M. L., and L. Pape. 1960. A total body irradiation chamber and its uses. *International Journal of Applied Radiationa and Isotopes* 8:141–3.

Janzen, L. A., and B. J. Spiegler. 2008. Neurodevelopmental sequelae of pediatric acute lymphoblastic leukemia and its treatment. *Developmental Disabilities Research Reviews* 14 (3):185–95.

Jursinic, P. A. 2007. Characterization of optically stimulated luminescent dosimeters, OSLDs, for clinical dosimetric measurements. *Medical Physics* 34 (12):4594–604.

Jursinic, P. A., and C. J. Yahnke. 2011. *In vivo* dosimetry with optically stimulated luminescent dosimeters, OSLDs, compared to diodes; the effects of buildup cap thickness and fabrication material. *Medical Physics* 38 (10):5432–40.

Kadan-Lottick, N. S., L. K. Zeltzer, Q. Liu, et al. 2010. Neurocognitive functioning in adult survivors of childhood non-central nervous system cancers. *Journal of the National Cancer Institute* 102 (12):881–93.

Kal, H. B., and M. L. van Kempen-Harteveld. 2009. Induction of severe cataract and late renal dysfunction following total body irradiation: Dose-effect relationships. *Anticancer Research* 29 (8):3305–9.

Kal, H. B., M. L. van Kempen-Harteveld, M. H. Heijenbrok-Kal, and H. Struikmans. 2006. Biologically effective dose in total-body irradiation and hematopoietic stem cell transplantation. *Strahlentherapie und Onkologie* 182 (11):672–9.

Keane, J. T., D. P. Fontenla, and C. S. Chui. 2000. Applications of IMAT to total body radiation (TBI). *International Journal of Radiation Oncology • Biology • Physics* 48 (3, Suppl. 1):239.

Keane, T. J., J. Van Dyk, and W. D. Rider. 1981. Idiopathic interstitial pneumonia following bone marrow transplantation: The relationship with total body irradiation. *International Journal of Radiation Oncology • Biology • Physics* 7 (10):1365–70.

Kelsey, C. R., M. E. Horwitz, J. P. Chino, et al. 2011. Severe pulmonary toxicity after myeloablative conditioning using total body irradiation: An assessment of risk factors. *International Journal of Radiation Oncology • Biology • Physics* 81 (3):812–8.

Kerns, J. R., S. F. Kry, N. Sahoo, D. S. Followill, and G. S. Ibbott. 2011. Angular dependence of the nanoDot OSL dosimeter. *Medical Physics* 38 (7):3955–62.

Kirby, T. H., W. F. Hanson, and D. A. Cates. 1988. Verification of total body photon irradiation dosimetry techniques. *Medical Physics* 15 (3):364–9.

Labar, B., V. Bogdanic, D. Nemet, et al. 1992. Total body irradiation with or without lung shielding for allogeneic bone marrow transplantation. *Bone Marrow Transplantation* 9 (5):343–7.

Lavallee, M. C., S. Aubin, M. Chretien, M. Larochelle, and L. Beaulieu. 2008. Attenuator design for organs at risk in total body irradiation using a translation technique. *Medical Physics* 35 (5):1663–9.

Lavallee, M. C., S. Aubin, M. Larochelle, I. Vallieres, and L. Beaulieu. 2011. 3D heterogeneous dose distributions for total body irradiation patients. *Journal of Applied Clinical Medical Physics* 12 (3):3416.

Lavallee, M. C., L. Gingras, M. Chretien, S. Aubin, C. Cote, and L. Beaulieu. 2009. Commissioning and evaluation of an extended SSD photon model for PINNACLE3: An application to total body irradiation. *Medical Physics* 36 (8):3844–55.

Laver, J., J. H. Kwon, and H. Castro-Malaspina. 1987. Effects of low dose rate irradiation on human marrow hematopoietic and microenvironmental cells: Sparing effect upon survival of stromal and leukemic cells. *Bone Marrow Transplantation* 2 (3):271–8.

Lawton, C. A., E. P. Cohen, K. J. Murray, et al. 1997. Long-term results of selective renal shielding in patients undergoing total body irradiation in preparation for bone marrow transplantation. *Bone Marrow Transplantation* 20 (12):1069–74.

Leung, W., M. M. Hudson, D. K. Strickland, et al. 2000. Late effects of treatment in survivors of childhood acute myeloid leukemia. *Journal of Clinical Oncology* 18 (18):3273–9.

Leung, P. M., W. D. Rider, H. P. Webb, H. Aget, and H. E. Johns. 1981. Cobalt-60 therapy unit for large field irradiation. *International Journal of Radiation Oncology • Biology • Physics* 7 (6):705–12.

Lin, J.-P., T.-C. Chu, and M.-T. Liu. 2001. Dose compensation of the total body irradiation therapy. *Applied Radiation and Isotopes* 55 (5):623–30.

Liu, J. C., E. T. Bacza, D. O. Findley, and B. W. Forell. 1983. Dosimetry of single fraction high dose total body irradiation as measured by thermoluminescent dosimeters. *International Journal of Radiation Oncology • Biology • Physics* 9 (9):1407–8.

Lutz, W. R., G. K. Svensson, and B. E. Bjarngard. 1983. Design and beam properties of a 4 MV total body irradiator. Abstract. *Medical Physics* 10:539.

Mangili, P., C. Fiorino, A. Rosso, et al. 1999. In-vivo dosimetry by diode semiconductors in combination with portal films during TBI: Reporting a 5-year clinical experience. *Radiotherapy and Oncology* 52 (3):269–76.

Marmont, A. M., M. M. Horowitz, R. P. Gale, et al. 1991. T-cell depletion of HLA-identical transplants in leukemia. *Blood* 78 (8):2120–30.

Miralbell, R., S. Bieri, B. Mermillod, et al. 1996. Renal toxicity after allogeneic bone marrow transplantation: The combined effects of total-body irradiation and graft-versus-host disease. *Journal of Clinical Oncology* 14 (2):579–85.

Miralbell, R., M. Rouzaud, E. Grob, et al. 1994. Can a total body irradiation technique be fast and reproducible? *International Journal of Radiation Oncology • Biology • Physics* 29 (5):1167–73.

Mody, R., S. Li, D. C. Dover, et al. 2008. Twenty-five-year follow-up among survivors of childhood acute lymphoblastic leukemia: A report from the Childhood Cancer Survivor Study. *Blood* 111 (12):5515–23.

Moleski, M. 2000. Neuropsychological, neuroanatomical, and neurophysiological consequences of CNS chemotherapy for acute lymphoblastic leukemia. *Archives of Clinicial Neuropsychology* 15 (7):603–30.

Molloy, J. A. 2010. Statistical analysis of dose heterogeneity in circulating blood: Implications for sequential methods of total body irradiation. *Medical Physics* 37 (11):5568–78.

Morgan, T. L., P. M. Falk, N. Kogut, K. H. Shah, M. Tome, and A. R. Kagan. 1996. A comparison of single-dose and fractionated total-body irradiation on the development of pneumonitis following bone marrow transplantation. *International Journal of Radiation Oncology • Biology • Physics* 36 (1):61–6.

Nathan, P. C., T. Whitcomb, P. L. Wolters, et al. 2006. Very high-dose methotrexate (33.6 g/m(2)) as central nervous system preventive therapy for childhood acute lymphoblastic leukemia: Results of National Cancer Institute/Children's Cancer Group trials CCG-191P, CCG-134P and CCG-144P. *Leukemia & Lymphoma* 47 (12):2488–504.

Niroomand-Rad, A. 1991. Physical aspects of total body irradiation of bone marrow transplant patients using 18 MV x rays. *International Journal of Radiation Oncology • Biology • Physics* 20 (3):605–11.

Oya, N., K. Sasai, S. Tachiiri, et al. 2006. Influence of radiation dose rate and lung dose on interstitial pneumonitis after fractionated total body irradiation: Acute parotitis may predict interstitial pneumonitis. *Int J Hematol* 83 (1):86–91.

Ozsahin, M., Y. Belkacemi, F. Pene, et al. 1994. Total-body irradiation and cataract incidence: A randomized comparison of two instantaneous dose rates. *International Journal of Radiation Oncology • Biology • Physics* 28 (2):343–7.

Ozsahin, M., F. Pene, E. Touboul, et al. 1992. Total-body irradiation before bone marrow transplantation. Results of two randomized instantaneous dose rates in 157 patients. *Cancer* 69 (11):2853–65.

Papież, L., J. Montebello, C. DesRosiers, and E. Papież. 1999. The clinical application of dynamic shielding and imaging in moving table total body irradiation. *Radiotherapy and Oncology* 51 (3):219–24.

Peters, L. 1980. Total Body Irradiation Conference: Discussion: The radiobiological bases of TBI. *International Journal of Radiation Oncology • Biology • Physics* 6 (6):785–7.

Peters, L. J. 1985. Relative effect of radiation dose rate on hemopoietic and nonhemopoietic lethality of total body irradiation. In *Life-Span Radiation Effects Studies in Animals: What Can They Tell Us? Proceedings of the Twenty-Second Hanford Life Sciences Symposium Held at Richland, Washington, September 27–29, 1983*, edited by R. C. Thompson and J. A. Mahaffey: Springfield, VA: National Technical Information Service.

Pino y Torres, J. L., D. S. Bross, W. C. Lam, M. D. Wharam, G. W. Santos, and S. E. Order. 1982. Risk factors in interstitial pneumonitis following allogenic bone marrow transplantation. *International Journal of Radiation Oncology • Biology • Physics* 8 (8):1301–7.

Pla, M., S. G. Chenery, and E. B. Podgorsak. 1983. Total body irradiation with a sweeping beam. *International Journal of Radiation Oncology * Biology * Physics* 9 (1):83–9.

Planskoy, B., A. M. Bedford, F. M. Davis, P. D. Tapper, and L. T. Loverock. 1996. Physical aspects of total-body irradiation at the Middlesex Hospital (UCL group of hospitals), London 1988–1993: I. Phantom measurements and planning methods. *Physics in Medicine and Biology* 41 (11):2307–26.

Planskoy, B., P. D. Tapper, A. M. Bedford, and F. M. Davis. 1996. Physical aspects of total-body irradiation at the Middlesex Hospital (UCL group of hospitals), London 1988-1993: II. *in vivo* planning and dosimetry. *Physics in Medicine and Biology* 41 (11):2327–43.

Pommier, P., M. P. Sunyach, C. Pasteuris, D. Frappaz, and C. Carrie. 2009. Second cancer after total-body irradiation (TBI) in childhood. *Strahlentherapie und Onkologie* 185 (Suppl. 2):13–16.

Quast, U. 1985. Physical treatment planning of total-body irradiation: Patient translation and beam-zone method. *Medical Physics* 12 (5):567–74.

Reiff, J. E., M. Werner-Wasik, R. K. Valicenti, and M. Saiful Huq. 1999. Changes in the size and location of kidneys from the supine to standing positions and the implications for block placement during total body irradiation. *International Journal of Radiation Oncology • Biology • Physics* 45 (2):447–9.

Rhee, J. G., C. W. Song, T. H. Kim, and S. H. Levitt. 1985. Effect of fractionation and rate of radiation dose on human leukemic cells, HL-60. *Radiation Research* 101 (3):519–27.

Ribas, M., N. Jornet, T. Eudaldo, et al. 1998. Midplane dose determination during total body irradiation using *in vivo* dosimetry. *Radiotherapy and Oncology* 49 (1):91–8.

Sahler, O. D. 1959. Development of a room specifically designed for total-body irradiation. *Radiology* 72 (2):266–7.

Sampath, S., T. E. Schultheiss, and J. Wong. 2005. Dose response and factors related to interstitial pneumonitis after bone marrow transplant. *International Journal of Radiation Oncology • Biology • Physics* 63 (3):876–84.

Sánchez-Doblado, F., J. A. Terrón, B. Sánchez-Nieto, et al. 1995. Verification of an on line *in vivo* semiconductor dosimetry system for TBI with two TLD procedures. *Radiotherapy and Oncology* 34 (1):73–7.

Sanders, J. E. 1990. Late effects in children receiving total body irradiation for bone marrow transplantation. *Radiotherapy and Oncology* 18 (Suppl. 1):82–7.

Sanders, J. E., J. Hawley, W. Levy, et al. 1996. Pregnancies following high-dose cyclophosphamide with or without high-dose busulfan or total-body irradiation and bone marrow transplantation. *Blood* 87 (7):3045–52.

Sarfaraz, M., C. Yu, D. J. Chen, and L. Der. 2001. A translational couch technique for total body irradiation. *Journal of Applied Clinical Medical Physics* 2 (4):201–9.

Scarpati, D., F. Frassoni, V. Vitale, et al. 1989. Total body irradiation in acute myeloid leukemia and chronic myelogenous leukemia: Influence of dose and dose-rate on leukemia relapse. *International Journal of Radiation Oncology • Biology • Physics* 17 (3):547–52.

Schaeken, B., S. Lelie, P. Meijnders, D. Van den Weyngaert, H. Janssens, and D. Verellen. 2010. Alanine/EPR dosimetry applied to the verification of a total body irradiation protocol and treatment planning dose calculation using a humanoid phantom. *Medical Physics* 37 (12):6292–9.

Schatz, J., J. H. Kramer, A. Ablin, and K. K. Matthay. 2000. Processing speed, working memory, and IQ: A developmental model of cognitive deficits following cranial radiation therapy. *Neuropsychology* 14 (2):189–200.

Schneider, R. A., J. Schultze, J. M. Jensen, et al. 2008. Long-term outcome after static intensity-modulated total body radiotherapy using compensators stratified by pediatric and adult cohorts. *International Journal of Radiation Oncology • Biology • Physics* 70 (1):194–202.

Schriock, E. A., M. J. Schell, M. Carter, O. Hustu, and J. J. Ochs. 1991. Abnormal growth patterns and adult short stature in 115 long-term survivors of childhood leukemia. *Journal of Clinical Oncology* 9 (3):400–5.

Schultheiss, T. E., J. Wong, A. Liu, G. Olivera, and G. Somlo. 2007. Image-guided total marrow and total lymphatic irradiation using helical TomoTherapy. *International Journal of Radiation Oncology • Biology • Physics* 67 (4):1259–67.

Shank, B. 1993. Hyperfractionation versus single dose irradiation in human acute lymphocytic leukemia cells: Application to TBI for marrow transplantation. *Radiotherapy and Oncology* 27 (1):30–5.

Shank, B., F. C. Chu, R. Dinsmore, et al. 1983. Hyperfractionated total body irradiation for bone marrow transplantation: Results in seventy leukemia patients with allogeneic transplants. *International Journal of Radiation Oncology • Biology • Physics* 9 (11):1607–11.

Shank, B., S. Hopfan, J. H. Kim, et al. 1981. Hyperfractionated total body irradiation for bone marrow transplantation: I. Early results in leukemia patients. *International Journal of Radiation Oncology • Biology • Physics* 7 (8):1109–15.

Shank, B., R. J. O'Reilly, I. Cunningham, et al. 1990. Total body irradiation for bone marrow transplantation: The Memorial Sloan-Kettering Cancer Center experience. *Radiotherapy and Oncology* 18 (Suppl 1):68–81.

Sharma, M., T. Dos Santos, N. P. Papanikolopoulos, and S. K. Hui. 2011. Feasibility of intrafraction whole-body motion tracking for total marrow irradiation. *Journal Biomedical Optics* 16 (5):058002.

Silber, J. H., J. Radcliffe, V. Peckham, et al. 1992. Whole-brain irradiation and decline in intelligence: The influence of dose and age on IQ score. *Journal of Clinical Oncology* 10 (9):1390–6.

Smith, C. L., W.-K. Chu, M. R. Goede, D.Granville, and G. V. Dalrymple. 1996. An analysis of the elements essential in the development of a customized TBI program. *Medical Dosimetry* 21 (2):49–56.

Socie, G., R. E. Curtis, H. J. Deeg, et al. 2000. New malignant diseases after allogeneic marrow transplantation for childhood acute leukemia. *Journal of Clinical Oncology* 18 (2):348–57.

Socie, G., A. Devergie, T. Girinsky, et al. 1991. Influence of the fractionation of total body irradiation on complications and relapse rate for chronic myelogenous leukemia. The Groupe d'Etude des greffes de moelle osseuse (GEGMO). *International Journal of Radiation Oncology • Biology • Physics* 20 (3):397–404.

Song, C. W., T. H. Kim, F. M. Khan, J. H. Kersey, and S. H. Levitt. 1981. Radiobiological basis of total body irradiation with different dose rate and fractionation: Repair capacity of hemopoietic cells. *International Journal of Radiation Oncology • Biology • Physics* 7 (12):1695–701.

Soule, B. P., N. L. Simone, B. N. Savani, et al. 2007. Pulmonary function following total body irradiation (with or without lung shielding) and allogeneic peripheral blood stem cell transplant. *Bone Marrow Transplantation* 40 (6):573–8.

Svensson, G. K., R. D. Larsen, and T. S. Chen. 1980. The use of a 4 MV linear accelerator for whole body irradiation. *International Journal of Radiation Oncology • Biology • Physics* 6 (6):761–5.

Tarbell, N. J., D. A. Amato, J. D. Down, P. Mauch, and S. Hellman. 1987. Fractionation and dose rate effects in mice: A model for bone marrow transplantation in man. *International Journal of Radiation Oncology • Biology • Physics* 13 (7):1065–9.

Tarbell, N. J., E. C. Guinan, L. Chin, P. Mauch, and H. J. Weinstein. 1990. Renal insufficiency after total body irradiation for pediatric bone marrow transplantation. *Radiotherapy and Oncology* 18 (Suppl. 1):139–42.

Teschendorf, W. 1925. Uber Bestrahlungen des ganzen menschlichen Korpers bei Blutkrankenheiten. *Strahlentherapie* 26:720–8.

Thomas, E. D., C. A. Ashley, H. L. Lochte, Jr., A. Jaretzki, 3rd, O. D. Sahler, and J. W. Ferrebee. 1959. Homografts of bone marrow in dogs after lethal total-body radiation. *Blood* 14 (6):720–36.

Thomas, E. D., R. A. Clift, J. Hersman, et al. 1982. Marrow transplantation for acute non-lymphoblastic leukemic in first remission using fractionated or single-dose irradiation. *International Journal of Radiation Oncology • Biology • Physics* 8 (5):817–21.

Thomas, E. D., H. L. Lochte, Jr., and J. W. Ferrebee. 1959. Irradiation of the entire body and marrow transplantation: Some observations and comments. *Blood* 14 (1):1–23.

Thomas, E. D., H. L. Lochte, Jr., W. C. Lu, and J. W. Ferrebee. 1957. Intravenous infusion of bone marrow in patients receiving radiation and chemotherapy. *New England Journal of Medicine* 257 (11):491–6.

Thomas, O., M. Mahe, L. Campion, et al. 2001. Long-term complications of total body irradiation in adults. *International Journal of Radiation Oncology • Biology • Physics* 49 (1):125–31.

Torrisi, J., S. Rustgi, T. Spitzer, et al. 1990. Total body irradiation: Comparison of direct *in vivo* dose measurement with calculated dose. *International Journal of Radiation Oncology • Biology • Physics* 19:250.

Travis, E. L., L. J. Peters, J. McNeill, H. D. Thames, Jr., and C. Karolis. 1985. Effect of dose-rate on total body irradiation: Lethality and pathologic findings. *Radiotherapy and Oncology* 4 (4):341–51.

Umek, B., M. Zwitter, and H. Habič. 1996. Total body irradiation with translation method. *Radiotherapy and Oncology* 38 (3):253–5.

Van Dyk, J., J. M. Galvin, G. P. Glasgow, and E. B. Podgorsak. 1986. *The Physical Aspects of Total and Half Body Photon Irradiation* (American Association of Physicists in Medicine Report No. 17). New York, NY: American Association of Physicists in Medicine.

Van Dyk, J., T. J. Keane, S. Kan, W. D. Rider, and C. J. Fryer. 1981. Radiation pneumonitis following large single dose irradiation: A re-evaluation based on absolute dose to lung. *International Journal of Radiation Oncology • Biology • Physics* 7 (4):461–7.

Van Dyk, J., T. J. Keane, and W. D. Rider. 1982. Lung density as measured by computerized tomography: Implications for radiotherapy. *International Journal of Radiation Oncology • Biology • Physics* 8 (8):1363–72.

van Kempen-Harteveld, M. L. R. Brand, H. B. Kal, et al. 2008. Results of hematopoietic stem cell transplantation after treatment with different high-dose total-body irradiation regimens in five Dutch centers. *International Journal of Radiation Oncology • Biology • Physics* 71 (5):1444–54.

van Kempen-Harteveld, M. L., H. Struikmans, H. B. Kal, et al. 2002. Cataract after total body irradiation and bone marrow transplantation: Degree of visual impairment. *International Journal of Radiation Oncology • Biology • Physics* 52 (5):1375–80.

van Kempen-Harteveld, M. L., M. H. van Weel-Sipman, C. Emmens, et al. 2003. Eye shielding during total body irradiation for bone marrow transplantation in children transplanted for a hematological disorder: Risks and benefits. *Bone Marrow Transplantation* 31 (12):1151–6.

van Weel-Sipman, M. H., E. T. van't Veer-Korthof, H. van den Berg, et al. 1990. Late effects of total body irradiation and cytostatic preparative regimen for bone marrow transplantation in children with hematological malignancies. *Radiotherapy and Oncology* 18 (Suppl. 1):155–7.

Vesterbacka, M., O. Ringden, M. Remberger, J. Huggare, and G. Dahllof. 2012. Disturbances in dental development and craniofacial growth in children treated with hematopoietic stem cell transplantation. *Orthodontics and Craniofacial Research* 15 (1):21–9.

Viamonte, A., L. A. R. da Rosa, L. A. Buckley, A. Cherpak, and J. E. Cygler. 2008. Radiotherapy dosimetry using a commercial OSL system. *Medical Physics* 35 (4):1261–6.

Vitale, V., A. Bacigalupo, M. T. Van Lint, et al. 1983. Fractionated total body irradiation in marrow transplantation for leukaemia. *Br J Haematol* 55 (3):547–54.

Waber, D. P., N. J. Tarbell, D. Fairclough, et al. 1995. Cognitive sequelae of treatment in childhood acute lymphoblastic leukemia: Cranial radiation requires an accomplice. *Journal of Clinical Oncology* 13 (10):2490–6.

Weichselbaum, R. R., J. S. Greenberger, A. Schmidt, A. Karpas, W. C. Moloney, and J. B. Little. 1981. *In vitro* radiosensitivity of human leukemia cell lines. *Radiology* 139 (2):485–7.

Weiner, R. S., M. M. Bortin, R. P. Gale, et al. 1986. Interstitial pneumonitis after bone marrow transplantation: Assessment of risk factors. *Annals of Internal Medicine* 104 (2):168–75.

Weshler, Z., R. Breuer, R. Or, et al. 1990. Interstitial pneumonitis after total body irradiation: Effect of partial lung shielding. *Br J Haematol* 74 (1):61–4.

Wilkie, J. R., H. Tiryaki, B. D. Smith, J. C. Roeske, J. A. Radosevich, and B. Aydogan. 2008. Feasibility study for linac-based intensity modulated total marrow irradiation. *Medical Physics* 35 (12):5609–18.

Wong, J. Y. C., A. Liu, T. Schultheiss, et al. 2006. Targeted total marrow irradiation using three-dimensional image-guided tomographic intensity-modulated radiation therapy: An alternative to standard total body irradiation. *Biology of Blood and Marrow Transplantation* 12 (3):306–15.

Wong, J. Y., J. Rosenthal, A. Liu, T. Schultheiss, S. Forman, and G. Somlo. 2009. Image-guided total-marrow irradiation using helical TomoTherapy in patients with multiple myeloma and acute leukemia undergoing hematopoietic cell transplantation. *International Journal of Radiation Oncology • Biology • Physics* 73 (1):273–9.

Yeginer, M., J. C. Roeske, J. A. Radosevich, and B. Aydogan. 2011. Linear accelerator-based intensity-modulated total marrow irradiation technique for treatment of hematologic malignancies: A dosimetric feasibility study. *International Journal of Radiation Oncology • Biology • Physics* 79 (4):1256–65.

Yukihara, E. G., G. Mardirossian, M. Mirzasadeghi, S. Guduru, and S. Ahmad. 2008. Evaluation of Al2O3:C optically stimulated luminescence (OSL) dosimeters for passive dosimetry of high-energy photon and electron beams in radiotherapy. *Medical Physics* 35 (1):260–9.

Zhuang, A. H., A. Liu, T. E. Schultheiss, and J. Y. Wong. 2010. Dosimetric study and verification of total body irradiation using helical TomoTherapy and its comparison to extended SSD technique. *Medical Dosimetry* 35 (4):243–9.

Zierhut, D., F. Lohr, P. Schraube, et al. 2000. Cataract incidence after total-body irradiation. *International Journal of Radiation Oncology • Biology • Physics* 46 (1):131–5.

Chapter **4**

Tumors of the Central Nervous System

4.1 Clinical Overview

Brain tumors are the leading cause of cancer-related death in children. Central nervous system (CNS) cancer as a group is the second most frequent malignancy of childhood, representing 16.6% of all malignancies and the most common of the solid tumors. Approximately 2500 children under the age of 20 develop a brain tumor each year in the United States, occurring in 3 out of every 100,000 children. Astrocytomas accounted for 52%, primitive neuroectodermal tumors (PNETs) 21%, other gliomas 15%, and ependymomas 9% of CNS malignancies. The incidence of CNS cancer is higher in children younger than 8 years of age than in older children or adolescents, and is slightly higher in males than in females, and in white children than in black children. There appears to have been a jump in incidence rate around 1984–1985, with a fairly constant rate both before and after that time period. Survival, which is dependent on the type and location of the CNS malignancy, tends to be worse in very young children than in older children. Primary brain tumors arise from normal cells in the brain, but little is known about how normal brain cells become malignant and grow into a brain tumor (Ries et al. 1999).

CNS tumors are usually treated with maximum feasible surgical resection, often followed by chemotherapy and radiation therapy. CNS tumors are the most common type of childhood tumor treated with radiation therapy, as the use of radiation is confined to a small proportion of children with leukemia. Radiation to the young, developing brain is well documented to be associated with significant toxicities, so its use on brain tumors is limited in children under 3 to 6 years of age to progressive tumors that are resistant to chemotherapy. In an age-dependent manner, these toxicities include lowered intelligence quotient (IQ) scores, impaired memory, impaired growth and abnormal hormone production, changes in behavior, hearing loss, and vision impairment. Studies have shown that the consequences of cognitive impairment during childhood are significantly lower educational attainment, less household income, less full time employment, and fewer marriages (Ellenberg et al. 2009). These effects are a result of radiation damage to specific structures that can be protected if identified during the planning process. In addition to these risks that can manifest in months to a few years, there is the chance of producing second malignancies much later in life (also see Chapter 2).

The increase in overall survival for all types of CNS malignancies from about 59% to 74% over the past several decades is due to improved diagnostic, surgical, and radiation technologies; new chemotherapeutic regimens; and improved supportive care measures.

4.1.1 Medulloblastoma and Supratentorial Primitive Neuroectodermal Tumor (sPNET)

Medulloblastoma is a highly malignant primary brain tumor that originates in the cerebellar vermis and accounts for approximately 20% of brain tumors in children. Medulloblastoma is the most frequently occurring CNS tumor in children under the age of 18. The greatest incidence for PNET is from infancy through age 3 (about 11 per million). Incidence steadily declines through childhood (Ries et al. 1999).

Medulloblastoma is categorized as one of the three types of PNETs. Medulloblastoma is the most common PNET originating in the posterior fossa. Another term for medulloblastoma is infratentorial PNET. These tumors that originate in the posterior fossa are referred to as infratentorial because they occur below the tentorium, a thick membrane that separates the cerebral hemispheres of the brain from the cerebellum. Medulloblastoma usually occupies a midline location and grows into the fourth ventricle, between the brainstem and the cerebellum, in 90% of childhood and 50% of adult cases (Figure 4.1). Common presenting symptoms are unsteadiness, headaches, and vomiting due to increased intracranial pressure (from blockage of cerebrospinal fluid flow). Supratentorial PNETs will be discussed in the next section.

All PNETs of the brain are invasive and rapidly growing tumors that, unlike most brain tumors, commonly spread through the cerebrospinal

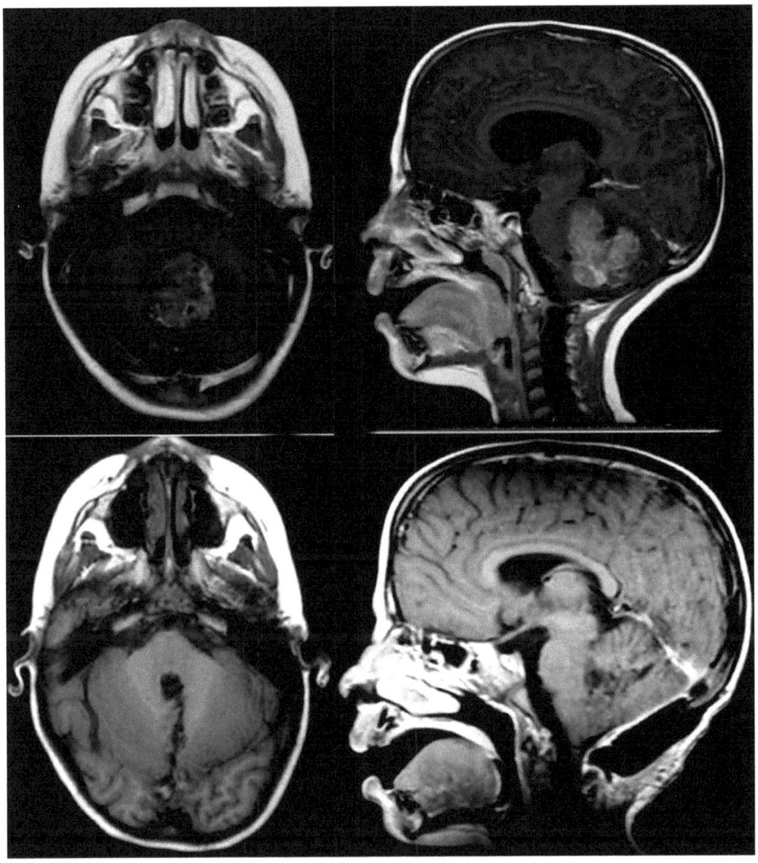

FIGURE 4.1 Axial and sagittal MRI of medulloblastoma preoperation (upper) and postoperation (lower).

fluid (CSF) and metastasize to different locations in the brain (including leptomeningeal spread) and spinal cord. Dissemination along the neuraxis through the cerebrospinal fluid is relatively common, with a reported incidence of 16% to 46% at diagnosis (Deutsch and Reigel 1981). As a result, craniospinal irradiation (CSI) has been the mainstay of postoperative treatment for medulloblastoma.

Medulloblastoma is classified as either average (also called standard) risk or high risk. Average-risk disease is defined as no evidence of metastasis, residual disease after surgical resection no more than 1.5 cm^2, and age at diagnosis of 3 years or above; otherwise the case is high risk. Conventional treatment for medulloblastoma consists of surgical resection, followed by chemotherapy both during and after radiotherapy producing cure rates of about 80% for average-risk (Packer et al. 2006) and 60% for high-risk disease (Zeltzer et al. 1999; Verlooy et al. 2006). Conventional treatment for average-risk cases is 23.4 Gy to the entire craniospinal axis plus a whole posterior fossa boost of 30.6 to 32.4 Gy (54 to 55.8 Gy total). High-risk cases are treated with 36 Gy to the entire craniospinal axis plus a boost to the whole posterior fossa to either 18.0 or 19.8 Gy (54 to 55.8 Gy total). The doses are given in 1.8 Gy fractions. Recent clinical trials have investigated reducing the volume of the boost to just the local site of the primary tumor as well as reducing the CSI dose to just 18 Gy in younger children.

Figure 4.1 shows an axial and sagittal MRI pre- and postsurgical resection of a posterior fossa tumor. For the CSI phase of treatment as well as whole posterior fossa boost treatments, the planning computed tomographic (CT) images with intravenous contrast may be sufficient for designing the fields. For surgical bed boost, MRI to CT image fusion can be helpful in delineating the gross tumor volume. MRI is also helpful in defining the location of the tentorium.

When PNET arises superior to the tentorium (supratentorial), it is distinctly different than medulloblastoma, which by definition occurs infratentorially. sPNET is considered a high-risk disease and requires 36 Gy CSI followed by a 19.8 Gy boost. Survival for sPNET is somewhat worse than that of high-risk posterior fossa medulloblastoma.

4.1.2 Ependymoma

Ependymomas make up about 9% of childhood tumors of the CNS and are the third most common brain tumor. Incidence peaks at age 2 (8.6 per million) and then levels off at about 1.4 per million from age 5 to 18. One-third develops in children less than 3 years of age (Ries et al. 1999). They develop from the ependymal cells that line both the ventricles of the brain and the spinal cord. Two-thirds arise from the floor of the fourth ventricle, situated in the posterior fossa, where they may produce headache, nausea, and vomiting by obstructing the flow of cerebrospinal fluid. Figure 4.2

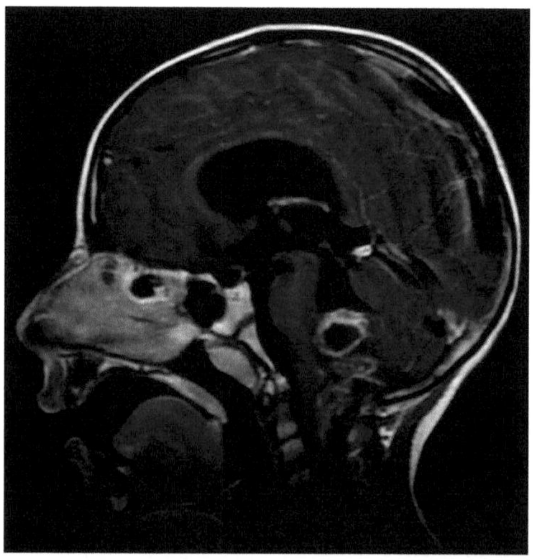

FIGURE 4.2 Sagittal MRI of ependymoma extending into the C-spine.

shows a sagittal MRI scan of unresected ependymoma extending into the cervical spine. Most intracranial ependymomas are classified as ordinary (differentiated) or anaplastic. Although prognosis may be better for ordinary ependymomas, the two types are treated the same. Maximal feasible resection is followed by involved field radiation. Chemotherapy is usually not given, except as a substitute for radiation therapy in very young children or for incompletely resected tumors. Chemotherapy has not been shown to increase survival of patients who have undergone complete resection and postoperative local radiotherapy.

In children older than 5, spread of disease throughout the brain or spinal cord occurs in about 10% of cases and when it occurs, CSI is given. The majority of cases does not require CSI and instead, if there is a complete resection, get 59.4 Gy in 1.8 Gy fractions (33 fractions) to the original tumor site plus a 0.5 to 1 cm margin. For children less than 2 years old with a gross total resection, the dose may be reduced to 54 Gy. Similar to medulloblastoma, ependymomas occurring infratentorially do better than those occurring supratentorially.

Patients with gross or near-total resection plus radiation therapy have reported survival rates up of 65% to 80% versus approximately 30% for patients receiving partial resection or biopsy only.

4.1.3 Glioma

Gliomas are a family of brain tumors that arise from the glial cells in the brain that are the nonneuronal cells that provide support, nutrition, maintain homeostasis, and participate in signal transmission in the nervous system. Incidence of astrocytoma and other gliomas is in a bimodal distribution with peaks at about age 6 and 13 (about 26 per million) (Ries et al. 1999). Figure 4.3 is an MRI showing an astrocytoma/glioblastoma in the supratentorial brain. Figure 4.4 and Figure 4.5 show a sagittal and axial MRI of a brainstem glioma. Gliomas are categorized according to their World Health Organization (WHO) grade:

Grade 1—Pilocytic astrocytoma (PCA)
Grade 2—Diffuse or low grade astrocytoma
Grade 3—Anaplastic astrocytoma
Grade 4—Glioblastoma multiforme (GBM) (most common in adults)

PCA, occurring in children but rarely adults, is considered a benign tumor and is usually cured by surgery alone. Radiation therapy is reserved for cases that recur after repeated resections and (in young children) progress despite chemotherapy. Grade 2 tumors are also treated with surgery alone, if resectable, and radiation therapy and chemotherapy are reserved for cases that cannot be resected and do not respond to chemotherapy or

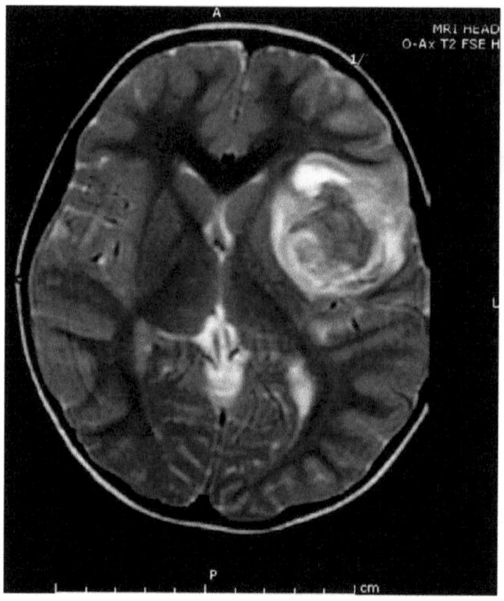

FIGURE 4.3 Axial MRI of a supratentorial astrocytoma.

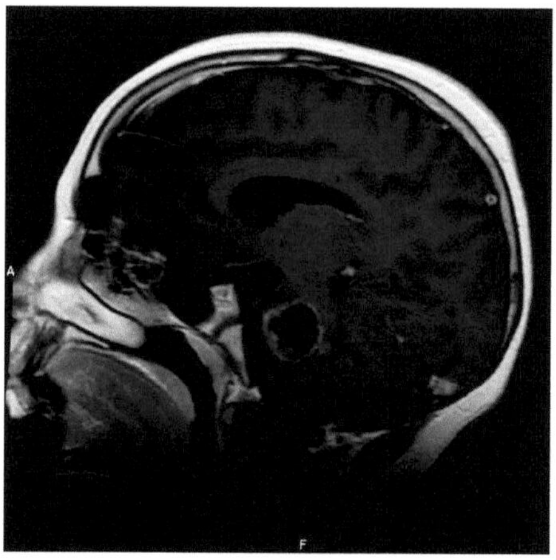

FIGURE 4.4 Sagittal MRI T1 FLAIR of a brainstem glioma.

recur after these therapies. Grade 3 and 4 gliomas require surgery (when feasible), chemotherapy, and radiation therapy. The prognosis is worst for grade 4, with average survival times of approximately 12 months, whereas for anaplastic glioma patients, survival is about 3 years.

The most devastating gliomas in children are brainstem gliomas. Because of the vital role of the brainstem, surgery or even biopsy is usually not attempted and diagnosis is made based on clinical signs and imaging studies. The anatomical location of the tumor provides three classifications: diffuse intrinsic pontine, tectal, and cervicomedullary. Intrinsic pontine gliomas have average survival times of less than 1 year. Longer survival and even cure is associated with the tectal and cervicomedullary gliomas. Tumors also are characterized on the basis of site of origin, focality, direction and extent of tumor growth, size, and presence or absence of cysts, necrosis,

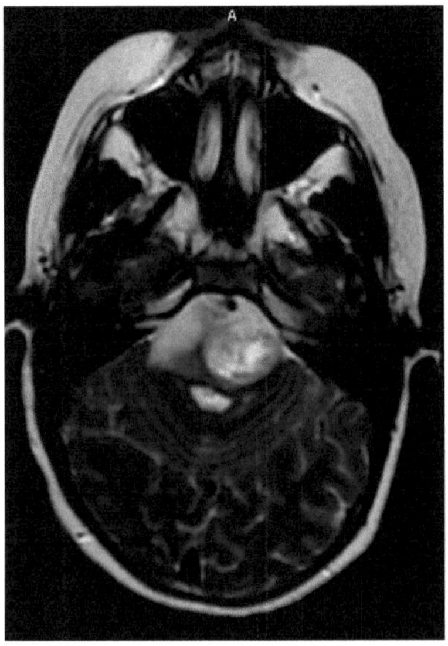

FIGURE 4.5 Axial MRI T2 of a brainstem glioma.

hemorrhage, and hydrocephalus. Because of the poor survival of this disease with 54 to 59.4 Gy with 1.8 Gy fractions and risk of brainstem injury with higher total doses, hyperfractionation has been tried by several centers using 1.0 to 1.2 Gy per fraction twice per day up to 78 Gy. Unfortunately, this regimen has not been shown to prolong survival compared to conventional regimens (Freeman and Farmer 1998).

In children, these tumors occur more frequently infratentorially compared to supratentorially for adults. Gliomas can also arise in the optic nerves or chiasm and account for 5% of childhood brain tumors. These are generally grade 2 and are more common in patients with neurofibromatosis. They are not resected so as not to destroy vision. Radiation or chemotherapy is used for progressive tumors causing vision loss.

Radiotherapy for low-grade astrocytoma is given as a focal treatment of 45 to 54 Gy. For the higher-grade gliomas, anaplastic and GBM, focal radiation therapy to doses of 54 to 59.4 Gy is used. For brainstem tumors, because of their extremely poor survival, doses of 80 Gy or more have been used without significant improvement in survival.

4.1.4 Central Nervous System (CNS) Germ Cell Tumors

CNS germ cell tumors arise from primitive germ cells in the embryo that were destined to travel from the brain down the notochord (embryonic spinal axis) to the gonads and produce the gametes. Instead, for unknown reasons, they become malignant and stay in their site of origin in the midline regions, pineal and suprasellar, within the brain. They are classified as pure germinoma or mixed malignant germ cell tumors (MMGCTs). Approximately

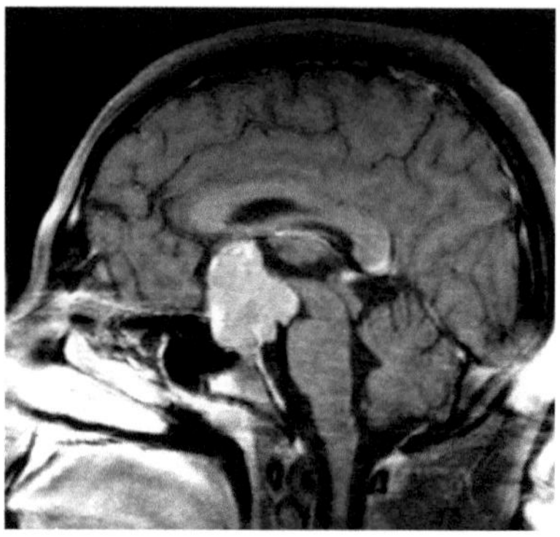

FIGURE 4.6 Sagittal MRI of a germinoma.

two-thirds of the germ cell tumors are pure germinoma. MMGCTs are comprised of four subtypes and often have multiple elements in the same tumor: choriocarcinoma, endodermal sinus (or yolk sac) tumor, embryonal carcinoma, or teratoma (mature or immature). Germinoma refers to tumors in the brain that have histology identical to dysgerminoma in the ovary and seminoma in the testes. Together these are germinomatous tumors. CNS germinoma is rare, occurring in about one in a million children, or 3% to 5% of all primary CNS tumors (Ries et al. 1999). They occur most often in the pineal or suprasellar region of the brain (Figure 4.6). Diagnosis is based on biopsy or tumor markers (alpha fetoprotein and human chorionic gonadotropin) present in the CSF and serum. Dissemination through the CSF is more common than for gliomas. Evaluation therefore involves MRI of the brain and spine, CSF cytology, and tumor marker levels in the CSF and serum. There is a classic calcification of the pineal region in about 15% of cases that can be seen on CT. The most frequent age at diagnosis is 10 to 12 years, which is older than ependymoma, PNETs, and brainstem gliomas. Unlike all the other tumor types discussed in this chapter, resection has not been an important component of treatment, although its role is being studied. Resection has been only important for mature or immature teratomas.

Radiotherapy has been the backbone of treatment for germinomas and nongerminomatous germ cell tumors, but the dose and volume needed is controversial, especially when chemotherapy is also given. Since the 1970s, the standard of care was to give radiation therapy alone, 30 to 36 Gy CSI with a boost to the primary site to a total of 45 to 50 Gy. More recently, smaller target volumes have been tried: whole brain or whole ventricles to a dose of 24 to 30 Gy plus a boost to a total of 45 to 50 Gy (Shibamoto et al. 2001). Similar high cure rates have been reported from several institutions using these smaller volumes. Local field radiotherapy alone appears to be an inadequate volume; a recent Children's Oncology Group (COG) study failed to show that in selected localized, good prognosis germinomas, two cycles of chemotherapy permits reduction of the treated volume to local only to a total dose of 30 Gy without compromising the high control rate achieved by standard radiation therapy alone. An alternative is to give either the whole brain or whole ventricles 21 to 24 Gy followed by a boost to the local tumor site to a total of 30 Gy. These doses can be given sequentially or concomitantly if intensity-modulated radiation therapy (IMRT) is used. With the concomitant boost, 15 daily doses of 2 Gy and 1.5 Gy have been given to the whole ventricles and local site, respectively (Khatua et al. 2010). Current (2012) clinical trials (COG ACNS1123) use chemotherapy response to determine the radiation dose in a randomized study of whole ventricular radiation therapy (18 or 24 Gy) plus local boost (12 Gy).

MMGCT, including mixed germ cell tumors and embryonal cell carcinomas (yolk sac tumors), have poorer prognosis of 40% to 70% (Matsutani 1997; Haas-Kogan 2003). CSI has been an essential component of treatment, usually to a dose of 36 Gy. A total of 50 to 55.8 Gy is given to the primary site after CSI. MMGCTs are treated with platinum-based chemotherapy followed by radiation therapy so ototoxicity is an issue similar to medulloblastoma treatment. COG trial ACNS1123 is testing the efficacy of reducing the initial target volume from craniospinal to whole ventricles (30.6 Gy) followed by the standard boost volume to a total of 54 Gy for patients with a complete or partial chemotherapy response.

4.2 Target Volume Definition

The planning CT scan should consist of contiguous slices no more than 2 to 3 mm thick. For noncoplanar beam treatments, they should extend from just above the top of the head down to at least C4. For coplanar treatments of the posterior fossa region, CT slices should be taken to at least C3 and at least 2 cm superior to the planning target volume (PTV) and for other locations, at least 2 cm superior and inferior to the PTV. A larger CT scan length may be needed to ensure the DRRs contain adequate bony anatomy. For CSI, the planning CT should include the entire head and spine to at least S5 (see Section 4.3.1). For each of the tumors described below, currently there is great interest in reducing the volume (and in some cases the dose) to be treated. Clinical trials are underway as of 2012 for medulloblastoma, ependymoma, and germinoma exploring this possibility. In each case below, after the gross tumor volume (GTV) is drawn in the planning system, a three-dimensional (3D) clinical target volume (CTV) and PTV margin is added to account for microscopic spread and daily setup variations. The CTV margin depends on the disease, while the PTV is usually between 3 and 5 mm, depending on the immobilization accuracy and reproducibility available. Noninvasive head immobilization systems are available that can provide 1 to 2 mm reproducibility and should be used whenever feasible to keep the PTV margin as small as possible. It is important to understand that the choice of

PTV margins should not be based on whether intensity-modulated radiation therapy (IMRT), 3D conformal, or anteroposterior–posteroanterior (AP–PA) treatment methods are used. PTV margins depend on the quality of immobilization and patient-specific reasons (including patient and organ motion) for anticipated daily setup variations. A study should be done by each institution for each treatment site and immobilization device to characterize the reproducibility obtained and define the needed PTV margin.

4.2.1 Medulloblastoma

The standard volume of the boost for medulloblastoma is the whole posterior fossa. However, the COG is currently conducting a randomized trial testing the hypothesis that for average-risk medulloblastoma, only the surgical resection bed and any residual disease (plus a margin) need the boost. Several published series demonstrated no increase in recurrence rate using a tumor bed boost. This is believed to be safe because failures rarely occur in the posterior fossa outside this volume (Merchant, Kun, et al. 2008). For high-risk cases, the boost volume still encompasses the whole posterior fossa. The whole posterior fossa is considered the CTV rather than the gross tumor volume (GTV). It includes C1, the entire cerebellum, and the entire brainstem superiorly. An MRI scan or contrast-enhanced CT scan is used to define the tentorium, which is one boundary of the posterior fossa (Figure 4.7).

For medulloblastoma, when a localized posterior fossa boost is treated or for sPNET boosts, the GTV is the surgical resection bed plus any gross residual. MRI–CT image fusion can be helpful in placing the presurgical tumor onto the postsurgical planning CT. For any of the brain tumors described in this chapter where a localized tumor bed is being treated, it is important to try to determine what tissues in the brain seen on the planning CT were in contact with the tumor before resection. This is often difficult and may require the radiation oncologist to consult with the surgeon to better define this margin.

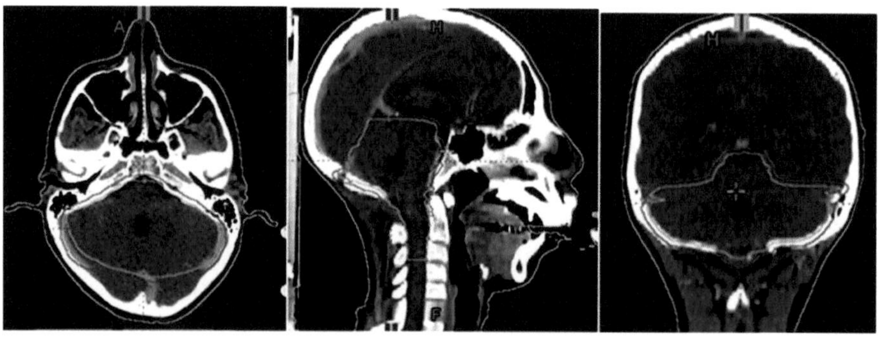

FIGURE 4.7 (See color insert.) Medulloblastoma (whole posterior fossa) CTV and PTV on axial, sagittal, and coronal images.

In the case of total resection, usually a volume smaller than the presurgical tumor volume can be defined as the GTV or CTV. Simply merging the presurgical tumor volume onto the planning CT will create unnecessarily large and misplaced target volumes. The CTV is made by uniformly expanding the GTV by about 1.5 cm. At this point, the CTV is inspected to see if it crosses into regions where tumor cells are unlikely to have spread, for example, in the cranial bones or superiorly across the tentorium. The CTV should be cropped to these barrier structures. The PTV is a 3 to 5 mm (depending on accuracy of immobilization) expansion of the cropped CTV. For medulloblastoma local boosts, it is not unusual for the volume of the PTV to be more than 60% the size of the whole posterior fossa PTV, but any reduction in volume is important and goes a long way toward reducing nearby critical structure dose. For average risk cases, some centers boost the whole posterior fossa to 36 Gy, then boost just the primary site to 54 Gy (Merchant, Kun, et al. 2008).

4.2.2 Ependymoma

For ependymoma, MRI–CT image fusion can be helpful in aligning the presurgical tumor onto the postsurgical CT. As mentioned earlier, the radiation oncologist should carefully consider what part of the fused volume actually remains in the planning CT. The GTV is expanded by about 1 cm to make the CTV. The CTV is cropped to avoid extension across bony or other anatomical barriers preventing the spread of tumor cells. Depending on the location of the target, a subset of the critical structures for brain tumors in Table 4.1 should be drawn.

TABLE 4.1 Organs at Risk and Total Dose Objectives (Percent of Prescribed Dose or Actual Total Dose [Gy])—Localized Targets in the Infratentorial Brain

PTV and Organ at Risk	Mean Dose	Maximum Dose	Dose–Volume
Infratentorial brain PTV		D2% ≤ 105%	D95% = 100%
Cochlea	35%		—
Pituitary/hypothalamus	27%		—
Optic chiasm	—	54 Gy	—
Optic nerves	—	54 Gy	—
Spinal cord (C-spine)		54 Gy	
Brainstem		D 1cc <62Gy	
Lenses	3 Gy		—
Temporal lobes	—	—	D50%<35% D10%<65%
Hippocampus	45%	—	—

Note: D50% means the dose that 50% of the volume receives, other percents are of the prescribed dose.

4.2.3 Glioma

Since low-grade astrocytomas are frequently totally resected, there may not be an abnormality on the planning CT. Image fusion is required to determine the CTV. Since these tumors are not infiltrative, a margin of 0.5 to 1.0 cm is used to expand the tumor bed region that was in contact with the tumor before surgery.

For the higher-grade gliomas, anaplastic and GBM that are resected, either partially or totally, MRI–CT image fusion can be useful in helping to delineate the GTV. For tumors that are not resected, MRI–CT fusion may be helpful if the tumor is not contrast enhancing on the planning CT. Margins of 1.5 to 2.0 cm may be necessary, depending on the degree of edema and areas of uncertainty for the GTV as seen on MRI.

4.2.4 Central Nervous System Germinoma

Localized CNS germinoma is sometimes treated postchemotherapy by radiation to the whole ventricles (WV) including the lateral, third, and fourth, which are expanded by 1 cm to create the WV CTV. This makes a rather large volume but still much smaller than treating the whole brain. Fortunately, the dose used for the WV is not more than 24 Gy in the current COG ACNS1123 study, which should not be neurotoxic for most children older than about 6 years old. The tumor bed GTV (for either pure germinoma or MMGCT) is identified as the prechemotherapy tumor extent plus any residual and is expanded by 0.5 cm to create the tumor bed CTV. In about 15% of the pure germinomas, both the pineal gland and suprasellar regions are involved with tumor, so there may be two boost volumes. The surfaces of these volumes are then cropped where they extend past barriers for tumor spread, such as the tentorium and bone (Figure 4.8).

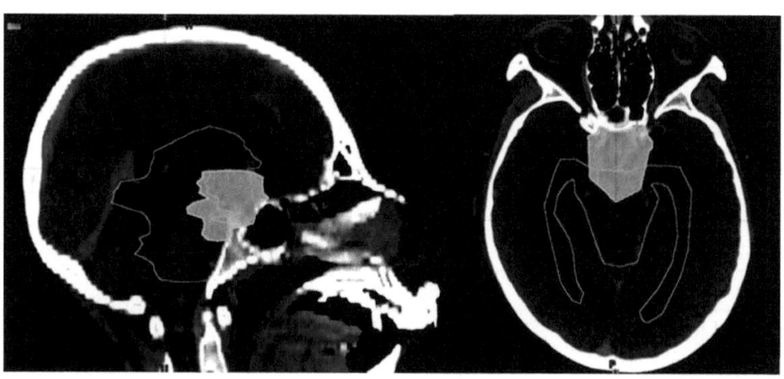

FIGURE 4.8 Germinoma whole ventricle and boost volumes: (a) sagittal, (b) axial.

4.3 Treatment Strategies

4.3.1 Craniospinal Axis Treatment

Whole craniospinal axis irradiation is one of the most challenging treatments to plan and deliver accurately. The most common tumors requiring CSI in pediatric radiation oncology are medulloblastoma, sPNET, central nervous system leukemia, and germ cell tumors.

Conventionally, it involves matching lateral opposed whole brain fields with one or more posterior spine fields. One important objective is to make the junction dose as uniform as possible to avoid a spinal cord myelopathy or tumor underdose, which is accomplished by leaving a small gap between fields, moving the junction, feathering field borders or using penumbra spoiling. Typically, a collimator rotation is used for the cranial fields to match the divergence of the spine field. In many institutions, a couch rotation is also used for the cranial fields to avoid divergence caudally into the spine field. There are numerous published treatment techniques and methods for assuring the accuracy of treatment delivery. More and more, the supine rather than prone patient position is used for increased patient comfort, positional stability and reproducibility, and access to the airway in the sedated child. The features of whole craniospinal treatment that make it challenging are the junction over the spinal cord of the whole brain fields with the whole spine field(s); the potential significant radiation dose to critical structures anterior to the spine, such as the intestine, heart, esophagus, thyroid, and trachea; and the coverage of the entire brain without giving the lenses of the eyes a cataract-forming dose.

Patient position, either supine or prone, is a major decision one must make with pros and cons for each. Historically, the prone position was used mainly so that the junction between the cranial and spinal fields could be visualized on the skin of the back of the neck. Lines separated by the desired gap were drawn on the posterior neck indicating where the inferior border of the cranial fields and superior border of the PA spinal field should be placed. Also, the PA spine field borders could be directly visualized on the posterior skin surface and aligned with the posterior processes of the patient's spine by palpation. The problems with the prone position are (1) the face is face-down in a mask that makes access to the airway difficult, a major concern for the anesthesiologist for the sedated child; (2) the prone position is not as stable as the supine position, so positional reproducibility is compromised, and (3) the boost treatment is usually given with the patient supine, necessitating repositioning the patient for this phase of treatment. Separate planning CTs will have to be done, and an accurate composite dose distribution will not be possible, and (4) for the awake child, the prone position is uncomfortable.

Many advocate the supine position because it does not suffer from any of the aforementioned problems. Although a thermoplastic facemask is typically used, a hole can be made at the location of the mouth and if needed the mask can be removed much more rapidly than if the patient were prone. The problems with the supine position are (1) inability to visualize the junction using the light field and (2) inability to visualize the alignment of the spine with the light field of the spine field. The first problem can be addressed by methods using radio-opaque markers at the junction that can be seen in the simulator and portal images. This method can be used both during the planning of the craniospinal fields and for verification on the treatment machine. Other methods to verify the junction include placing small pieces of film under the patient's neck, which remain in place during the treatment of all fields or film dosimetry in phantoms. The second problem can be managed by the use of an AP setup field aligned to lines drawn on the skin.

For supine patients, it is efficient to make a head immobilization device that can be used for both the CSI and boost phases of treatment. Body immobilization for a single PA spine field for a cooperative (or sedated) patient is not necessary if the patient is marked with horizontal leveling lines, a sagittal line, and a center line, and these lines are carefully matched to the lasers daily. To keep the chin out of the exit from the PA spine field, the neck should be flexed so the chin is extended superiorly as far as patient comfort and reproducibility will allow. The patient's head and body should be aligned such that the sagittal laser runs through the patient's sagittal midline from the head to the pelvis. A CT scout or fluoroscopic image can be used to make corrections to the patient's body position, which closely aligns the longitudinal beam axis with the vertebral column. For patients more likely to move or more complex beam arrangements, custom immobilization for the body is recommended. Thomadsen et al. (2003) described his use of a plastic-mesh facemask and either a foam-based or vacuum bead body mold with a connecting bracket for CSI treatment in the supine position, and we have used a carbon fiber head fixation system abutted to a body-sized vac-lock bag (Figure 4.9).

Slampa et al. (2007) reported that for 33 patients, overall and disease-free survival as well as side effects of CSI in the supine position was comparable with results from treatment in the prone position.

4.3.1.1 Conventional Three-Field Craniospinal Irradiation (CSI) Treatment Planning Methods

The CSI treatment can either be planned on a conventional simulator or virtually planned using CT simulation. If the patient's position will not change between the CSI and boost phases of treatment, then the same CT scan can be used for the planning of the boost. Slices separated by 1.5 to

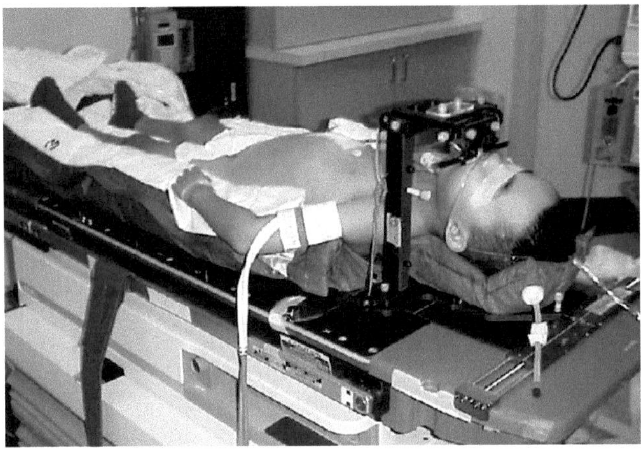

FIGURE 4.9 Head to foot immobilization for supine CSI treatment.

3 mm should be obtained in the head and neck (to just below the inferior-most junction level) and no more than 3 mm for the rest of the body down to about the inferior aspect of the sacrum. This slice spacing makes for very readable digitally reconstructed radiographs (DRRs). For conventional simulation, radio-opaque markers can be placed at the lateral canthi to represent the posterior aspect of the lenses. The following sections generally apply to patient treatment in the supine position with a junction in the inferior aspect of the cervical spine just superior to the shoulders.

4.3.1.1.1 The Craniospinal Junction An important objective of the treatment technique used for craniospinal irradiation is to create a "comfortable" junction between the whole brain and the whole spine treatments. A comfortable junction may include a small gap but no overlap. This junction is usually shifted by 1 cm cephalad every 9 to 12.6 Gy to smear out the low dose strip over several centimeters. Although the clinical significance of the dose in the gap region has not been found to increase the incidence of residual or recurrent disease, most recommend feathering based on their dosimetric analysis of small junction gaps (Lees et al. 1988; Tatcher and Glicksman 1989; Holupka et al. 1993; Cheng et al. 1994; Kiltie et al. 2000). Tinkler and Lucraft (1995) suggested that well planned and executed stationary junctions are as good as feathering and presented data showing that disease-free survival was comparable to other centers that use feathering. The junction can be created either as a single plane (compensating for the beam divergences) containing the inferior border of the brain fields and the superior border of the spine field(s), or by retaining the beam divergences. It should be noted that some centers choose to avoid the couch rotation that produces the coplanar junction and instead introduce an additional 2 to 3 mm skin gap to account for the fact that the lateral brain fields diverge inferiorly into the spine field. If there is perfect abutment of cranial and spinal fields at the center of the cord without a table rotation, there is a wedge-shaped overlap on the exit side of the cord that can be up to 1.5 mm from center to cord edge. There is a similar gap on the entrance side of the cord. By introducing a small additional gap, this wedge-shaped overlap can be avoided. Many centers that employ the fully coplanar junction method also introduce a small gap as a safety margin against a systematic error.

We now examine the impact of various gaps on the dose in the junction region for a conventional PA spine field and opposed lateral cranial field plan with or without applying junction shifts. Figure 4.10 and Figure 4.11 show the junction region for the abutted fields without and with two junction shifts with a 2-mm and 5-mm gap, respectively. Figure 4.12 compares the junction dose with and without two shifts (three junction positions, i.e., treatment of high-risk medulloblastoma) for the situation where there is either an abutment, or a 2-mm or 5-mm gap. With a perfect abutment, as one might expect, the dose is highly uniform (within 5%) even without employing any shifts. But with a 2-mm or 5-mm gap, there is a 27% and 67% dose deficit in the junction, respectively, if no shift is used. By shifting the junction twice (the junction exists at three locations 1 cm apart), the maximum

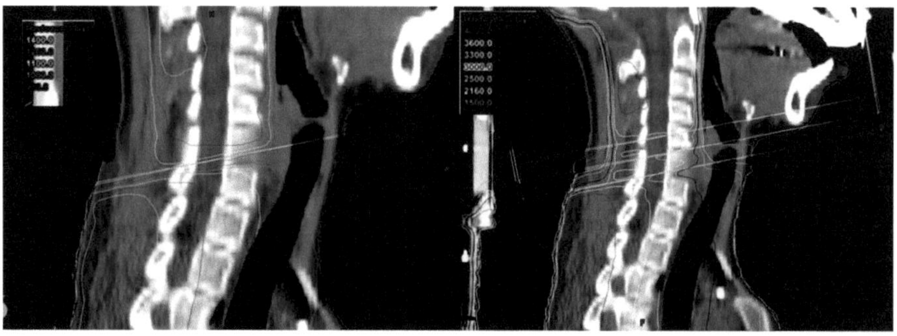

FIGURE 4.10 (See color insert.) Two-millimeter gap between PA spine and whole brain fields, without (left panel) and with (right panel) two junction shifts.

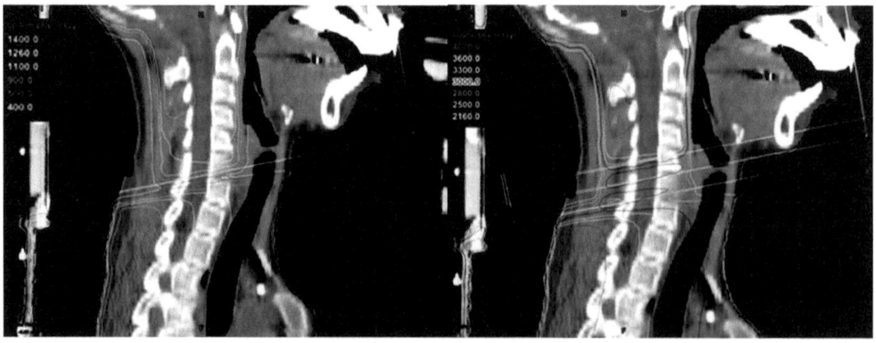

FIGURE 4.11 (See color insert.) Five-millimeter gap between PA spine and whole brain fields, without (left panel) and with (right panel) two junction shifts.

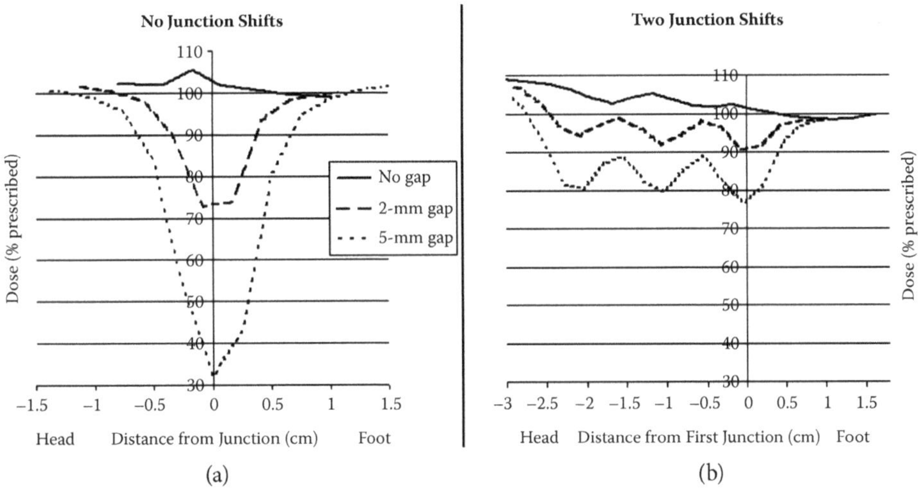

FIGURE 4.12 Dose across junction, either abutted, 2-mm or 5-mm gap: (a) without shifts, (b) with two 1-cm shifts.

dose decrease is effectively reduced to just one-third of these values, 9% and 23%, respectively. Dose deficits are spread over a 2 cm long region and range from 1% to 9% for the 2-mm gap and 12% to 23% for the 5-mm gap. These values are similar to what others have reported in the literature (Lees et al. 1988; Tatcher and Glicksman 1989; Holupka et al. 1993; Cheng et al. 1994; Kiltie et al. 2000). Based on these data, most have recommended feathering the junction. It is important to note that these values are highly dependent on the accuracy of the beam penumbral model used in the planning system. In addition to dose heterogeneity at the junction due to summation of doses from the three fields, one should be aware of additional decreases in dose due to attenuation by immobilization devices present at the level of the junction. A good rule of thumb is that the dose is attenuated by about 4% per centimeter of intervening solid plastic-type material.

As discussed earlier, methods of addressing the junction dose conventionally include using a small gap or an abutment that is shifted periodically. Because of the added complexity of creating and moving a junction during treatment, several methods have been devised that avoid junctions altogether, including TomoTherapy or by using a beam spoiler or IMRT to feather the dose daily across a 2 to 5 cm long "junction" region producing a uniform dose and eliminating the need for junction shifts.

The junction level is also an important consideration. Two locations for the junction have been widely used, one high up in the neck (around C2) and the other much lower in the neck (between C5 and C7). One motivation for using a junction higher in the neck is that the dose to the neck is typically higher than at central axis of the cranial fields. There are three reasons for this higher dose: (1) the lateral thickness of the lower neck is much less than that of the brain, (2) the source to surface distance (SSD) along ray lines inferior to the central axis will be smaller than at central axis when the couch is rotated, and (3) if the inferior border of the field is away from the central axis, the spinal cord will receive about a 3% higher dose due to the "horns" of the beam. Using the spine field instead of the cranial fields to treat the upper cervical spinal cord gives a dose to the spinal cord that is close to the prescribed dose. Another motivation for a high junction is that split beam whole brain fields can be used and still cover the entire brain with the 20 cm field length. This avoids using a couch rotation during the whole brain fields to compensate for the divergence of symmetric fields. A persuasive reason not to use the high junction is that the pathway of the exit dose of the spine field will traverse the thyroid, airway, larynx, pharynx, mouth, mandible, salivary glands, and other structures that can be blocked with lateral cranial fields when the junction is lower. Narayana et al. (1999) reported on junction dose for either a high junction at C1-C2 or a low junction at C5-C7 for patients receiving 36 Gy CSI. The average dose to the cervical spinal cord was 11.9% higher (40.3 Gy) than the prescribed dose with the low junction, and 6.7% (38.4 Gy) higher with the high junction. However, doses to the thyroid gland, mandible, pharynx, and larynx were increased by an average of 29.6%, 75.8%, 70.6%, and 227.7%, respectively, by the use of the high junction compared to the low junction. Most of the thyroid gland was irradiated with either junction, getting a mean dose of at least 20 Gy. The thyroid gland is highly radiosensitive and susceptible to the carcinogenic action of ionizing radiation and every precaution should be made to minimize dose to this structure.

Note that if the spine is treated at extended distance and a 1 cm junction shift is desired, the inferior jaw of the isocentric brain field will reduce by 1 cm, but the superior jaw of the spine field will increase by 1 cm × (100/SSD). Thus, if treating at 140 cm SSD, the superior spine field jaw will increase by only 0.7 cm so that on the skin at 140 cm, it increases by the same 1 cm as the brain field decreases.

4.3.1.1.2 The Spine Field The clinical target volume for the whole spine is the entire spinal cord and thecal sac and therefore usually involves field lengths on the skin longer than 40 cm in older and taller patients. Traditionally, a single PA field is used. This length can be achieved by either using extended SSD when necessary or by placing the superior end of a 40 cm long field at the inferior edge of the cranial fields and then abutting a second PA spine field to the inferior edge of the upper spine field (Liu et al. 2009).

For the case where split-beam cranial fields are not used, it is easier to plan the spine field first because the couch and collimator rotations for the brain fields depend on the jaw setting for the superior portion of the spine field. For a low junction, the inferiormost junction lies about 1 cm superior to the top of the shoulders, which always should be pulled down to achieve the lowest

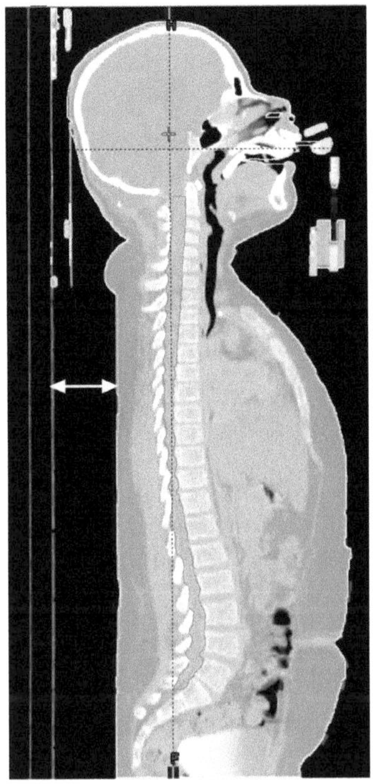

FIGURE 4.13 Supine craniospinal positioning using 5 cm thickness (arrow) of Styrofoam under patient to level the spine.

possible junction, frequently achievable at C6. To assure that the lateral brain fields do not irradiate the shoulders, pull the arms down and use an immobilization device to reproduce this position. Head immobilization with a hyperextended neck will avoid the PA beam from exiting through the mouth. For the awake child, holding on to posts can accomplish this. For sedated children, a device to preserve the planned shoulder position may be needed. To straighten the T- and L-spine relative to the C-spine, one can either place a 4- to 8-cm spacer using Styrofoam sheets under the patient from below the neck to the sacrum (Figure 4.13) or one can position the head below the level of the body using a special headrest (Liu et al. 2009). The following simulation procedures have steps for 2D planning as well as CT-based planning.

First, the patient should be straightened so that the sagittal laser goes through the midline of the head and truck. The vertical side lasers can be used to mark the location for the first junction just above the shoulders. BBs should be attached to the skin, mask, or immobilization device on this laser line at the level of the spinal cord. It is highly recommended that the patient be in an indexed immobilization system to provide the most accurate junction match each day as well as accurate field positioning. If the treatment will be indexed, then it is preferable that the BBs be placed on the mask or immobilization device rather than on the patient. The method described here provides a geometrically reproducible junction at a point that corresponds to C5-C6 but may vary on a daily basis due to differences in patient position. This variation is advantageous as it serves to further feather the dose across the junction. The gantry is set to a PA beam angle, and the field size set as necessary to cover the distance between the top of the field during final shift and S2 (or wherever the inferior border of the field should be). Ensure that the central axis of the spine field was placed such that the superior jaw setting of the spine field will never exceed 20 cm. The inferior border can be found from a recent MRI sagittal image of the spine and is commonly at S2-S3 and may be as high as S1 and as low as S4. Between 10% and 33% of the time, placement of the inferior border at the bottom of the S2 vertebral body would miss the bottom of the thecal sac (Dunbar et al. 1993; Scharf et al. 1998).

It is usually found simpler to use extended SSD to cover lengths greater than 40 cm, but isocentric abutting PA spine fields can also be used. Koshy et al. (2004) compared using a single PA spine field at 140 cm SSD to two 100 SSD spine fields abutted at the anterior edge of the spinal cord. They noted that the penumbra of the 140 SSD beam is .77 cm versus .58 cm for 100 SSD, due to increased geometric penumbra at the larger SSD. The dose was more uniform over the cord for the 140 cm SSD beam but, due to the increased PDD with increased SSD, there was increased normal tissue exit dose to intestine, heart, lungs, and ovaries (typically about a 2 Gy dose increase in absolute terms for 36 Gy prescribed dose). Considering the trade-offs, they recommended two gapped spine fields instead of one extended SSD field to lessen the chance of thyroid dysfunction, carcinogenesis, and female sterility.

If a lower spine field is required, then one way to match the two spine fields is to rotate the couch 90 degrees for the lower field and rotate the gantry so that the superior border of the lower spine field matches the divergence of the inferior border of the upper spine field (see Figure 4.11). For long spine fields, place the junction inferior to the conus of the spinal cord, generally located between the T12 and L2 level to avoid the risk of cord damage in the advent of an overlap of fields. The conus is easily identified on a sagittal MRI of the spine. The inferior field is planned by one of two methods: (1) a couch rotation to create a common junction plane or (2) by introducing a skin gap with a field abutment at the anterior aspect of the spinal cord. In the first method, the couch is rotated 90 degrees for the inferior field and then the gantry is rotated by an amount equal to the sum of the divergence of the lower half of the upper spine field plus the upper half of the lower spine field. The formula to calculate the gantry angle is

Gantry angle = Divergence of the inferior jaw from upper spine field (IJUS) + Divergence of superior jaw from lower spine field (SJLS)

= Arctan (IJUS/SAD) + Arctan (SJLS/SAD)

The SAD (source–axis distance) is usually 100 cm. For example, if the upper spine field is symmetric 40 cm long, the lower spine field is asymmetric; the superior jaw is 15 cm; and the inferior jaw is 0,

Gantry angle = Arctan (20/100) + Arctan (15/100) = 11.3° + 8.5° = 19.8° (beam angled from superior to inferior along patient's long axis)

Note that the divergence of the inferior jaw of the lower field will cause irradiation of anterior tissues that are more inferior than the target volume to the same degree whether or not you use a split beam for this field. This is because either the inferior jaw is diverging or the jaw is nondivergent (if jaw is on central axis) but the gantry is rotated superoinferiorly based on the length of the superior jaw. Usually the junction between the upper and lower spine fields is shifted at the same time as the C-spine junction.

Figure 4.14 shows the dosimetric result of using a couch rotation for the lower spine field and perfectly abutting it to the upper spine field with the proper gantry angle and central axis position compared to not using couch or gantry rotations and simply using a skin gap with the field edges crossing at the anterior aspect of the spinal cord. With the perfect coplanar abutment, the prescription isodose line runs smoothly parallel to the skin surface anterior to the spinal cord.

In the second method, a simple gap calculation is used to position the intersection of the borders of the upper and lower spine fields near the anterior edge of the spinal cord. A 1 cm junction shift is usually made at this junction as well as the neck junction, adding to the complexity of the treatment. With intersecting spine fields, there is an approximately 20% low dose

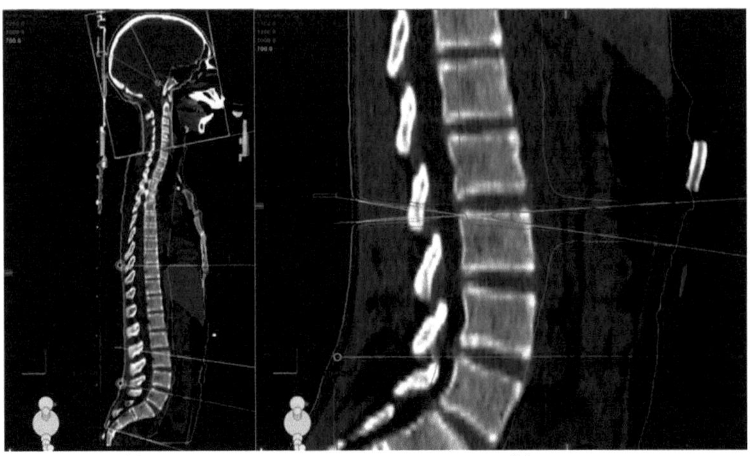

(a) (b)

FIGURE 4.14 Dose across superior and inferior PA spine fields: (a) abutted using couch rotation for inferior spine field, (b) skin gap without using couch rotation. (12.6 Gy given for first junction location.)

triangle in the spinal PTV as well as a 30% high dose triangle in the tissues anterior to the spine. These values will of course be reduced and distributed over 1 to 2 cm by shifting the junction.

The divergence of the lower spinal field into the upper pelvis in patients with two spinal fields could result in unintentional irradiation of the ovarian tissue. Although the ovaries are typically inferior to the bottom of the spine field, at the coccyx level, they potentially could be irradiated depending on the individual patient's anatomy. Mitchell et al. (2007) reported using MRI to find the ovaries and then surgically transposing them if needed, which lowered their dose from about 10 Gy to just 1 Gy for 23.4 Gy target dose with 6 MV x-rays. In one study, 14% of the girls receiving a mean ovarian dose of 2.9 Gy experienced ovarian failure compared to none for those receiving a mean dose of 0.54 Gy (Wallace et al. 1989). Because doses <5 Gy can cause sterility, care should be taken to minimize any dose to the ovaries.

Once the full extent of the spinal target volume has been determined, adjustment of the SSD, central axis position, and field size should be made to obtain the correct upper and lower borders of the PA spine field. The DRR (or fluoroscopic) image of this spine field should show that the superior edge of the field goes through the markers placed at the junction level. The inferior border should extend to the predetermined sacral level. For supine treatment, once the PA spine field is set, an AP setup field can be established by bringing the gantry to the AP position and the field edges marked on the patient using the light field. If an extended SSD is used, the couch can be brought to a convenient height and then the AP setup field borders and horizontal leveling lines can be drawn (Figure 4.15). These lines together with the couch digital readouts provides for a reproducible setup.

Figure 4.16 shows the spinal cord CTV and PTV in regions within and between vertebral bodies. The PTV margin laterally is covered by placing field edges 1 cm outside the recesses of the vertebral bodies to include the spinal ganglia. The depth of calculation is typically taken as the average depth to the anterior aspect of the spinal cord across the entire length of the field. This approach can result in ±15% dose deviations due to the more superficial cord depth at the level of the neck and thoracic spine and deeper depth in the lumbar region. The field width is typically about 4 to 5 cm wide until it widens at about the level of L4 to ensure coverage of the nerve roots and sacral foramina. The jaw setting for the PA spine field is typically about 8 cm wide to allow for this expansion at the inferior portion of the field (Figure 4.17). Halperin (1993) found that the inferior aspect of the spine field should widen by 1.2 to 1.8 cm but does not need to include the sacroiliac joints, which others have routinely included. Multileaf collimator (MLC) leaves should be opened for at least 2 cm above the top of the field to allow for the increase in field size by 1 cm for one or more junction shifts that occur every 5 to 7 fractions.

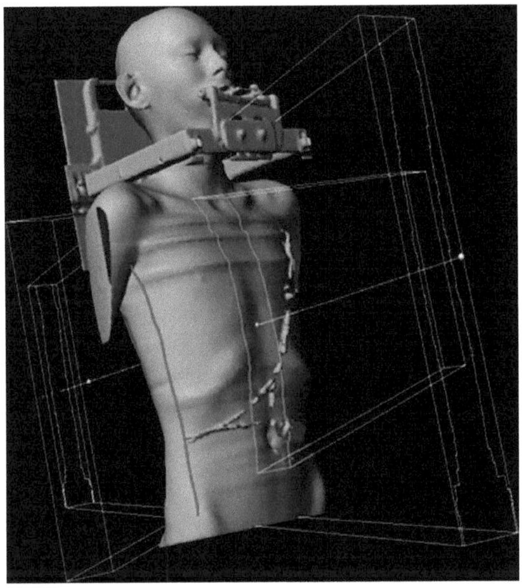

FIGURE 4.15 (See color insert.) PA spine treatment field with AP setup field and horizontal laser leveling line.

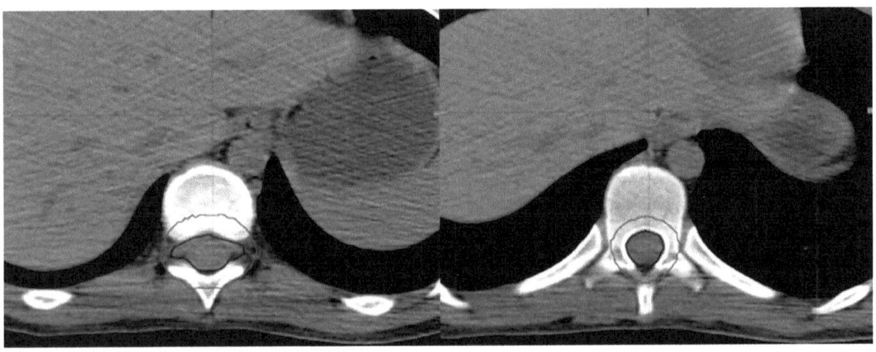

(a) (b)

FIGURE 4.16 Spinal CTV and PTV for (a) intravertebral region containing spinal ganglia and (b) within vertebral body.

Parker and Freeman (2006) described a time-saving CT-based planning technique for supine treatment in which the position of the isocenter of the spine field is always 20 cm inferior to the isocenter of the half-beam brain fields. Thus, the inferior jaw of the brain fields is set to 0 cm, the superior jaw of the spine field is set to 20 cm, and the collimator angle for the brain fields is 11 degrees for every patient. If multiple posterior fields are required, the isocenter of the second spine field is always a fixed longitudinal distance from that of the first. Due to the need to include the entire brain with flash in a 20 cm field, this technique relies on the use of a high junction.

It should be noted that children's spines are rather flexible, and that one can see more than 5 mm changes in the position of the skin surface at the inferiormost aspect of the field and S1 (radiographically) for the same

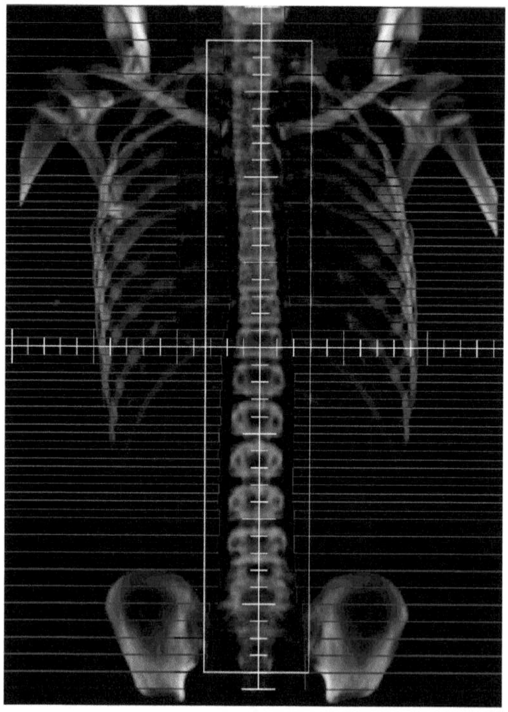

FIGURE 4.17 CSI PA spine field MLC shape over DRR.

position of the central axis or superior border of the field. The AP setup field can help the therapist orient the patient's spine so that the central axis as well as both the upper and lower borders of the setup field match the lines on the skin (see Figure 4.15).

When using a single PA spine field, doses along the spinal PTV can vary by more than 10% as the spinal depth varies. Methods to homogenize this dose are discussed later in this chapter.

4.3.1.1.3 The Cranial Fields The clinical target volume for this set of fields is the whole brain and the spinal cord down to the junction with the spine field. Whole brain fields are typically treated with a simple set of right and left lateral photon beams that are designed to meet the following objectives: (1) flash the skin superiorly, anteriorly, and posteriorly by about 1 cm (this gives about 1.5 cm at the brain surface) at levels above the base of the skull to give full dose to the brain; (2) protect the lenses; (3) give full dose to the cribriform region; (4) spare the skin and tissues at level of the neck posterior to the vertebral bodies; (5) block the face, oral cavity, pharynx, larynx, and thyroid gland; and (6) avoid passing through shoulders by setting the inferior border above them.

A good starting point for the central axis of the lateral beam is a point 3 cm posterior to the sella (for a C5 junction). Typically, the central axis cannot be used as the inferior border for a low junction plan because the total length of the brain fields is more than 20 cm from C5 to the level that flashes the top of the head. A good starting field length is 22 cm. Pulling the arms down can move the shoulders inferiorly by 1 to 3 cm, which allows a more inferiorly-placed junction. Along with having a hyperextended neck, the junction can then be shifted superiorly at least twice without having the PA spine field exit through the mouth.

To create a perfect junction plane with the PA spine field, one method is to rotate the collimator of each lateral brain field to match the divergence of the spine field and rotate the couch so that the inferior planes of the brain fields are coplanar with each other and with the plane of the superior border of the spine field. These angles can be easily calculated as follows:

Let L_{Brain} be the length of the inferior part of the brain fields and L_{Spine} be the length of the superior part of the spine field at 100 cm from the source. Then,
Brain field collimator angle = Arctan(L_{Spine}/SAD) and,
Brain field couch angle = Arctan(L_{Brain}/SAD). The SAD is usually 100 cm.

If the fields are symmetric, then L_{Spine} and L_{Brain} are one-half of each field length.

Note that these angles are not affected by the SSD in an extended SSD treatment. In other words, the divergence angle of the spine field borders is the same for any SSD.

After the couch and collimator angle is set for the first lateral brain field, the superior, anterior, and posterior borders can be set to extend past the inner aspect of the skull by about 1.5 cm. The inferior border should be set to go through the BBs placed during the spine field planning. With CT-based planning, the junction can be matched graphically. If the inferior jaw size is changed after the couch angle is determined, the couch angle should be redetermined and the field position adjusted to again place the inferior border through the BBs or a few millimeters away to create the desired gap. As the inferior border of the brain fields decrease by 1 cm for each shift, the couch angle should decrease by about 0.5 degrees to maintain a coplanar border. In the same way, the collimator angle should increase by about 0.5 degrees for each time the superior jaw of the spine field increases by 1 cm. These small changes, amounting to at most about 1 to 1.5 degrees during the final junction position, may be ignored or addressed depending on the junction gap chosen and the degree of junction dose uniformity required. If these angular adjustments are made, note that the MLC leaf shape may have to be adjusted.

The gantry should now be rotated so that the lenses (or markers on the canthi) are aligned with the longitudinal beam axis. This will allow MLC edges to block both lenses rather than diverging anteriorly into the

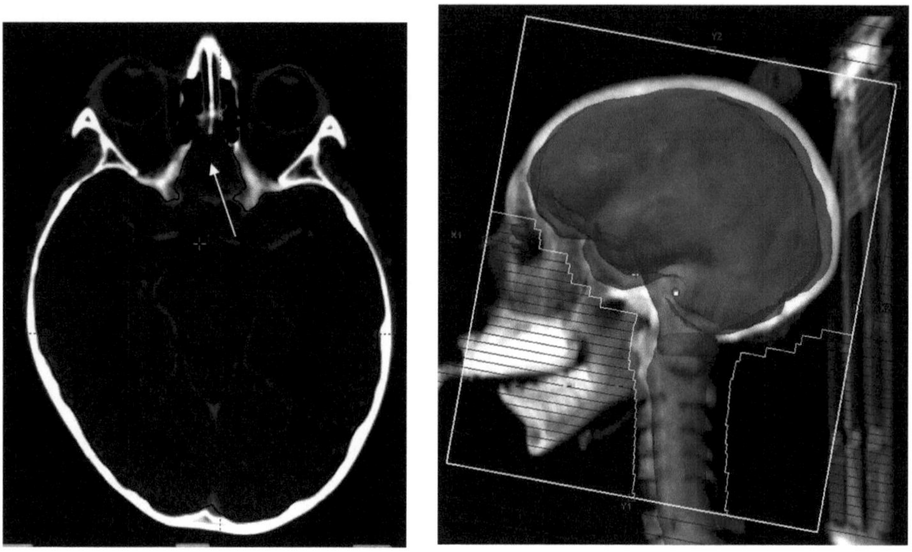

(a) (b)

FIGURE 4.18 (a) Axial CT with contour for the brain including the cribriform region (arrow) and lenses. (b) Left lateral cranial field for CSI, MLC shape, lenses, and brain and C-spine PTV contours.

contralateral lens. Some centers place the isocenter just behind the lenses rather than at the midplane, block the anterior half of the beam, and use a 90 and 270 degree gantry angle to achieve bilateral lens sparing (Woo et al. 1989). Figure 4.18 shows a CT slice through the lenses and cribriform region of the brain and the field shaping for a typical lateral whole brain field.

One of the more challenging objectives of CSI treatment is to fully treat the cribriform plate without producing a cataract. If one is planning the brain fields without the benefit of CT scans, then the locations of the cribriform plate and lenses are uncertain. For conventional simulation, radio-opaque markers placed on the bony orbital canthi are used to represent the posterior aspect of the lenses. Although the lateral canthi can locate a reasonable position for the anterior edge of the brain field in most children, for children under about 5 years old, this point is often not the dosimetrically optimal location for this field border (Figure 4.19). As one can see from this CT scan of a 3-year-old, there is no space between the posterior aspect of the lenses and the cribriform plate. One must irradiate the lenses of this child to completely treat the cribriform region. Thus, CT-based planning is encouraged.

Lateral brain field shaping can be performed by contouring the brain from the superior extent of the cribriform to the inferior extent of the base of the skull (as a minimum), and also both lenses. Modern treatment planning systems can autocontour the brain with one command. Also contour the lenses with a 0.4 cm margin posteriorly. If there is at least 0.5 cm from the lens margin contour to the brain contour, there should be enough to adequately treat

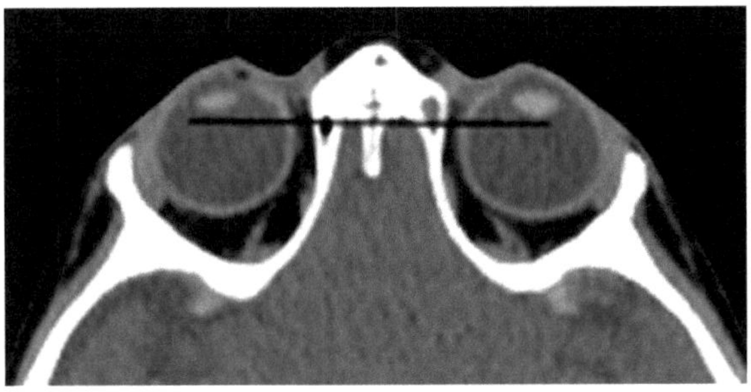

FIGURE 4.19 Three-year-old showing no separation between lenses and cribriform.

the brain and spare the lenses. Viewing each lateral brain DRR (or simulator image), one can create a field shape that has at least a 0.5 cm margin on this brain contour in the region of the lenses. Insufficient coverage of the cribriform plate and the use of overgenerous lens blocks have been given as causes for frontal area relapse (Jereb et al. 1981; Carrie et al. 1992; Mah et al. 1998). Others have not found an association (Taylor et al. 2004). Karlsson et al. (1995) in a CT study of 66 adult patients observed that the cribriform plate was located at the same level as the lenses in half of the cases. Thus, shielding the lens may lead to underdosage of the cribriform plate in patients treated with conventional cranial irradiation. Miralbell et al. (1997) documented a case of cribriform relapse due to wrong field design and showed a significant correlation between inadequate field margins and recurrence in the region of the cribriform. The relationship between the lenses and the cribriform plate cannot be accurately determined on a simulator film. Assuming one has an accurate beam model of the penumbra across the MLC leaves, a dose calculation can show if the brain is receiving adequate treatment while keeping the lenses below about 20% of the prescribed dose (5–7 Gy). Iterative MLC leaf adjustments can be made to optimize these dosimetric relationships. Figure 4.20 shows the isodose distribution at the level of the cribriform and lenses in which the cribriform is within the 100% isodose while at the same time the lenses are at about 15% of the prescribed dose. In cases where it is not possible to achieve both lens sparing and full dose to the cribriform region, full target coverage is usually chosen over sparing the lenses, as cataracts can be surgically repaired but recurrent PNET is often fatal. MLC leaf width is a factor in obtaining adequate dose coverage and sparing. Using 0.5 cm leaf width or smaller is recommended. Hood (2005) found that 0.5 cm wide MLC leaves and 6 MV photons gave 20% lower lens dose than 1 cm wide MLC leaves and 4 MV due to less scattering and higher-resolution field shaping with the former. Those who only have 1 cm wide leaves should use cerrobend shields in the cribriform-lens region (Kalapurakal et al. 2000).

All MLC leaves outside the contoured cribriform and facial region will be pulled out of the field (superiorly, anteriorly, and posteriorly where flash is

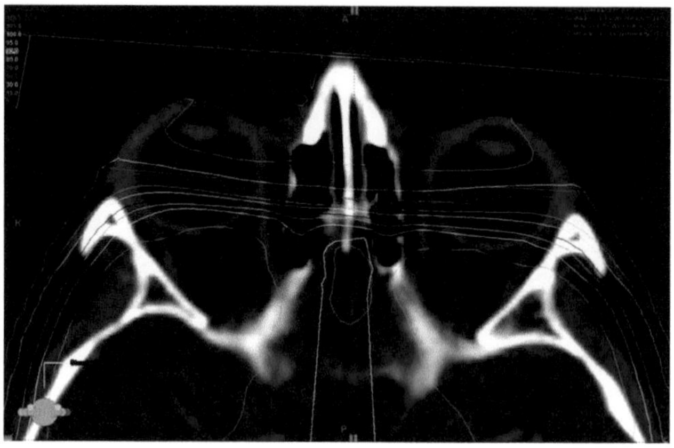

FIGURE 4.20 Dose in the cribriform and lens region from opposed lateral brain fields. Lens dose is below 15%, cribriform region is within 100% dose in this older child.

desired) except to protect the superficial tissues of the posterior aspect of the neck while including the spinous processes.

Doses in the cranial field will exceed the central axis dose by up to 10% in the periphery where there is tangential penetration and in the neck where in both cases, the separation is less than on the central axis.

4.3.1.1.4 Treatment Setup and Verification Figure 4.21 shows a three-dimensional representation of the classic three-field CSI beam arrangement for the patient in the supine position. With the prone position, the junction of the brain and spine fields can be checked daily by drawing a line on the back of the neck that corresponds to the light field from the inferior border of the brain fields and then matching the light field from the superior border of the spine field to that line. With the supine position, this method must be replaced. An effective method utilizes the same principles used to plan the fields for the supine position. At treatment, the brain fields are treated first. A radio-opaque marker is placed on the mask or immobilization device at the inferior border of each field at the level of the spinal cord. Prior to treating the PA spine field, a double exposure is taken. On this image, the two markers should be apparent. A line can be drawn between the markers and this line should be the planned distance from the top of the spine field (either a perfect match or a deliberate gap) (Figure 4.22) (Lau et al. 2006; Munshi and Jalali 2008). Lau et al. (2006) used thermoplastic mask immobilization and an indexed couch system to set up the spine and brain fields. By using an electronic portal imaging device (EPID), they found a 1.4 mm average displacement of the fields. Hawkins (2001) described a geometrical procedure at treatment of determining the position of the PA spine field based on using a predetermined AP setup field and its relationship to the cranial fields. Although many centers have reported on their successful

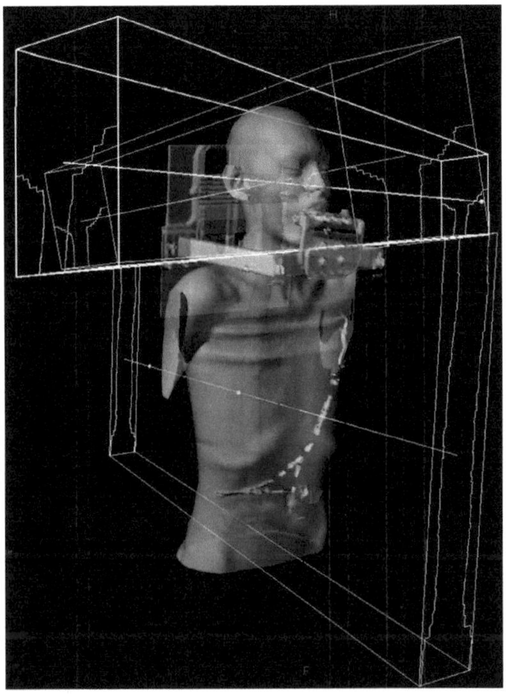

FIGURE 4.21 Three-field CSI shown in 3D.

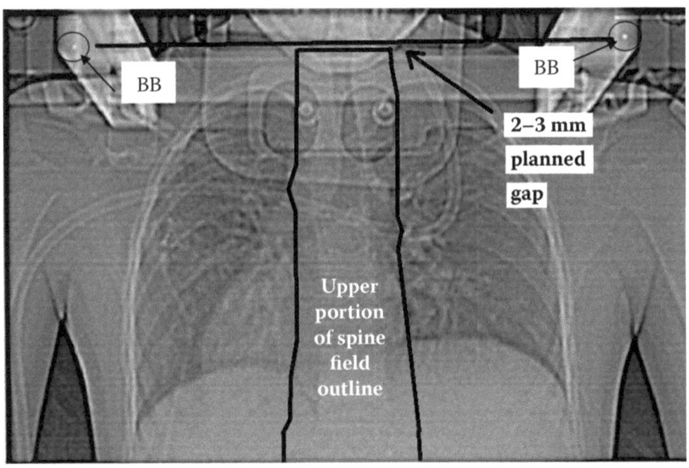

FIGURE 4.22 Portal image of PA spine field showing BB markers, outline of upper part of spine field, and graphical method of verifying the gap.

methods of verifying the matching of fields at the junction, Chang et al. (2003) found a 5 mm random error for the supine technique that was larger than for the prone technique reportedly due to the inability to visualize the junction on skin.

Other methods of verification have been proposed that measure the dose at the junction on film either in a phantom or with the patient present. Rades et al. (2001) and Michalski et al. (2002) describe systems of exactly

FIGURE 4.23 Verification film showing match of PA spine and cranial fields. (From Panandiker, A. P., et al., *International Journal of Radiation Oncology • Biology • Physics* 68 (5):1402–9, 2007. With permission.)

matching the three fields by using the record and verify system to assure the correct collimator settings and rotation, digital couch readouts, and gantry parameters. A small piece of film (Kodak XV) was placed behind the patient's neck (Michalski et al. made a slot in a foam rubber headrest) and was exposed by all treatment fields (posterior flash from the lateral cranial fields and entrance from the PA spine field) to assess field placement accuracy at the junction of these three fields. Also, placement of radio-opaque markers at the junction was visualized in each portal radiograph. Others have described film-in-phantom methods of verification of the three-field match (Figure 4.23) (Panandiker et al. 2007). These methods have shown that the craniospinal junction for a supine patient can be verified and is generally at least as accurate as for the prone position. Several reports on long-term treatment clinical results have shown that CSI in the supine position results in similar tumor control and lack of spinal cord injury as was historically seen for prone CSI (Miemietz et al. 2007; Slampa et al. 2007; South et al. 2007).

Verification of the entire field location should include both visual and radiographic methods. It is a good practice to look at the light field regularly at the anterior aspect of the lateral brain fields in the region of the lenses to be sure that the lens blocks are shielding the intended tissues. Also, when verifying the whole spine field with electronic portal imagers, it is usually not possible to capture the entire field length within one image. Typically, one must image an upper and lower half of the field, each split at the central axis, and compare these separately to the reference images. For spinal fields that are treated at extended distance, it may be necessary to either take images of the upper, middle, and lower portions of the field, or just the upper and lowermost portion that fits on the imager.

4.3.1.2 Variations on Conventional Methods

If a high junction is used or one is treating a young child, then one may be able to use a split beam for the lateral brain fields as long as the field size is less than 20 cm (Thomadsen et al. 2003). In this case, the central axis of the field is the inferior border, placed at the location of the first junction. Because the inferior field edge does not diverge, couch rotation is not necessary. In Thomadsen's technique, the couch is rotated 90 degrees and the gantry is rotated by an angle equal to its divergence to get a vertical plane for the cephalic edge of the spine field (Figure 4.24). Collimator rotation is still needed to compensate for the divergence of the PA spine field.

If the child is very young, it may be possible to use a split beam for both the brain and spine fields. In this case, neither a couch or collimator rotation is needed for the brain fields, making for a vertical junction plane. Since the length of the spine is longer than 20 cm, a junctioned lower spine field is added using the methods discussed earlier.

If maintaining a low neck junction is desired, another variation is to use the full field for the whole brain treatment but use an SAD split beam upper spine field (superior jaw set to 0) with an abutted lower spine field (either with a couch rotation for a coplanar match or with just a skin gap). This enables one to avoid the collimator rotation for the brain fields but the couch rotation is still necessary.

Another method that avoids both couch and collimator angulations for the brain fields is to use a split beam for the brain fields (with a high junction

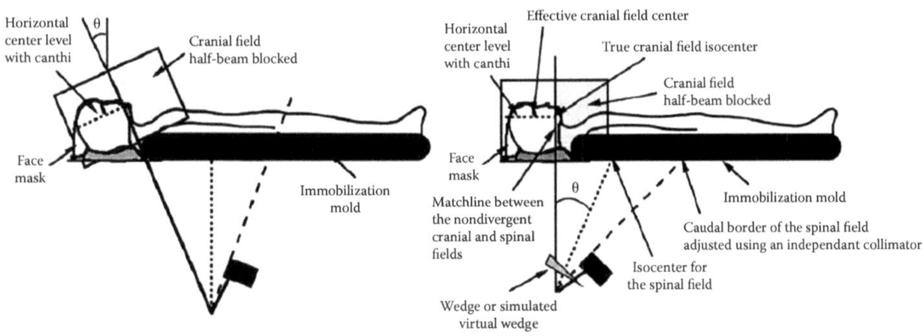

FIGURE 4.24 Variations on CSI setups: (a) cranial field half-beam block, (b) cranial field half-beam block, spinal field gantry, and couch rotated to match cranial field with wedge used to even the dose. (From Thomadsen, B., et al., *Medical Dosimetry* 28 (1):35–38, 2003. With permission.)

or young child) without a couch or collimator rotation but a 90 degree couch rotation and a gantry rotation for the spine field, either using an SAD or extended distance SSD beam. The gantry rotation for the PA spine field is equal to the divergence angle of that beam, making a vertical junction plane in the neck. A wedge (or field-in-field) should be applied to the spine field with the thick end toward the patient's head to compensate for the fact that the distance from the source to the skin is smallest at the superior field edge and increases toward the inferior field edge (Figure 4.24b). Either extended SSD or a second matched inferior spine field can be used for field lengths that exceed 40 cm (Thomadsen et al. 2003).

4.3.1.2.1 Unconventional Craniospinal Irradiation (CSI) Treatment Methods

For all of the unconventional treatment methods described next, the objectives are to reduce exit dose, simplify the field junctions, or to homogenize the dose.

The simplest and most widely achievable modification of the conventional three-field CSI treatment is to add oblique fields to the PA spine field. Here one can both reduce exit dose and homogenize the dose to the spinal cord along its entire length by taking advantage of the oblique beams to compensate for the changing depth in the superior–inferior direction. Either a right and left posterior oblique pair angled 30 to 45 degrees from the vertical is added or a total of five (or more) fields are used, adding two oblique fields to the right and to the left, one 20–30 degrees and the other 40–60 degrees from vertical on each side. Wedges (or field-in-field compensation) should be used for each of the oblique fields. The dose uniformity across the spinal target volume will be increased compared to just a single PA field and if field-in-field or IMRT techniques are used, this increased dose homogeneity can be realized at all levels of the spine. The exit dose will be dramatically reduced, decreasing the chances that the patient will experience uncontrolled nausea that interrupts treatment (Figure 4.25). Also, the dose to the heart, esophagus, and trachea are greatly reduced (Figure 4.26). The junction dose can also be improved because the superior borders of each of the three or more spine fields can be independently set a few millimeters apart to in effect feather the junction. The disadvantage of this method is that dose will be given to structures that normally would not receive any, such as the kidneys, lungs, and liver. By judicious weighting of the beams, these normal structures will receive only a fraction of their tolerance dose even for 36 Gy prescribed dose, as seen in Figure 4.27. Caution should be exercised for females when posterior oblique beams exit through breast tissue due to the small increased risk of radiation-induced carcinogenesis. Note that for long spine fields, an isocentric technique for the spine fields may not be possible. In that case, extended SSD may be employed that requires the couch to be moved both vertically and laterally between each oblique spine field. A

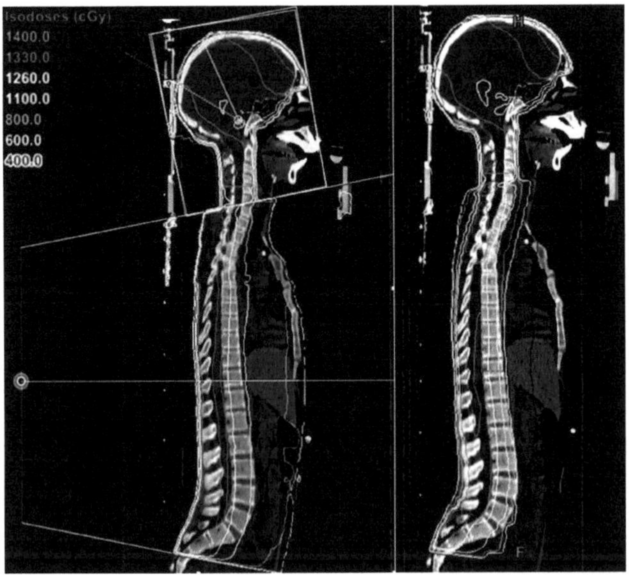

FIGURE 4.25 (See color insert.) Three-field spine (right panel) versus single PA spine field isodoses. Note that the 86 cGy/fx (for 180 cGy/fx target dose) isodose line is at the anterior skin surface for the single PA field but at the anterior edge of the vertebral body for the three-field spine plan.

less attractive alternative is to use an upper and lower version of each of the multiple isocentric spine fields.

IMRT techniques have been reported for whole spine treatment, using either a single field or multiple posterior oblique fields. When a single PA IMRT spine field is used, the objective is to reduce the dose variation with changing depth of the spinal cord along the cephalad–caudad direction. When multiple spine fields are employed, the exit dose can also be reduced (as with non-IMRT multifield spine treatment). Panandiker et al. (2007), using a single IMRT PA spine field, reported a 7% reduction in the target volume receiving >110% of the prescribed dose and an 8% increase in the target volume receiving >95% of the prescribed dose. Although target homogeneity was improved, the maximum dose delivered in the posterior superficial tissues was 13% higher with IMRT. Dose coverage improved for the IMRT plan in the cervical and lumbosacral regions where the depth to the cord was deeper than in the middle of the field. The IMRT single spine field required up to 40% more monitor units (MUs) than the conventional field. Treatment planning treatment time for the complete craniospinal plan was 30 to 45 min.

Parker et al. (2007) compared 23.4 Gy CSI using 2D, 3D, or IMRT methods all using 6 MV x-rays. Organs at risk (OARs) in this study included the thyroid, esophagus, heart, lungs, liver, kidneys, small bowel, and ovaries. The brain–spine junction was moved twice. The 2D plan used conventional field borders, whereas the 3D and IMRT plan used the contoured spine

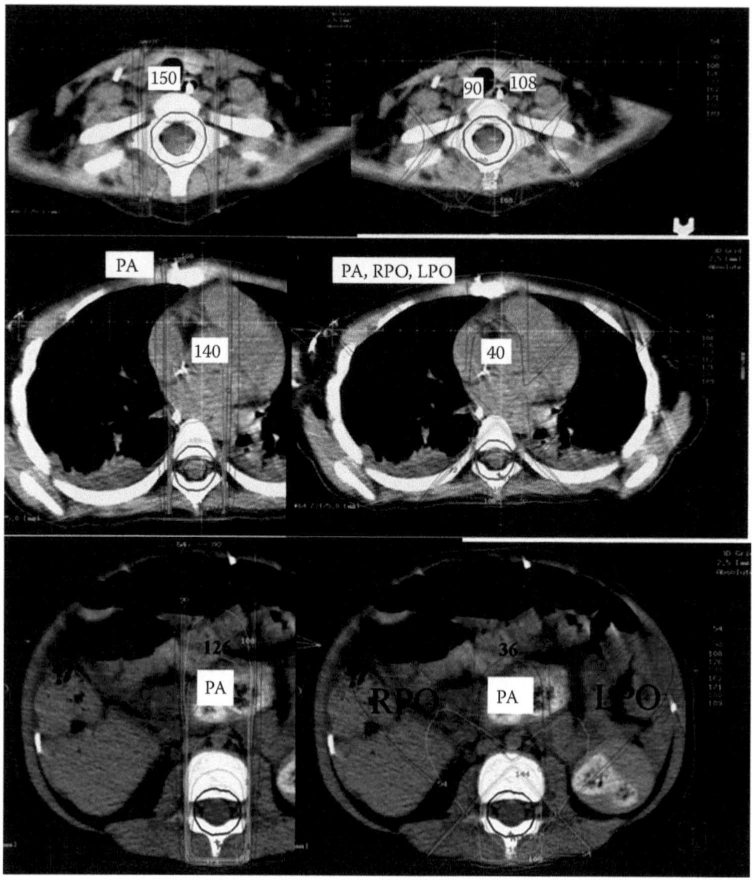

FIGURE 4.26 (Right) Three-field spine versus (left) single PA spine treatment: upper panel, esophagus and trachea doses, middle panel, heart dose, lower panel, gut dose. Inset values are daily doses.

(narrower spine field) and brain with a specified margin. The 2D and 3D plans used a single PA spine field, whereas the IMRT plan used a PA and four posterior oblique fields. IMRT plans gave 3% to 4% better prescribed dose coverage, 2% to 4% better V95% coverage, 3% versus 37%–38% V107%, and 114% versus 118%–119% maximum dose compared to a 2D or 3D plan. For the OARs, the IMRT plan gave higher V5 Gy than the 3D plan but generally less than for the 2D plan. IMRT gave a larger volume dose less than 5 Gy than the 3D plan but less volume above 10 Gy. Less than 1% of the heart was irradiated to doses above 15 Gy with IMRT. He also found a lower nontarget tissue integral dose for the IMRT than 3D plan for doses greater than or equal to 5 Gy.

Others have reported on the use of field-in-field techniques in place of customized compensators for the spine field or all fields. Autosequencing the subfields allowed the junction to be moved during each fraction. This moving junction is achieved by creating three copies of the PA spine field

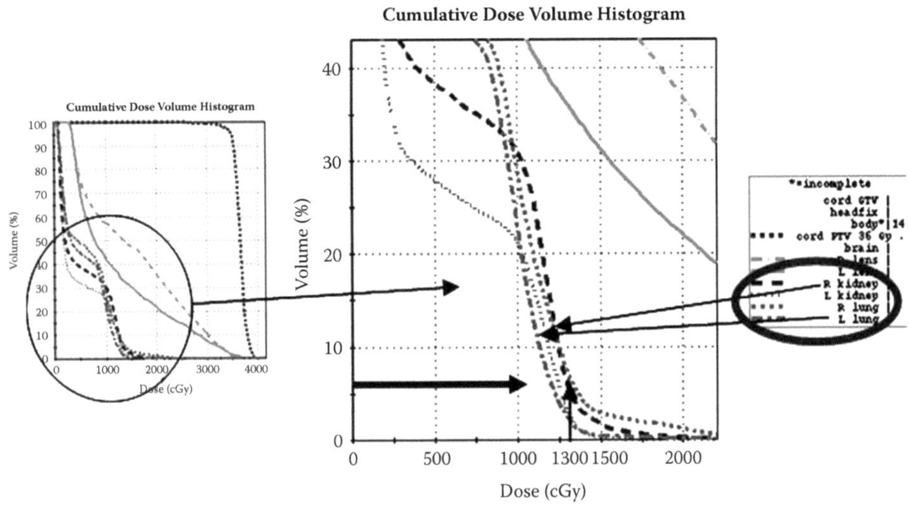

FIGURE 4.27 DVH for three-field spine treatment showing doses to kidney and lung are below tolerance.

and then extending the superior jaws by .75 and 1.5 cm for the second and third field, respectively. After dose calculation, regions that exceed a 4% dose decrease from the normalization point are corrected by adding segments with a minimum segment size of 4 cm and a beam weight of approximately 5%. Segments are routinely added in the upper cervical and lower lumbar regions. A field-in-field approach for the upper half of the spine field can often serve to homogenize the dose to be within 7% to 10% of the prescription. To accomplish the feathering in the junction, three sets of opposed lateral cranial fields are also planned with subfields to compensate the 105% and 110% dose regions that occur with open fields. The inferior borders are aligned with those of the three spinal fields (Happersett et al. 2004; Wilkinson et al. 2007; Yom et al. 2007; South et al. 2008).

In an attempt to simplify the usual process of shifting the field edges at the brain–spine junction, Happersett et al. (2004) used one posterior spine field and five noncoplanar IMRT brain fields. The inferior border of the five brain fields overlapped with the spine field by varying amounts such that the junction region extended over a 5 cm length. IMRT for the brain fields feathered the dose across this region producing a uniform dose and eliminating the need for junction shifts. Dose–volume histogram (DVH) statistics for PTV and OARs including lenses, parotid, cochlea, mandible, and thyroid were 10% to 63% less than with conventional treatment.

Helical TomoTherapy (HT) (Accuray, Sunnyvale, California) has been used as a way to avoid the junction issue altogether due to its ability to continuously translate the patient through the beam throughout the craniospinal target volume. Megavoltage CT imaging is used prior to treatment to assure correct patient positioning. Both conventional and unconventional target volumes and doses (including simultaneous integrated boosts) have been treated with normal tissue sparing for doses >10 Gy at the expense of larger volumes of normal tissues receiving lower doses (Parker et al. 2010).

Bauman et al. (2005) used HT to deliver 36 Gy CSI and compared the dosimetry to that of conventional three-field treatment. OARs in the brain, abdomen, and thorax were protected. They found that to adequately treat the cribriform while protecting the lenses, 10 mm pitch was needed, which increases treatment time to 40 min. It was also noted that large volumes get 3 to 10 Gy, which caused them to be concerned about future production of second malignancy.

Penagaricano et al. (2005) also reported on their use of HT to deliver 36 Gy CSI. They found that HT produced lower doses for the heart, lungs, kidneys, thyroid for both D10% and D50%. Only the eyes had lower D50% for conventional radiotherapy. Total body integral dose was higher by 6.5% for HT over conventional radiotherapy. In their center, beam on time was 27 min for HT versus 3 min for conventional radiotherapy.

Swanson et al. (2005) compared standard three-dimensional conformal radiation therapy (3DCRT) to HT. They defined nontarget organs to include the heart, right and left lungs, esophagus, stomach, right and left kidneys, transverse colon, and liver. Comparison of the percent of prescribed dose received by 5%, 50%, and 90% of the defined organs showed an overall statistically significant difference ($p = 0.0001$) favoring treatment with HT over 3DCRT (Table 4.2).

Volumetric modulated arc therapy (VMAT) techniques can also be used for CSI and produces a dose distribution very similar to HT. Fogliata et al. (2011) provided planning and dosimetric data for five patients treated with RapidArc (Varian Medical systems, Palo Alto, California). All patients were treated with two to three isocenters; the first isocenter was in the brain region

TABLE 4.2 Organ-at-Risk Doses for CSI Treatment with TomoTherapy versus 3DCRT

	3DCRT			TomoTherapy		
	5%	50%	90%	5%	50%	90%
Heart	100	67	7	14	7	3
R and L Lung	68	6	3	23	9	5
Esophagus	103	83	81	22	13	8
Stomach	11	5	3	10	6	3
R and L Kidney	69	4	3	18	8	4
Colon	70	40	2	14	11	8
Liver	85	9	4	17	11	3

Source: Swanson, E., et al., *International Journal of Radiation Oncology • Biology • Physics* 63 (2):S265, 2005.

or at the level of C2, the second isocenter was located in the lumbar region (L1 to L3), and in some cases a third isocenter was needed in the upper thorax (T3-T5). For each isocenter, one to three arcs were used. Adjacent arcs were intentionally overlapped by at least 2 cm so the optimization program could provide a uniform dose for the summation of the plans. It was found that setup errors of 5 mm in the distance between isocenters resulted in 5% junction dose errors. PTV coverage and OAR sparing was at least as good as for HT. For all arc therapy modalities, it is important to carefully define the anterior aspect of the spinal PTV to avoid differential dosing to the vertebral bodies, which can cause scoliosis or kyphosis. Inclusion of the entire vertebral body is generally the safest approach for spinal cord doses of greater than 18 Gy. Another consideration is the consequences of giving a low dose to a large volume of normal tissue. There have been reports of correlation of total lung V5 with pulmonary toxicity (Wang et al. 2006), although that dose level is also highly correlated with V20 and higher. In females, breast dose could also be a concern. Figure 4.28 shows a dosimetric comparison of a two-isocenter arc plan with a conventional opposed lateral brain with extended distance PA spine plan for 36 Gy prescribed CSI dose. Significant sparing of normal structures at the 12 to 25 Gy level is achieved at the expense of greater volumes receiving 5 Gy.

More recently, helical VMAT using a conventional linear accelerator and manual synchronization of the couch longitudinal and gantry motion has been described. An alternating gantry direction was used to avoid the need for the gantry to pass through 180 degrees. Using a 40 cm × 40 cm MLC and a pitch of 10 to 15 cm provided uniform PTV coverage and good sparing of critical structures with a reasonable number of MUs and delivery time (Bedford et al. 2011).

Another method that has been used to reduce exit dose for the PA spine field is to use electrons instead of photons. By carefully choosing the energy of the beam to cover the deepest extent of the spinal cord at any level of the field, good coverage of the spinal target volume can be obtained with no exit dose. Complications of this method are that generally a second electron field needs to be abutted to the first to cover the entire length of the field, even if the collimator is rotated 45 degrees to take advantage of the diagonal of the field. Due to the nature of the isodose distribution of an electron field, it is not possible to obtain a perfect match plane by rotating the couch and gantry for the inferior field. Instead, a skin gap is used, if possible, at a level caudal to the conus of the spinal cord. The junction in the neck is also complicated by the difference in the shape of the dose distribution between photons and electrons. Dosimetric measurements, which include shifting the junction every five to seven fractions, should be made to establish the optimal junction dose.

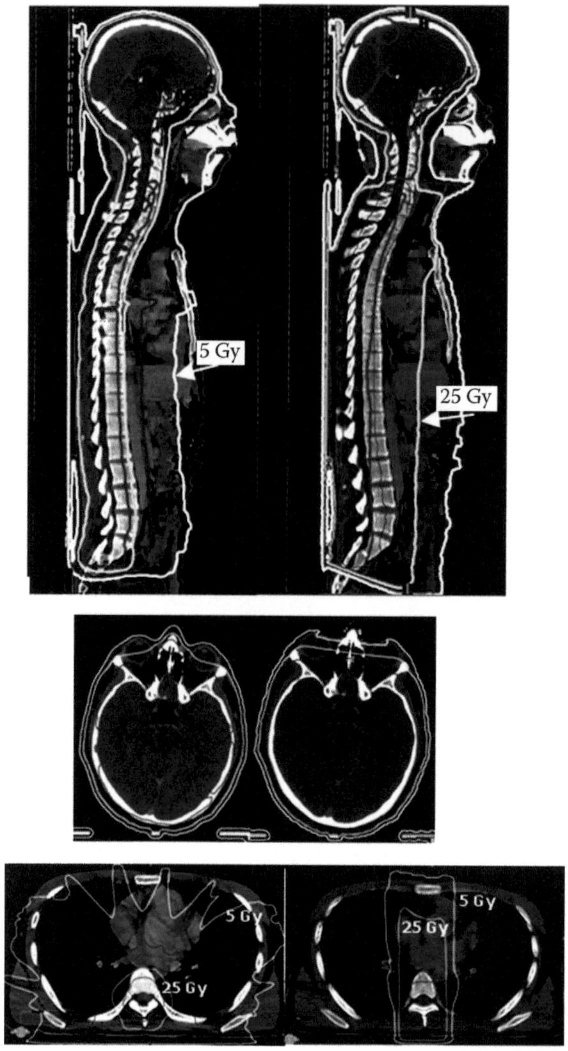

FIGURE 4.28 Left side is 2-isocenter arc plan, right side is opposed lateral brain plus PA spine field. Lens dose is 7 Gy for arcs, 6 Gy for conventional plan. PTV dose is 36 Gy. 5 Gy and 25 Gy isodose lines are indicated showing dramatic differences between two plans.

Li et al. (1994) reported on a technique that involves moving the photon beam in three steps to degrade its penumbra to match that of the electron field. Couch rotations for the brain fields were used to minimize divergence inferiorly. Dosimetric measurements showed that the use of electrons for the spinal field leads to significant sparing of deep-seated normal structures except for an increased dose to lung due to lateral scattering. Customized bolus was used to compensate for the varying spinal cord depths.

Chapek et al. (2002) presented dosimetric results from a study that simulated dynamic couch motion with a single electron field with varying electron energies depending on spinal cord depth. This technique could produce a very conformal dose distribution across the entire spine. However, brain–spine junction matching was not addressed. The current generation of linear accelerators is not capable of continuously moving the couch and varying beam energy.

Maor et al. (1985) advocated the use of electron beams for spine treatments where at least the 90% isodose line can be used. It was noted that this treatment method could only be done with patients prone. The inferior border of the brain field was feathered by shifting it 0.9 cm every five fractions in an attempt to match the penumbral spread of the PA spine electron beam. They reported that none of the five patients receiving 30 Gy to the brain and spine with this treatment technique reported odynophagia or nausea. They note that the doses in the junction region were within –15% and +10%.

Roback et al. (1997) used an extended SSD electron technique. They pointed out four dosimetric problems to address (1) a mismatch in geometric penumbras between the lateral photon brain and posterior electron spine fields; (2) increased electron beam penumbra and decreased width of higher isodose lines due to geometric scattering over the extended SSD, requiring larger widths for PTV coverage but also giving a larger irradiated volume; (3) varying spinal cord depth; and (4) absorption and scattering by bone in the posterior spinal processes requires increasing the beam range by 5 to 7 mm. Roback et al. advocate the use of tertiary lead collimation placed 1 to 5 cm from the skin surface to tighten the penumbra, widen the 90% isodose at the spinal cord level, and reduce the hot spots in the brain–spine junction by 10%.

A variation of this method is to use a mixed photon-electron treatment for the spine field. Patients were positioned either prone if awake or supine if anesthetized, in a custom foam body cradle and plastic facemask. Six MV x-rays plus the lowest electron energy for which the 90% isodose fell 1 cm anterior to the spinal cord were used. The electron field was 1 cm wider than the photon field. In very small children, electrons alone were used. The junction was moved 1 cm daily over a 3-day cycle with a 0.8 cm gap between the brain and photon spine field. When two spine fields were required, they were separated by a gap of 0.5 to 1 cm, which also moved 1 cm on a 3-day cycle. Different energies could be used for the two electron fields, for example, in the thoracic region 12 MeV and in the lumbosacral region, 15 MeV. The upper level of the electron field was matched on the skin to the upper level of the photon spine field. The usual weighting between electron and photon fields was 1:1. Phillips et al. (2004) found that 50% of the heart gets 16 Gy with the mixed beam plan versus 27 Gy for the all photon plan. The mixed beam plan produced a maximum dose per fraction to the upper abdomen of 0.75 Gy compared to 1.5 Gy for the photon plan while large parts of the abdomen received no dose at all. The advantage of the mixed beam plan over all electrons was that a lower skin dose was obtained.

Other methods of increasing junction dose uniformity have been described. Sohn et al. (1994) reported a technique where a single isocenter is used at the junction of the cranial and upper spinal fields. From the isocenter, these fields are extended 1 cm inferiorly and superiorly, respectively. A modifier attached to the wedge tray provides a 2 cm wide penumbra in the overlapping region for each beam, providing a dose across the junction uniform within 10%. Glasgow and Marks (1983) designed an unusual hockey-stick-shaped PA field that encompassed the entire brain and spine by using the diagonal of the field and having patients turn their head first to the right and then to the left side to achieve a uniform dose through the brain.

4.3.1.2.2 Quality Assurance Using immobilization devices indexed to the linear accelerator couch top along with digital couch position readouts for daily setup can reduce the chance of human error and increase the reproducibility of both field positioning and junctions. Choosing an indexed head fixation system that works for both the whole brain and the boost is advisable. For treatments that employ a single PA spine field, indexed immobilization systems may not be necessary if a set of skin marks are drawn on the torso that include an AP setup field rectangle including the central axis and horizontal leveling lines on both sides of the patient from just inferior to the armpit to the level of the bottom of the field.

When relying on digital couch positions, checks of the accuracy and linearity of the longitudinal, lateral, and vertical digital values should be made monthly and after any servicing work on the couch or recalibration of digital values. Linearity in the horizontal and longitudinal direction can be checked by placing a long ruler on the couch, either horizontally or longitudinally, setting the couch to an extreme but clinically relevant position, positioning the ruler so the laser strikes one end of the scale, then moving the couch across its range of travel and noting the agreement between the changing digital readout value and the new location of the laser across the ruler. The distance moved according to the digital readout should be within 2 mm of that measured by the ruler. For the vertical dimension, the ruler is taped to the edge of the couch and the closest side laser is used to indicate the actual distance moved. The measured distance is again compared to the change in the vertical couch digital readout. Linearity of both the couch lateral and vertical is tested for the case of extended SSD oblique spine fields, where the couch may well need to be moved close to its limit of travel laterally at the same time that it is extended 30 cm or more vertically away from the isocenter. It is prudent to create a monthly check where the side and sagittal lasers are set to a reference point on the couch (at a notch or other obvious edge) and the digital couch readouts compared to the baseline values.

For treatments with a single PA field with the patient supine, especially for high-risk medulloblastoma that requires 36 Gy, skin dose must be a consideration in the choice of immobilization and couch devices. Skin dose is minimized with the use of a "tennis racket" (open mesh) couch support and no other material between the support and the patient. Use of a carbon fiber couch top with a vacuum bag immobilization device will significantly increase skin dose and can result in grade 2 or 3 skin reactions. Even a table pad on the tennis racket couch insert can increase surface dose to over 90% depending on its composition. When multiple oblique spine fields are used, skin dose is less of an issue because each beam entrance receives only a fraction of the dose. For this method of treating the spine, indexed immobilization is highly recommended. Couch and immobilization devices also attenuate radiation and must be investigated for this effect. Although tennis racket couch tops have negligible attenuation effects, sandwich-type carbon fiber couches attenuate the beam by at least 1% and often more than 2% (Seppälä and Kulmala 2011). Immobilization devices attenuate the beam by amounts that may be clinically significant, especially for plans with few beams. Solid plastic frames and base plates attenuate about 4% per cm of thickness. Usually vacuum bags by themselves introduce negligible attenuation although they do increase skin dose. Measurements, or at least estimates based on thickness and composition, should be made of attenuation in the structural sections of immobilization devices so that highly attenuating areas can be avoided. These devices should be included in treatment planning system dose calculations if possible.

Multiple oblique fields combined with the PA spine field will greatly increase dose homogeneity if wedges or field-in-field techniques are used. If one is using wedges, then one should ensure the accuracy of the wedged beam model in the treatment planning system not only in the wedged direction but also with respect to the shape of the profile in the unwedged direction. This requires that profiles are measured with each wedge, usually in a water phantom, along the unwedged direction for a 40 cm long field.

Accurate monitor unit calculations for the long and narrow spine field requires the understanding of backscatter into the monitor chamber within the head of the linear accelerator, either by the treatment planning system or by use of correction factors to hand calculations of monitor units. This effect is caused by the radiation being able to more readily scatter back upstream to the monitor chamber from the top surface of the upper jaws of a narrow field than from the lower jaws. Thus, the monitoring system of the linear accelerator behaves like more dose has been delivered than it really has, causing the beam to be turned off earlier than if a square field of the same equivalent square as the long narrow field was used. This effect can result in about a 2% dose error if not taken into account.

When performing second check hand calculations of the treatment planning system MU calculations for the PA spine field(s), the following less commonly used factors are needed: collimator backscatter factor because of the highly elongated field shape and inverse square factor for extended SSD. For wedged fields, wedge factors and wedge off-axis factors if the normalization point is not on the central axis. For field-in field methods, off axis factors and off-axis percent depth doses are needed. For complex field-in-field treatments, measurements or calculation by secondary systems may be the most accurate verification of beam-on time.

For accurate estimates of junction doses, treatment planning system (TPS) beam models must be accurate in the penumbral region. Typically, the 20% to 80% penumbra distance of a 6 MV x-ray beam is about 4 to 5 mm for 5 to 10 cm depth. For each 1 mm error in this parameter, dose errors of 10% to 20% can occur in junction dose calculations. Profiles in water should be taken using diodes or other high-resolution devices to maximize accuracy of the data given to the TPS during beam model creation, and the doses calculated by the model should be compared to measurements. The accuracy of the calculations of dose gradients at the edges of the MLCs will also be affected by the accuracy of the profile data in the beam model. Errors in MLC penumbral modeling will be important when evaluating the dose gradient between the cribriform plate and the lenses, and, of course, for any IMRT treatment. If the spine is treated with electron beams, a dosimetric study using the treatment geometry, and film or computed radiography (CR) plates should be conducted to determine the best junction method and validate the planning system calculated dose distribution. Other factors influencing the accuracy of dose calculation especially in the junction region is the TPS dose grid resolution as well as the CT scan slice spacing. Dose grid resolution and CT slice spacing should be less than or equal to 3 mm, the smaller the better. Dose will be calculated fairly accurately during interpolation across the CT slices and dose grid points to the extent that it is changing linearly in that region.

Six MV x-rays are typically used for the whole brain treatment, but energies from Co-60 up to 10 MV are suitable. Higher energies will produce low dose regions near the more superficial aspects of the brain due to skin sparing.

4.3.2 Treatment Planning for Localized Brain Tumors

In this section, we discuss treatment planning strategies for medulloblastoma and sPNET boost treatments, either for the whole posterior fossa (medulloblastoma) or only the tumor bed (medulloblastoma or sPNET), entirely focal treatments for ependymoma and glioma, as well as whole ventricular irradiation with a tumor bed boost for germinoma.

Good head immobilization will allow PTV margins to be reduced to about 3 mm, important for reducing the irradiated volume of normal brain and other critical structures when more focal treatments are being given.

CT planning should be performed using slices 2 to 3 mm apart from the top of the head down to at least C2 and is preferably done with the same CT dataset used to perform CT planning of the craniospinal treatment (if performed), assuming the same patient position is used. A composite dose distribution of the CSI plus boost can be useful in accurately understanding the total doses in the brain and in the cervical spinal cord included in the whole brain fields and the boost. This part of the spinal cord usually receives more than the prescribed dose to the brain due to being at a lesser depth and shorter SSD than the brain. The composite dose is often required to be included with the clinical trial data submission.

4.3.2.1 General Treatment Planning Guidelines

Historically, opposed lateral fields were used to treat the medulloblastoma boost volume. However, delivering 54 Gy to such a large volume of normal brain (including the temporal lobes) has been associated with severe cognitive dysfunction as mentioned earlier. Because the whole brain has already received a dose of at least 18 to 23.4 Gy for average risk medulloblastoma, a dose that by itself is enough to cause cognitive deficits in young children, the boost should be planned to minimize the volume of normal brain that gets more than about 40% of the boost dose along with minimizing dose to specific normal structures. For high-risk medulloblastoma and sPNET, where the whole brain receives 36 Gy, this objective is especially important. It is important to calculate the dose to the C1-C2 spinal cord from the whole brain cranial fields and add this dose to the boost dose to this same volume. It is likely that for 36 Gy to the whole brain, an additional 2 Gy is delivered to the upper spinal cord, which is significant if one is nearing spinal cord tolerance with the boost treatment. Three-dimensional conformal techniques have been recognized as clinically beneficial for treatment of pediatric brain tumors (Smitt et al. 1998; Loeffler et al. 1999; Merchant, Zhu, et al. 2002). Conformal posterior oblique fields have been used, which avoid the pituitary and supratentorial brain and reduce the dose to cochlea by about 35% (Fukunaga-Johnson et al. 1998; Breen et al. 2004). Although 3D conformal planning techniques are superior to opposed lateral beams, IMRT will nearly always provide better normal tissue sparing. At least 8 noncoplanar beams should be used to treat the boost volume, and 10 beams can be demonstrated to further improve both conformality of target coverage as well as normal tissue sparing. More than 10 beams provide only minimal incremental improvement in the dose distribution. Table 4.3 shows a dosimetric comparison of opposed lateral beams, 6 and 10 noncoplanar

TABLE 4.3 Dosimetric Comparison of Organ-at-Risk Doses for Whole Posterior Fossa versus Local Boost, IMRT versus 3DCRT, 6 Beams versus 10 Beams

Technique	Cochlea Mean Dose Gy	Pituitary Mean Dose Gy	% STB Brain 10 Gy	% STB Brain 20 Gy	% TL 10 Gy	% TL 20 Gy
Opposed Laterals WPF	31	4	20	17	74	69
Opposed Laterals Local	31	3	17	14	68	61
IMRT, 6 Beam WPF	13	8	25	6	59	17
IMRT, 6 Beam Local	8	7	24	4	43	8
IMRT, 10 Beam WPF	10	7	27	6	36	6
IMRT, 10 Beam Local	6	6	18	3	30	4
Conformal, 10 Beam WPF	24	16	33	11	81	27
Conformal, 10 Beam Local	20	9	21	5	67	12

Notes: STB, supratentorial brain; TL, temporal lobe.

IMRT beams, and the same 6 and 10 noncoplanar 3-D conformal beams for either whole posterior fossa or local boost medulloblastoma treatments. Although the pituitary dose is the lowest for opposed lateral beams because these beams are posterior to this structure where the other beam arrangements enter or exit through the pituitary, all other metrics generally favor the 10-beam IMRT plan for either boost volume.

As described earlier, potential OARs are the normal brain, pituitary/hypothalamic complex, the optic chiasm and nerves, the cochlea, temporal lobes, lenses, spinal cord (C1-C2), and the hippocampus. As a general reference for contouring organ-at-risk structures in the brain, a color CT atlas can be found at the following Web site: http://www.crcpress.com/product/isbn/9781420085099. The spinal cord and optic chiasm are allowed to receive a maximum of 54 Gy in recent COG brain tumor protocols. The brainstem is not usually at risk of exceeding its tolerance dose of about 62 Gy during treatment of CNS tumors except in the case of posterior fossa ependymoma, brainstem, and other high-grade gliomas near the brainstem, where hot spots of over 105% of 59.4 Gy begin to exceed the tolerance dose. For boost doses of 30.6 Gy after 23.4 Gy CSI, one would have to have hot spots of over 38 Gy (125%) in the brainstem to approach brainstem tolerance.

Noncoplanar beam plans will provide better target conformality and normal tissue sparing compared to the same number of coplanar beams. This is because with coplanar beams, there is a larger volume receiving dose from overlapping fields than for noncoplanar beams. However, with noncoplanar beams, there is a larger volume receiving dose from some beam, causing larger volumes of the brain to receive low doses. Thus, noncoplanar

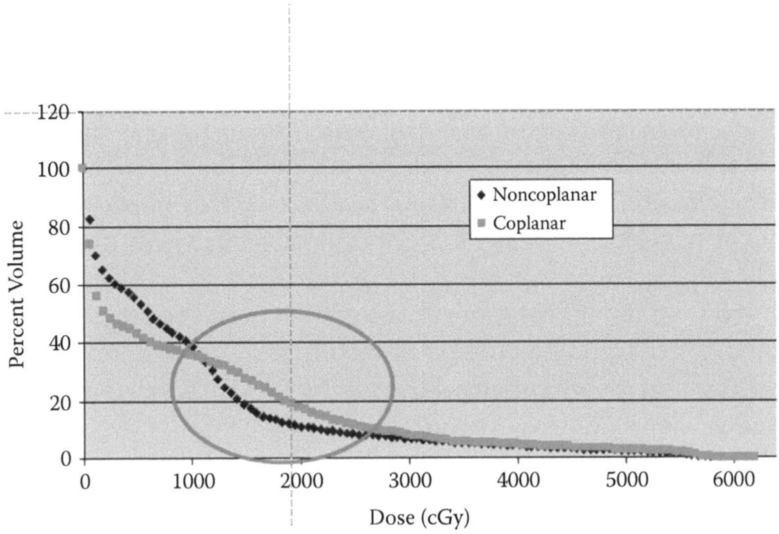

FIGURE 4.29 DVH for noncoplanar versus coplanar beams used to treat brain tumor. Noncoplanar beams treat less normal brain tissue to a clinically important dose than do coplanar beams.

beams treat smaller volumes of brain to medium to high doses but larger volumes of brain to low doses compared to coplanar beams (Figure 4.29). This difference leads to a decreased risk of cognitive and auditory deficits and may even reduce the risk of SM or at least not increase it (see chapter 2). However, the larger the target volume, the less advantage noncoplanar beams have over coplanar beams. As an extreme example, if the PTV fills the cranium from anterior to posterior on the right side, treating with conformal AP–PA beams would be better than either coplanar or noncoplanar multibeam arrangements because with the latter, fairly large doses would be given to the left side of the brain with no better coverage of the target. The Appendix provides noncoplanar beam arrangements that I have found useful for a range of tumor locations in the brain.

After one has decided on whether to use coplanar or noncoplanar beams, we focus on beam angle selection, which is based on the principle that the best plan is the one for which beams (1) avoid critical structures, (2) beam entrances and exits are well separated (by at least 20 degrees), and (3) have the shortest path possible to the tumor. These rules suggest that for eccentrically positioned targets, eccentric rather than uniformly spaced beams may be best.

If nonintensity modulated beams are used, wedges (or field-in-field) will likely be needed to improve dose homogeneity, especially if noncoplanar beams are used. For noncoplanar beams, the wedged direction of the beams do not act in a single plane, so that the wedged effect is diminished. A solution to this problem is to rotate the collimator of each wedged beam to align the wedged direction to the necessary axis. This process can be accomplished by looking at one pair of wedged beams at a time in 3D visualization mode and rotating the collimator until the wedged axes are seen to be collinear. In some cases, it may be advantageous to use a collimator angle that distributes the wedged effect to two different directions at the same time if needed. Here, a higher angle wedge is usually required to achieve adequate dual-plane wedging.

For 3D conformal plans, if the OAR is adjacent to or inside the PTV, one must use the MLC leaves to block the critical structure in as many beams as it takes to achieve the needed dose reduction. This will reduce the dose to the required level but will also create a cold region in the target.

Although sPNET and medulloblastoma are planned similarly for the CSI phase of treatment, treatment planning for the sPNET boost is different than for medulloblastoma due to the more superior location of sPNET. The normal brain is still an important critical structure to avoid cognitive deficits later in life. However, the spinal cord is far from the target and the cochlea, optic nerves, and retina are usually far enough away that they will not get a significant dose. However, the pituitary, optic chiasm, lenses, temporal lobes, and hippocampus are potentially close enough to receive doses exceeding tolerance. Planning for other localized tumors of the posterior fossa, such as ependymoma or gliomas, can be approached similarly, whereas supratentorial tumors need more individualization due to the diversity of potential locations and nearby critical structures. MMGCTs, on the other hand, require 36 Gy CSI plus a boost to 54 to 55.8 Gy, so all of the guidelines for sPNET apply to this disease. The PTV may overlap the top of the brainstem. As long as the dose to the brainstem from the boost is less than 24 Gy (over 120% of the 19.8 Gy boost dose), this structure is safe. Generally, the brainstem dose can be kept to much less than 110% of the boost dose.

The most radio-responsive tumor in this group of CNS diseases is the pure germinomas, which need only about 30 Gy after good response to chemotherapy. Since these tumors typically involve the sellar and pineal region, the pituitary and hypothalamus are frequently damaged before the patient ever gets to the radiotherapy department. The other OARs listed earlier all can handle doses of 30 Gy and will typically receive much less due to their distance from the PTV. The WV irradiation will require larger fields than the other tumors and IMRT is definitely useful to conform the high dose region to a very irregular volume and requires a planning strategy that minimizes normal brain tissue dosing. IMRT is also needed if a simultaneous integrated boost is to be used.

In general, a planning-at-risk volume (PRV) should be created for any OAR that could get close to its tolerance dose (including the lenses of the eye). The PRV margin should be at least as large as the PTV margin. One should plan the limiting dose to fall on or outside the PRVs, which will ensure that the actual structure does not exceed this dose due to daily setup and other structure position variations.

Protecting OARs for the brain tumors that are only given local treat (i.e., without CSI) is the most difficult for ependymoma because the dose is the highest, 59.4 Gy. This higher dose presents a special challenge for posterior fossa ependymoma where the upper part of the spinal cord is both part of the target and also an OAR. Generally, a maximum of 54 Gy can safely be given to this part of the spinal cord, however, this means that either a second plan needs to be made that omits the spinal cord for the last 5.4 Gy or if IMRT is being used, 1.64 Gy per fraction (54 Gy over 33 fractions) can be given daily to the included spinal cord target while 1.8 Gy per fraction is given to the rest of the PTV. This reduced daily dose also lessens the biological effect of the 54 Gy given to the spinal cord. The radiation oncologist will decide whether all target volumes are given 1.8 Gy per fraction or if the somewhat lower daily dose to the spinal cord target is acceptable. Due to the higher prescribed dose, all the OARs are just a little harder to protect when the target is nearby. However, since CSI is not given, we can realize the full protective effects of our advanced technology planning and delivery methods. For localized treatments, Table 4.1 gives achievable OAR doses as percentages of the prescribed dose. Note that a higher cochlear dose (45 Gy) is generally tolerated when platinum-based chemotherapy is not part of the treatment. Therefore, 50% of 59.4 Gy or about 30 Gy should be very safe and allows better PTV coverage in the vicinity of the cochlea compared to limiting the cochlear dose to the otherwise achievable 35% of the target dose.

In almost every case, 8 to 10 noncoplanar intensity-modulated beams will do the best job of conformally treating the PTV while minimizing dose to OARs. It should be noted that IMRT is not always required to treat localized brain tumors. Those occurring in the supratentorial brain or small volume targets in the posterior fossa that are at least 1 cm away from OARs can be treated with multibeam 3DCRT if care is taken to choose beam angles and weights that optimize the dose distribution. Sagittal beams that are aligned with the patient's long axis or exit through the thyroid should be avoided. Multiple noncoplanar VMAT beams can also be used for these localized treatments.

Isocenter verification can be accomplished using an orthogonal pair of kilo- or megavoltage images (i.e., AP and a left lateral) or cone beam CT. Much has been written about minimization of imaging dose in children, including for radiotherapy verification (Olch et al. 2007). One should minimize the dose to nontarget tissue without jeopardizing the quality of image information by using field shaping of these two portals (Figure 4.30). Also, kilovoltage (kV) imaging will generally give less dose than megavoltage (MV) portal imaging. Daily cone beam CT can reduce the setup margin component of the PTV to 2 mm compared to 3.5 mm for weekly cone beam CT (Beltran, Krasin, et al. 2011).

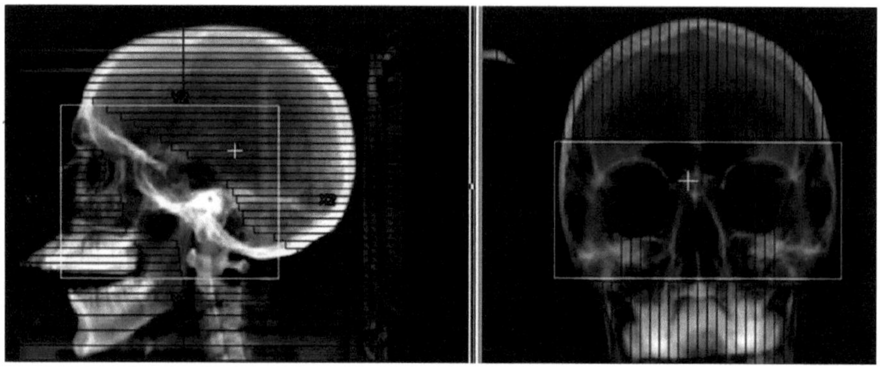

FIGURE 4.30 AP and left lateral images for isocenter verification showing blocking of the brain and lenses.

For patients requiring corticosteroids to reduce inflammation in the brain, weight gain and facial tissue expansion can compromise the fit of facemask immobilization devices and cause reduced target dose coverage. In a study of brainstem glioma patients receiving steroids, an 8% mean weight gain was found, but dosimetry based on conebeam CTs performed after steroid use found less than a 1.5% target dose decrease from the planned dose, indicating that adaptive replanning was not warranted (Beltran, Sharma, et al. 2011).

4.3.2.2 Intensity-Modulated Radiation Therapy (IMRT) Planning

IMRT is recognized as providing more conformal dose to the brain tumor while better sparing normal tissues (Cardinale et al. 1998; Paulino and Skwarchuk 2002; Chan et al. 2003; Penagaricano et al. 2004; Rembielak and Woo 2005). It is especially attractive for treatment of childhood brain tumors because of the increased need to reduce the dose to normal adjacent brain as well as other sensitive structures. It is most often used in children for conformal avoidance of critical structures or for simultaneous integrated boost (SIB) (Paulino and Skwarchuk 2002; Chan et al. 2003; Penagaricano et al. 2004; Schroeder et al. 2008). Much of the information that follows is applicable to IMRT treatment of any site. Subsequent chapters will describe the use of IMRT as it specifically relates to the other disease sites.

There are many sources of information in the literature on commissioning, quality assurance (QA), and generalized treatment planning so these will not be discussed here. This section will focus strictly on methods to optimize IMRT plans in children with brain tumors. The examples shown are for 6 MV x-rays, 0.5 cm MLC leaf width, and 10 intensity-level step-and-shoot IMRT. Sliding window IMRT and VMAT methods can also produce good plans. Sliding window plans generally produce plans with more MUs and VMAT with less MUs than step-and-shoot IMRT.

In general, the dose gradient that one can achieve with IMRT is limited by a number of factors that include the number of beams, the beam energy, target depth, and the MLC leaf width. The maximum limit to the dose gradient is the penumbral width of a single beam, which is about 5 mm for the 20% to 80% dose range for a 6 MV beam at a depth of 7 cm, or about 12% per millimeter. In a multibeam brain tumor treatment plan, because of scatter and complex dosimetric summation, one can expect about an 8% decrease in dose per millimeter of distance between the PTV and the organ at risk for a convex target. For example, an 8 mm separation between the PTV and the center of the cochlea results in about a 64% dose reduction for that OAR. If one is willing to accept the 100% isodose line indenting into the PTV by 2 mm just at the location of the cochlea, for example, an additional 16% dose reduction can be achieved.

For average-risk medulloblastoma, the boost dose is usually 30.6 to 32.4 Gy, whereas for high-risk medulloblastoma or sPNET, the boost dose is 18 to 19.8 Gy in 1.8 Gy fractions. The percentage reduction of prescribed dose to each OAR is based on the beam arrangement and geometry of the target and OARs. Due to the CSI component of treatment, one must accept the fact that all the OARs in the brain will get a higher total dose than for the localized tumors not needing CSI no matter how skillful one is in planning the boost.

The dose objectives for the boost PTV relative to the prescribed dose are that at least 95% of the PTV should receive 100% and no more than 2% of the PTV should receive more than about 112%. Much more homogeneous PTV coverage can be obtained with smaller PTVs. Usually high priorities are given to the PTV for both upper and lower dose limits, which are set close to each other to force the optimizer to make the dose as uniform as possible. The Appendix (Figures A4.1, A4.6, and A4.7) gives beam arrangements suitable for treatments in the posterior fossa, either whole or limited volume.

Table 4.4 along with the following discussion describe the OARs and total dose objectives for IMRT planning for the whole posterior fossa. Dose, volume, and priority values one would enter into a planning system optimization program are not provided since these are dependent on the particular user interface and algorithm of each system. The doses listed are both achievable and thought to be significantly sparing of each structure. Depending on the distance from the PTV to OAR, one may be able to achieve lower doses than stated in the table. For small structures like the cochlea, pituitary, and lenses, the mean dose is more meaningful than maximum dose or dose-volume limits.

The cochlea is within 0.8 cm of the posterior fossa brain tissue (Figure 4.31). The cochlea can usually be kept below a mean dose of 35% of

TABLE 4.4 Organs at Risk and Total Dose (CSI + Boost) Objectives (Gy)-Whole Posterior Fossa (WPF)

PTV and Organ at Risk	Mean Dose (Gy) AR,HR	Maximum Dose	Dose–Volume AR,HR
WPF total boost PTV		D2% = 58 Gy	D98% = 54 Gy
Cochlea	34,43	–	–
Pituitary/hypothalamus	35,43	–	–
Optic chiasm	–	54 Gy	–
Optic nerves	–	54 Gy	–
Spinal cord (C-spine)		54 Gy	
Brainstem		D 1cc <62Gy	
Lenses	8	–	–
Temporal lobes	–	–	D50%<34,42 Gy; D10%<43,48 Gy
Hippocampus	37,44	–	–

Notes: D50% means the dose that 50% of the volume receives. AR, average risk, HR, high risk.

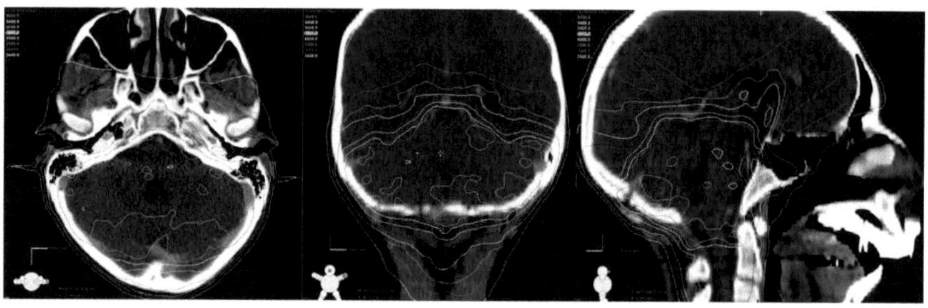

FIGURE 4.31 (See color insert.) Average-risk medulloblastoma whole posterior fossa composite doses, 23.4 Gy CSI plus 30.6 Gy boost. Isodose lines range from 24–56 Gy.

the prescribed dose for whole posterior fossa boost and much less than that for local boosts if the CTV does not abut the cranial bone adjacent to the cochlea. This means that the mean cochlear dose should not be more than between 7 and 11 Gy from the boost, and not more than 43 Gy and 34 Gy in total for either the high risk or average risk cases, respectively. Thirty-four Gy should be sparing even with platinum-based chemotherapy but much less so for 43 Gy. A maximum cochlear dose and its priority value are sufficient for dose optimization.

The pituitary and hypothalamus are structures that are usually taken to have the same tolerance dose and are located close to each other, the first just inferior and the second just superior to the optic chiasm. These structures are second most sensitive after the lenses and for cases where 36 Gy is given to the whole brain, may have already used up their tolerance dose before the boost is given. Nevertheless, the dose to this structure should be limited as much as possible. The whole posterior fossa boost PTV usually nearly abuts the sella so is only about 7 mm posterior to the center of the pituitary. Mean boost doses of between 7 and 12 Gy (40%) can be expected for the pituitary for high- and average-risk cases, respectively. Again, every millimeter of additional space allows one to further reduce the dose by 8%, so local boosts that are more posterior will result in much lower doses to the pituitary and hypothalamus. A maximum pituitary/hypothalamic dose and priority value are sufficient for dose optimization.

The brainstem is part of the target volume for medulloblastoma and therefore routinely gets at least 54 Gy. Small volume hot spots up to about 60 Gy do not appear to cause brainstem toxicity but the number of surgical procedures in the area before radiotherapy can reduce the radiation tolerance (Debus et al. 1997, 1999; Merchant et al. 2010).

The three structures in the tables with only maximum dose limits are the optic chiasm and nerves and the spinal cord in the C-spine. All three structures can maximally tolerate 54 Gy. Especially for these structures, a 2 to 3 mm margin expansion, creating the PRV, should be used for optimization, which ensures that even with daily patient setup variations, the maximum tolerated dose is not exceeded to the actual structure. Generally, the optic nerves are well away from the PTV and so are easy to keep below 54 Gy. In some cases, the chiasm or the spinal cord are overlapping or inside the PTV. Here, a nonoverlapping structure should be made (using Boolean operators if available) that creates a new structure that does not share voxels with the PTV. For many IMRT planning systems, this reduces the confusion of competing dose constraints on voxels that are assigned to more than one structure. This new OAR structure should be given a maximum dose limit of about 5% less than the tolerance dose as well as high priority for dose uniformity to ensure that hot spots are minimized and if they do occur, they will not take the total dose in the cord, optic chiasm, or optic nerves above 54 Gy (Figure 4.32).

The lenses of the eye can best be protected by missing them altogether by each beam in the plan. With careful CT-based planning, the lenses will receive about 15% of the whole brain dose, which is between 3 and 5 Gy. Ideally, they will not receive more than another 3 to 5 Gy from the boost. This is also a very low percentage of the boost dose. Depending on the number of beams, it may be acceptable for one or more of the beams to enter or exit through each lens.

The temporal lobes abut the superior aspect of the posterior fossa medially and extend laterally to the inside of the cranium. The most effective way to minimize dose to these structures is by defining DVH points in the optimizer. For whole posterior fossa boosts, one should allow for about 10% of each temporal lobe to get at least 65% of the prescribed dose so as not to

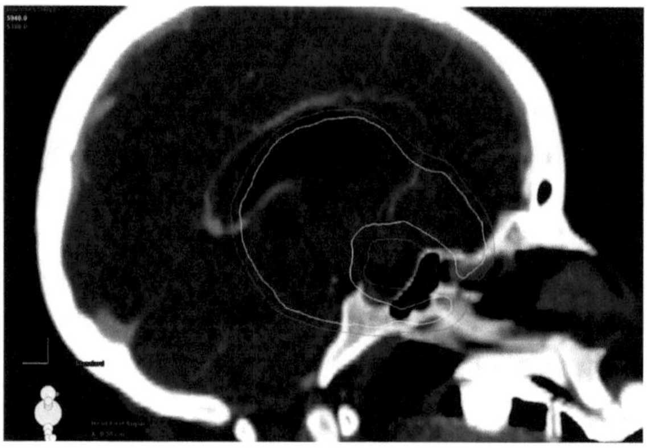

FIGURE 4.32 IMRT isodose distribution with 59.4 Gy to target but 53.8 Gy carved out around optic chiasm.

indent the prescribed dose surface into the PTV. At least 50% of the temporal lobe volume should get less than 35% of the prescribed dose. The dose has to come from somewhere to fully enclose the PTV so decreasing the dose lateral to the PTV increases dose anteriorly and posteriorly. For local tumor bed boosts, generally there is more distance between the PTV edges to the temporal lobes and the PTV is smaller, so it is easier to further reduce the volume of temporal lobes getting 50% of the dose. The hippocampi are located in the medial aspect of the temporal lobes. The dose–effect relationship for the hippocampus is not known, so one can only minimize its dose to the extent it does not negatively impact the PTV coverage or dose uniformity.

Table 4.1 shows dose objectives for localized targets in the infratentorial brain. This table can be used for local medulloblastoma boosts, posterior fossa ependymomas, or other tumors in the posterior fossa. The doses in the table serve as an upper limit and most structures can be spared further depending on the distance from the PTV to the structure. Note that as one pushes the dose down to all of these OARs, dose inhomogeneity increases in the PTV as well as in the undefined volume. One must balance the magnitude of the hot and cold spots created with the tissue sparing objectives of the radiation oncologist.

One good approach to IMRT planning for any site is to start with few constraints and obtain excellent target coverage and uniformity. Then one can add higher OAR constraints in a systematic fashion, one structure at a time, until PTV coverage begins to suffer. At that point, one can use OAR constraints that reflect the physician's priorities and optimize the DVH for both targets and OARs. The distance between OAR and PTV, the ratio of OAR to PTV dose, and the number of OARs will determine the PTV dose uniformity. Other features of one's IMRT planning system will influence the quality of the plan, including number of beam segments, beam model accuracy, and optimization algorithm. Also note that judicious use of fluence smoothing can greatly improve dose homogeneity without significant reduction in OAR sparing. An added benefit is that the total MUs will be decreased as well. Once a good set of optimization parameters have been established for a particular disease, these can be reused as a starting point for subsequent similar cases.

Figure 4.31 and Figure 4.33 show the composite dose distribution for CSI plus whole posterior fossa boost treatment for average-risk and high-risk medulloblastoma, and Figure 4.34 for average-risk medulloblastoma with local boost for the same 10-beam IMRT plan.

Posterior fossa ependymoma treatment planning is essentially the same as for localized boost medulloblastoma treatment. For ependymoma or gliomas in other locations in the brain, one can use an appropriate beam arrangement from the Appendix. For a regularly shaped supratentorial PNET, where there is a greater distance between the PTV and OARs, eight beams suffice. When germinoma is treated by giving whole ventricular irradiation followed by a localized boost, we have used 8 to 10 beams to optimally cover the complex shape of the whole ventricles and have included a SIB (Figure 4.35). Raggi et al. (2008) reported on their method of using seven noncoplanar IMRT fields for the complex-shaped whole ventricular volume, and six noncoplanar non-IMRT fields for the subsequent semispherically

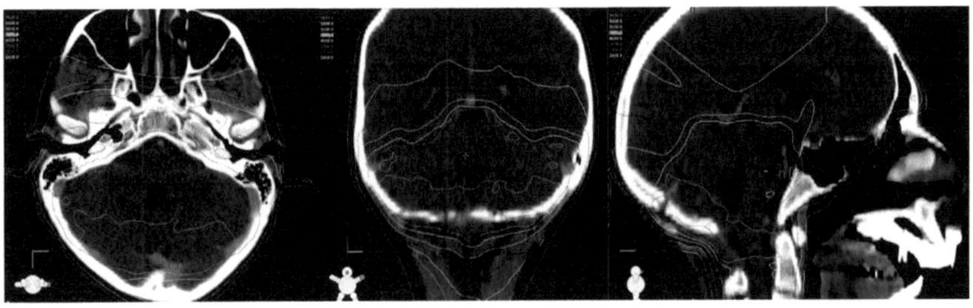

FIGURE 4.33 High-risk medulloblastoma whole posterior fossa boost composite doses, 36 Gy CSI plus 18 Gy boost.

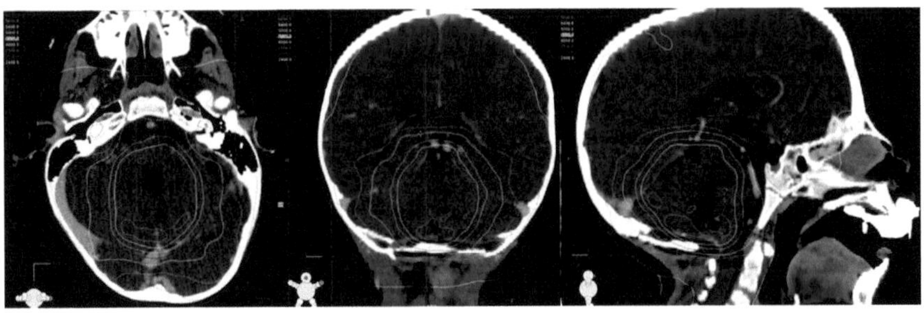

FIGURE 4.34 Average-risk medulloblastoma with local boost. Composite dose for 23.4 Gy CSI plus 30.6 Gy boost.

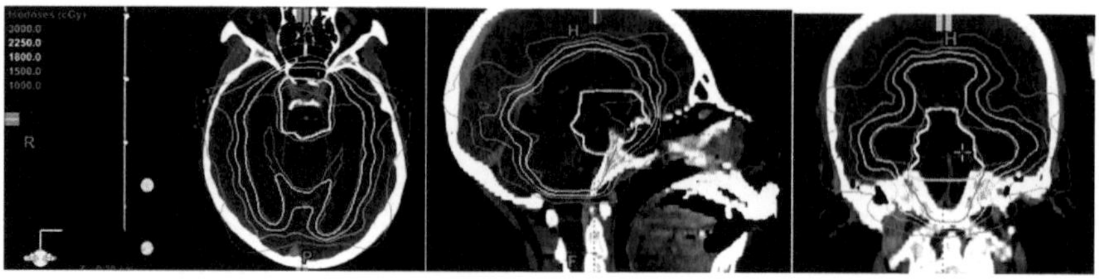

FIGURE 4.35 Germinoma treatment using SIB doses of 22.5 Gy to WV and 30 Gy to primary.

shaped boost. Similar to our process, Raggi et al. was able to conform the 95% isodose line around at least 95% of the WV PTV and found significant sparing of normal brain with this multibeam plan compared to simpler plans. Since the dose is low, OAR sparing is not difficult, especially since the pituitary is part of the target volume and therefore not a candidate for sparing. Generally, more beams are needed as the number of critical structures increases, the distance between PTV and OARs decreases, and the shape becomes more complex. Several investigators have reported on the use and clinical efficacy of IMRT in various pediatric brain tumors (Huang et al. 2002; Rembielak and Woo 2005; Raggi et al. 2008; Schroeder et al. 2008).

IMRT is required for SIBs, which is a technique whereby a higher daily dose (inner or adjacent) volume is treated at the same time as a lower daily dose volume. This requires separate PTVs and dose constraints for each. For some planning systems, it may be necessary to subtract the inner from the outer volume so they do not have voxels in common. Shared voxels can confuse the optimizer because each would have two different dose objectives. The rationale for treating with SIB is that one can hit the core of the tumor harder each day while synchronizing the dose falloff with the anticipated density of tumor cells. One must be careful not to use SIB in situations where the two doses are more than about 30% different because otherwise, either too large or small daily doses may be obtained. We allow prescribed daily doses up to about 200 cGy, knowing that small hot spots of up to 220 cGy can occur. Where this dose would exceed a critical structure's tolerance dose, ensure that the maximum dose does not occur in that structure. Brainstem glioma and germinoma have been treated this way (Lavey et al. 2006). For brainstem glioma, 30 fractions with 2 Gy, 1.8 Gy, and 1.5 Gy per fraction have been given to the GTV without margin, GTV with a 0.5 cm margin, and GTV with a 1.5 cm margin, respectively. Total doses amount to 60 Gy, 54 Gy, and 45 Gy, respectively. For germinoma, 15 fractions of 2 Gy per fraction have been given to the boost PTV, while the WV PTV gets 1.5 Gy per fraction. Others have reported on the use of SIB for ependymoma and pediatric GBM (Chan et al. 2003; Gutierrez et al. 2007; Schroeder et al. 2008).

4.3.2.3 Proton Planning

Proton beam treatment has been described for CSI and the posterior fossa boost for medulloblastoma as well as for most other pediatric brain tumors. Protons are particularly attractive for whole spine treatment because of the total lack of dose anterior to the vertebral bodies. This both reduces acute and late toxicity and greatly reduces the integral dose, which theoretically could reduce the risk of secondary malignancy formation. By one calculation, the lifetime risk of secondary cancers was estimated to be 30% for IMRT, 20% for conventional photon CSI, and just 4% for intensity-modulated proton treatment (IMPT) (Mu et al. 2005) (see Chapter 2 for commentary). In some cases, it may be possible to fully treat the spinal PTV yet spare the vertebral bodies from growth retardation, but in general it is safer to include the whole vertebral body (St. Clair et al. 2004).

Publications by the Loma Linda University (California), Massachusetts General Hospital (Boston), and the Paul Sherrer Institute (Switzerland) proton therapy centers describe their techniques for planning proton therapy for CSI (St. Clair et al. 2004; Yuh et al. 2004; Timmermann et al. 2007). Multiple posterior fields are abutted craniocaudally to cover the full length of the spine while opposed lateral oblique fields are used for the whole cranium. The posterior fossa boost is delivered with two to four oblique fields. The junctions are shifted as with photon plans. Dosimetric comparisons of proton plans to conventional or IMRT plans for the spinal PTV show that protons give much lower doses to the heart, lung, stomach, kidney, and colon, but the esophagus was not better spared.

Comparing protons to IMRT for the posterior fossa boost, protons gave virtually no dose to the pituitary, hypothalamus, tempromandibular joint, parotid, and pharynx compared to a mean dose of 6% to 30% for IMRT; protons gave just 10% to 26% of the prescribed dose to 50% of the cochlea compared to 45% for IMRT (Lin et al. 2000; St. Clair et al. 2004). As mentioned earlier, more sparing IMRT plans can be created that give just 30% of the prescribed dose to the cochlea, but this is still greater than for a proton plan. Further improvements in hearing preservation over what has been demonstrated with IMRT may be difficult to achieve in medulloblastoma because of the use of ototoxic chemotherapy that is standard of care.

Proton treatment of the whole ventricles (23.4 Gy) plus a suprasellar and pineal boost (45 Gy total) for intracranial germinoma has been described, including dosimetric comparisons of six-beam coplanar IMRT, three-field scattered, and spot-scanned proton beams (3–8 mm pencil beams). For the structures where these doses might cause late effects, such as the brain outside the ventricles and the temporal lobes, protons reduced the volume receiving 20 Gy from about 50% to about 25% (MacDonald and Yock 2010).

With the potential for normal brain tissue sparing with protons comes the possibility of reduced neurocognitive deficits. Merchant, Hua, et al. (2008) calculated changes in IQ scores based on mathematical models for children with medulloblastoma, infratentorial ependymoma, cranipharyngioma, and optic pathway glioma treated with unmodulated 3D conformal photons versus IMPT with 3 mm pencil beams. Using these extremes, IQ retention was predicted to be about 10 points better for proton-treated patients 5 years after therapy.

4.4 Organ-at-Risk Doses and Late Effects

The consequences of treating a child's brain with curative doses of radiation is far more serious than in adults due to the potential for arrest of development, which for adults is not a consideration, and permanent cognitive deterioration which can cause a reduction in quality of life ranging from lifelong dependence on others to total incapacitation. Some structures in the young brain are generally taken to have the same tolerance dose as for adults, such as optic nerves and lenses of the eye, but we do not really know if these are the same or not. Many of the tolerance doses are taken from adult data, but wherever possible, pediatric-specific tolerance doses have been included. These tolerance doses must be taken in the context of the total patient's condition, surgical, and chemotherapeutic history, as these factors can greatly alter the radiation sensitivity of a particular organ or system.

4.4.1 Cognitive Dysfunction

Treatment techniques that reduce the dose to the temporal lobes and other nontarget brain tissues are critically important in preserving cognitive function. In 2004, the Children's Oncology Group (COG) began accruing patients to a protocol (ACNS0331) for average risk patients that tests whether 18 Gy to the craniospinal axis is sufficient for children under 8 years of age and also whether the boost dose can be limited to just the tumor bed rather than the whole posterior fossa. The objective of this randomized trial is to see if by lowering the doses in the brain, neurocognitive morbidity can be reduced without sacrificing tumor control. Silber (1992) has estimated an approximate 5-point decline in full-scale IQ for a 6-year-old treated to a CSI dose of 24 Gy compared to an estimated decline of 17-points in those treated with a dose of 36 Gy. In a recent Childhood Cancer Survivor Study of neurocognitive outcomes in long-term survivors of childhood CNS tumors, 36 Gy to the whole brain was only slightly worse than 24 Gy for memory loss but much worse for task efficiency. Younger age at diagnosis was not correlated to increased cognitive dysfunction, but this may have been because these patients selectively received a lower dose of radiation (Ellenberg et al. 2009). Other studies have demonstrated that intellectual decline is greater in patients younger than about 3 years of age (estimated drop of 21 points at 4 years after radiotherapy) (Silber 1992; Fouladi et al. 2005). Because of this finding, these very young children are treated with high-dose chemotherapy and autologous bone marrow transplant or peripheral stem cell rescue instead of craniospinal irradiation. In contrast to these younger children, in a review of 22 germinoma patients with a mean age of 16.9 years at diagnosis, a quality-of-life survey on average 10 years after successfully completing treatment showed all patients had completed high school, nine completed or were in college, and five had advanced degrees. Eighteen out of 22 were treated with 36 Gy of craniospinal irradiation (Sutton et al. 1999). In another study, germinoma patients with a mean age of 12 years at diagnosis who received craniospinal doses of up to 23 to 32 Gy did not demonstrate significant differences between pre- and postirradiation full-scale, verbal, and performance IQ scores (Merchant et al. 2000).

The temporal lobes are associated with memory and sensory processing. The dose–response relationship is not known, but it is generally accepted that reducing the dose to these structures without sacrificing target coverage should be a goal of the planning process. The hippocampus is a structure that has been implicated in cases of neurocognitive deficits caused by radiation. The hippocampi are small supratentorial structures located bilaterally in the medial aspect of the temporal lobes near the superior aspect of the posterior fossa. There is evidence that the hippocampi and periventricular zones are the niche for neural stem cells that can facilitate repopulation of normal brain cells after radiation damage. There has been keen interest in reducing the dose that the hippocampi receive during brain irradiation as a way to minimize cognitive deficits (Barani, Benedict, et al. 2007; Barani, Cuttino, et al. 2007; Gutierrez et al. 2007; Gondi et al. 2011). In a study of adults receiving whole brain irradiation for metastatic disease, sparing of the hippocampi to doses less than 6 Gy for a whole brain dose of 30 Gy was technically achieved using HT or linac-based IMRT. In another adult study by the same group, memory scores were significantly correlated to doses of greater than 7.3 Gy to more than 40% of the bilateral hippocampi (Gondi et al. 2010, 2011). Whether it is safe to spare this portion of the brain during whole brain irradiation in childhood medulloblastoma or other diseases is not known. Hippocampi sparing in children receiving localized brain irradiation, when there is no suspicion of tumor involvement, is theoretically beneficial but not proven.

Armstrong correlated a number of neurocognitive outcomes with the region of the brain receiving radiation, including the posterior fossa, temporal, parieto-occipital, and frontal lobes. They failed to find a correlation with dose for any of the four brain regions for emotional regulation or organization but did find a significant dose correlation for working memory for doses greater than 30 Gy to the frontal, temporal, and parieto-occipital regions and for task efficiency for doses greater than 30 Gy to the temporal lobe. Other quality of life measures showed increased risk with dose for each of the four brain regions (Armstrong et al. 2010).

When IMRT is used for brain boosts, it has been hypothesized that sparing certain structures such as the cochlea may result in a redistribution of dose to the normal brain, increasing the dose to the medial temporal lobes and therefore the chance for cognitive deficits. In a study comparing the cognitive outcomes of 25 medulloblastoma patients treated with either 3D conformal or IMRT brain boosts techniques, the enhanced cochlear sparing of IMRT was not associated with worse cognitive outcomes (Jain et al. 2008).

4.4.2 Hearing Loss

Sensor neural hearing loss is also a concern for PNET patients because they receive ototoxic platinum-based chemotherapy following their radiation therapy. The combined effect of platinum-based chemotherapy and 54 Gy to the cochlear hair cells leads to a high probability of permanent high-frequency hearing loss requiring the use of hearing aids as well as necessitating a reduction in the dose of chemotherapy the patient can receive (Williams et al. 2005). Thus, opposed lateral beams for the boost phase of treatment is discouraged in favor of IMRT or other methods, which should be used to lower the cochlear dose (Huang et al. 2002). Although comprehensive dose–response data is not available, the severity of hearing loss correlates to the cochlear dose. The data that exist suggest that cochlear doses should be restricted to below about 37 Gy if platinum-based chemotherapy is used (Huang et al. 2002; Paulino et al. 2010). In a study of 78 brain tumor patients receiving 3DCRT or IMRT for brain tumors but without ototoxic chemotherapy, a tolerance dose of 35 to 45 Gy was determined. However, of 30 cochleae (for 23 patients) that received greater than 50 Gy, 14 developed hearing loss but 16 remained normal (Hua et al. 2008).

4.4.3 Endocrine Dysfunction

Sklar and Constine (1995) give an excellent review of the anatomy and function of the hypothalamic–pituitary axis (HPA) as well as the relationship of radiation dose and its effects on neuroendocrine function. According to research examining the effect of cranial irradiation on endocrine function, 56% of children were reported to have growth hormone deficiency after 24 Gy and 75% after 45 Gy. The effect of radiation on endocrine function is influenced by the tumor itself, surgery, and chemotherapy. Pituitary–hypothalamic doses above 30 Gy reduces adult height in most children, with more severe reductions in height if spinal radiation is also given (Rappaport and Brauner 1989). Merchant developed a mathematical model to predict growth hormone (GH) levels as a function of hypothalamic dose and concluded that a cumulative dose of 16 Gy to the hypothalamus was the mean radiation dose required to achieve a 50% risk of growth hormone deficiency (GHD) at 5 years (TD(50/5)) (Merchant, Goloubeva, et al. 2002; Merchant et al. 2011). Thus, high-risk PNET that requires 36 Gy to the whole brain brings with it a very high risk of pituitary–hypothalamic dysfunction.

Thyroid abnormalities after CSI radiation include hypothyroidism, hyperthyroidism, goiter, and thyroid carcinoma. Several small series report hypothyroidism in 32% to 69% of patients after CSI (Hirsch et al. 1979; Maor et al. 1985). Constine et al. (1984) found thyroid abnormalities in 17% of children who received <26 Gy to the thyroid and in 78% of those who received >26 Gy. Laughton et al. (2008) found in a group of 88 average-risk and high-risk medulloblastoma patients that the cumulative incidence of GHD, thyroid-stimulating hormone (TSH) deficiency, adrenocorticotropic hormone deficiency, and primary hypothyroidism at 4 years from diagnosis was 93%, 23%, 38%, and 65%, respectively. TSH deficiency was the only measure that was found to be dose dependent, with irradiation of the pituitary and hypothalamus to doses greater than about 42 Gy resulting in an increased effect. In a study on acute lymphoblastic leukemia (ALL) patients, those who received less than 20 Gy whole cranial irradiation alone were not at increased risk of subsequent hypothyroidism. Those receiving pituitary doses greater than 20 Gy combined with thyroid doses greater than 10 Gy were at a significantly increased risk of hypothyroidism (hazard ratio = 9.9) (Chow et al. 2009).

4.4.4 Spinal Growth Changes

We are concerned about dose to the vertebral bodies of children for two reasons: (1) possible decrease in sitting height, and (2) possible differential growth resulting in scoliosis or kyphosis. The first problem is relevant in the context of spinal treatment during CSI, the second for treatments of other sites near the spine. Hartley et al. (2008) studied the growth of eight vertebral body sections after CSI. With a minimum of 3 years follow-up, using MRI measurements, they found that vertebral bodies continued to grow after CSI but grew faster following 23.4 Gy CSI than 36 Gy CSI. Radiation seems to slow lumbar more than thoracic or cervical spine growth. Final height of boys was more affected by CSI than girls, perhaps because boys have more growth remaining than do girls.

4.4.5 Spinal Cord Myelopathy

Since the spinal cord is both the target and an important critical structure, methods to accurately address the junction between the posterior spine and lateral cranial fields have been developed to avoid the possibility of spinal cord over- or underdose. In addition, the C1-C2 spinal cord is part of the PTV for medulloblastoma. Adamus-Gorka (2008) reviewed myelitis data for the cervical and thoracic regions, and studied a radiobiological model to explain the clinical data. The data suggests that the cervical cord is more sensitive and is more dependent on irradiated length than the thoracic spinal cord. All of the cases of myelitis in the cervical region occurred above 55 Gy and all but one thoracic myelitis occurred above 60 Gy with conventional fractionation. Schultheiss (2008) reviewed 18 radiation myelopathy series and concluded that the probability of cervical myelopathy is 0.2% at 50 Gy and 5% at 59.3 Gy. His analysis also indicates that thoracic cord is less sensitive than cervical cord. The literature suggests that very few cases of myelopathy have occurred due to treatment for medulloblastoma. COG medulloblastoma and ependymoma protocols allow 54 Gy to be given to the upper spinal cord.

4.4.6 Acute Toxicities of the Upper Aerodigestive and Lower Gastrointestinal Tract

Acute toxicities of irradiation of the upper aerodigestive and lower gastrointestinal tract from the spinal treatment include sore throat, dysphagia, nausea, vomiting, hoarseness, and cough. Marrow and gastrointestinal toxicity are often severe enough to cause treatment interruption. Bone marrow suppression can result from the large volume of bone marrow irradiated in the cranium and spine. Nausea is often daily-dose related, so treatment methods that can reduce the exit dose from the spinal treatment are beneficial. Other significant late effects include cardiovascular disease, reduced density and growth of the irradiated vertebrae, pulmonary fibrosis, and thyroid cancer 8 to 18 years after treatment (Roggli et al. 1979; Xu et al. 2004). Despite the small volume of lung irradiated by the relatively narrow whole spine field, one study found that restrictive lung disease was four times more likely in these patients than in those not receiving spinal irradiation (Jakacki et al. 1995).

4.4.7 Cardiac Abnormalities

The exit dose from the PA spine field irradiates a portion of the heart. Jakacki et al. (1993) described a number of cardiac function abnormalities after spinal irradiation in childhood and propose that irradiation of the posterior cardiac wall might be responsible. Milano et al.'s (2007) review of the literature showed that the dose to produce a 5% complication probability is 30 to 42 Gy.

4.4.8 Brainstem Toxicity

The volume of brainstem getting a high dose and not just the maximum dose appears to be important for producing brainstem toxicity. Doses of over 60 Gy to small volumes (about 1 cc) appear to be tolerated, but the number of surgical procedures in the area before radiotherapy and very young patient age can reduce the radiation tolerance (Debus et al. 1997, 1999; Merchant et al. 2010). In an infratentorial ependymoma study, no relationship between dose, volume, and recovery of brainstem function was found up to brainstem doses greater than 60 Gy (Merchant et al. 2010). Although the edge of the brainstem may be in the high dose gradient for treatments of head and neck tumors, the brainstem itself is the target in brainstem glioma, which has a dismal prognosis at the highest doses given. Thus, in the context of brainstem gliomas, the prescribed dose is pushed above what is otherwise taken as brainstem tolerance and may be higher than 60 Gy.

4.4.9 Visual Impairment

The optic nerves and chiasm are usually taken to have a similar dose tolerance. Data suggest that at 50 Gy or less one would not expect visual neuropathy, but between 50 and 60 Gy there is about a 10% incidence. There also are some data showing that patients under 20 years old seem to have a higher tolerance than those over 20 years. Daily dose higher than 1.8 Gy was suggestive of being more toxic than for lower daily doses (Bhandare et al. 2005). Parsons et al. (1994a) found no optic neuropathy below doses of 59 Gy. Monroe et al. (2005) reported that the incidence of retinopathy is low below 50 Gy with a statistically significant increase above 60 Gy, but a pure threshold dose has not been defined. Parsons et al. (1994b) found a dose–response curve for retinopathy with no incidence at 45 Gy but a large increase between 45 and 55 Gy. There are reports that lowering the dose per fraction below 1.9 Gy in the range of 45 to 55 Gy to the retina and optic nerves and chiasm may lower the risk of late visual toxicity (Parsons et al. 1994a,b). Monroe et al. (2005) and Bhandare et al. (2005) reported that BID (twice-a-day) treatment to 1.1–1.2 Gy/fraction resulted in less injury to the optic apparatus.

Cataract formation is a concern for brain tumor patients. Normal lens epithelial cells lie beneath the anterior capsule and equator only and are the only ones that undergo mitosis, making them the most sensitive to irradiation. The anteriorly located cells undergo mitotic division and migrate posteriorly to the center of the lens making up the lens fibers. The posterior pole of the lens is free of dividing cells, which maintains an optically clear zone.

Radiation damage that causes migration and proliferation of lens epithelial cells posteriorly increases lens opacity, causing a cataract. Radiation-induced cataracts began to develop 14 to 20 months after irradiation. Cataract formation has been reported to occur above doses of 2 Gy but reports indicate that doses of 15 Gy don't always produce a cataract (Letschert et al. 1991; Henk et al. 1993). The dose to the center of the lens appears to be more relevant than the maximum dose posteriorly in the context of opposed lateral whole brain fields. Reviews articles by Gordon et al. (1995) and Letschert et al. (1991) presented dose–effect data for these and other orbital structures.

Appendix

Noncoplanar Beam Arrangements for Brain Tumors of Various Locations (IEC Gantry and Couch Angles)

Field Name	Gantry	Couch
RAIO	315	340
LAIO	45	10
RPIO	255	340
LPIO	105	10
RPSO	215	32
LPSO	145	328
RASO	300	40
LASO	60	320
ANTSAG	320	90
POSTSAG	240	90

FIGURE A4.1 Ten beams for posterior fossa tumors.

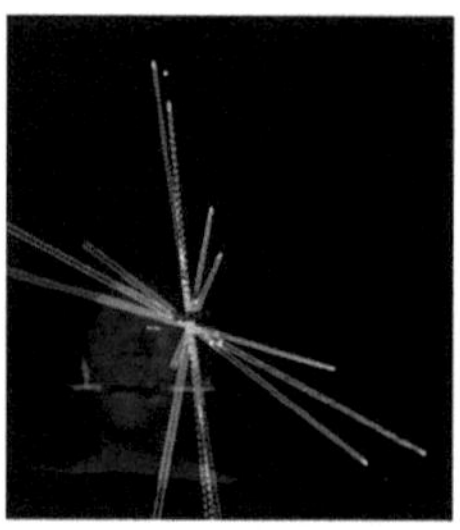

Field Name	Gantry	Couch
LPIO	95	24
POSTSAG	230	90
RSO	270	80
LAIO2	50	40
LAIO	10	10
LPO	140	0
RASO	310	80

FIGURE A4.2 Seven beams for left-sided tumors.

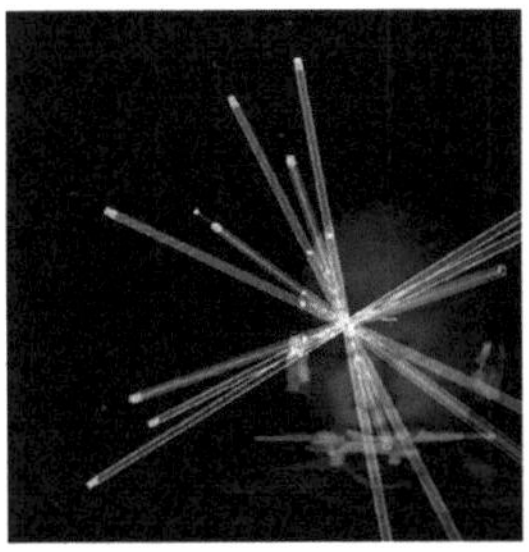

Field Name	Gantry	Couch
RAO	335	0
RASO1	290	30
RAIO	295	340
RPSO1	240	20
RPIO	255	340
RPSO2	260	60
ANTSAG	320	90
RASO2	290	85

FIGURE A4.3 Eight beams for right-sided tumor.

Field Name	Gantry	Couch
RAIO	295	340
LAIO	65	20
RPO	235	0
LPO	125	0
RPSO	240	55
LPSO	120	305
RASO	300	45
LASO	60	315

FIGURE A4.4 Eight beams for superior or midbrain tumor.

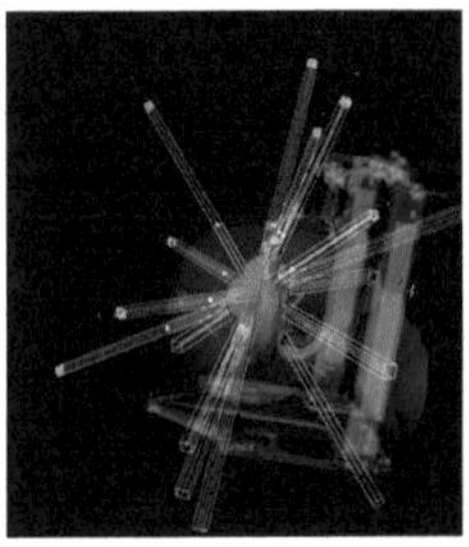

Field Name	Gantry	Couch
ANTSAG	320	90
POSTSAG	240	90
RSO	270	40
RAIO	300	345
RAIO2	330	345
LAIO	50	20
LAIO2	10	30
LSO	90	340

FIGURE A4.5 Eight beams for anterior brain tumor.

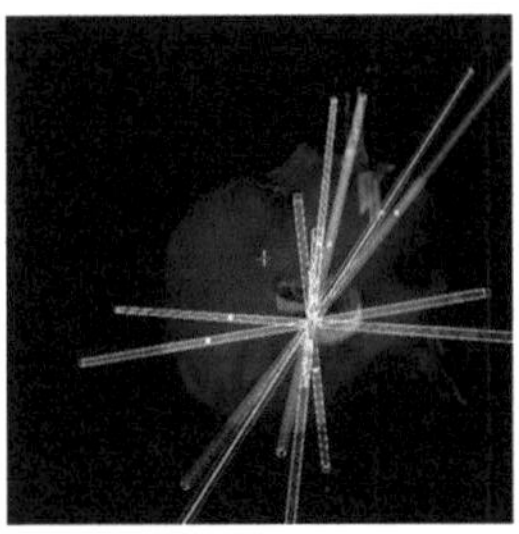

Field Name	Gantry	Couch
RAIO1	315	308
LAIO1	45	52
RPO	250	0
LPO	110	0
RAIO2	310	350
LAIO2	50	10
RSO	270	45
LSO	90	315

FIGURE A4.6 Eight beams for inferior tumors.

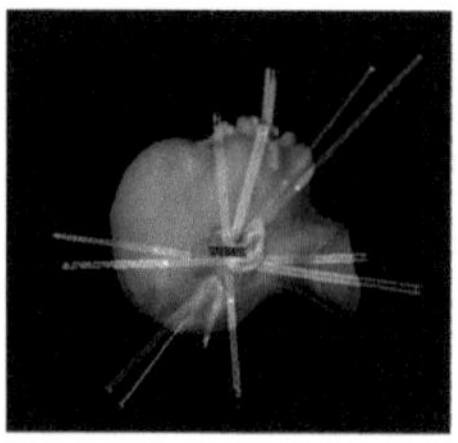

Field Name	Gantry	Couch
RAIO	318	303
LAIO	45	50
RPO	240	0
LPO	120	0
RSO	270	45
LSO	90	315

FIGURE A4.7 Six beams for inferior tumors.

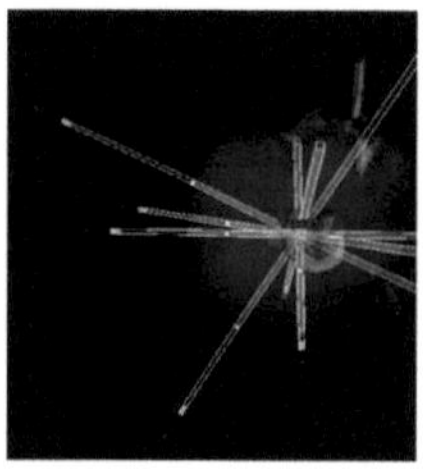

Field Name	Gantry	Couch
LPO	130	0
RPO	250	0
LSO	90	315
RSO	270	45
ANTSAG	290	90
POSTSAG	210	90

FIGURE A4.8 Six beams, all from superior hemisphere.

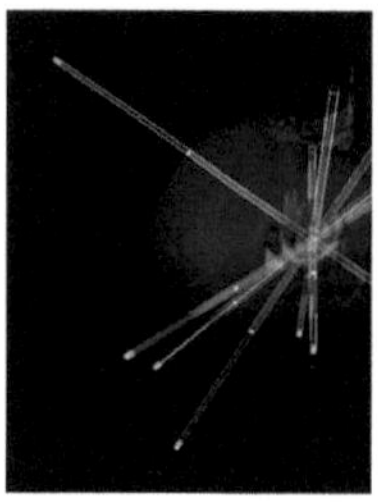

Field Name	Gantry	Couch
RPO	240	0
LPO	110	0
ANTSAG	300	90
POSTSAG	210	90
RPSO	240	45
LPSO	120	315

FIGURE A4.9 Six beams, five out of six are posterior for posterior tumor.

References

Adamus-Gorka, M., A. Brahme, P. Mavroidis, and B. K. Lind. 2008. Variation in radiation sensitivity and repair kinetics in different parts of the spinal cord. *Acta Oncologica* 47 (5):928–36.

Armstrong, G. T., N. Jain, W. Liu, et al. 2010. Region-specific radiotherapy and neuropsychological outcomes in adult survivors of childhood CNS malignancies. *Neuro-Oncology* 12 (11):1173–86.

Barani, I. J., S. H. Benedict, and P. S. Lin. 2007. Neural stem cells: Implications for the conventional radiotherapy of central nervous system malignancies. *International Journal of Radiation Oncology • Biology • Physics* 68 (2):324–33.

Barani, I. J., L. W. Cuttino, S. H. Benedict, et al. 2007. Neural stem cell-preserving external-beam radiotherapy of central nervous system malignancies. *International Journal of Radiation Oncology • Biology • Physics* 68 (4):978–85.

Bauman, G., S. Yartsev, T. Coad, B. Fisher, and T. Kron. 2005. Helical TomoTherapy for craniospinal radiation. *British Journal of Radiology* 78 (930):548–52.

Bedford, J. L., Y. K. Lee, F. H. Saran, and A. P. Warrington. 2011. Helical volumetric modulated arc therapy for treatment of craniospinal axis. *International Journal of Radiation Oncology • Biology • Physics* 83 (3):1047–54.

Beltran, C., M. J. Krasin, and T. E. Merchant. 2011. Inter- and intrafractional positional uncertainties in pediatric radiotherapy patients with brain and head and neck tumors. *International Journal of Radiation Oncology • Biology • Physics* 79 (4):1266–74.

Beltran, C., S. Sharma, and T. E. Merchant. 2011. Role of adaptive radiation therapy for pediatric patients with diffuse pontine glioma. *Journal of Applied Clinical Medical Physics* 12 (2):3421.

Bhandare, N., A. T. Monroe, C. G. Morris, et al. 2005. Does altered fractionation influence the risk of radiation-induced optic neuropathy? *International Journal of Radiation Oncology • Biology • Physics* 62 (4):1070–7.

Breen, S. L., P. Kehagioglou, C. Usher, and P. N. Plowman. 2004. A comparison of conventional, conformal and intensity-modulated coplanar radiotherapy plans for posterior fossa treatment. *British Journal of Radiology* 77 (921):768–74.

Cardinale, R. M., S. H. Benedict, Q. Wu, R. D. Zwicker, H. E. Gaballa, and R. Mohan. 1998. A comparison of three stereotactic radiotherapy techniques; ARCS vs. noncoplanar fixed fields vs. intensity modulation. *International Journal of Radiation Oncology • Biology • Physics* 42 (2):431–6.

Carrie, C., C. Alapetite, P. Mere, et al. 1992. Quality control of radiotherapeutic treatment of medulloblastoma in a multicentric study: The contribution of radiotherapy technique to tumour relapse. The French Medulloblastoma Group. *Radiotherapy and Oncology* 24 (2):77–81.

Chan, M. F., K. Schupak, C. Burman, C.-S. Chui, and C. C. Ling. 2003. Comparison of intensity-modulated radiotherapy with three-dimensional conformal radiation therapy planning for glioblastoma multiforme. *Medical Dosimetry* 28 (4):261–5.

Chang, E. L., P. F. Wong, K. M. Forster, M. D. Petru, A. V. Kowalski, and M. H. Maor. 2003. Verification techniques and dose distribution for computed tomographic planned supine craniospinal radiation therapy. *Medical Dosimetry* 28 (2):127–31.

Chapek, J., G. Watson, L. M. Smith, and D. D. Leavitt. 2002. The application of electron beam delivery using dose rate variation and dynamic couch motion in conformal treatment of the cranial-spinal axis. *Medical Dosimetry* 27 (4):265–8.

Cheng, C. W., I. J. Das, and D. J. Chen. 1994. Technical note: Dosimetry in the moving gap region in craniospinal irradiation. *British Journal of Radiology* 67 (802):1017–22.

Chow, E. J., D. L. Friedman, M. Stovall, et al. 2009. Risk of thyroid dysfunction and subsequent thyroid cancer among survivors of acute lymphoblastic leukemia: A report from the Childhood Cancer Survivor Study. *Pediatric Blood & Cancer* 53 (3):432–7.

Constine, L. S., S. S. Donaldson, I. R. McDougall, R. S. Cox, M. P. Link, and H. S. Kaplan. 1984. Thyroid dysfunction after radiotherapy in children with Hodgkin's disease. *Cancer* 53 (4):878–83.

Debus, J., E. B. Hug, N. J. Liebsch, et al. 1997. Brainstem tolerance to conformal radiotherapy of skull base tumors. *International Journal of Radiation Oncology • Biology • Physics* 39 (5):967–75.

———. 1999. Dose-volume tolerance of the brainstem after high-dose radiotherapy. *Frontiers of Radiation Therapy and Oncology* 33:305–14.

Deutsch, M., and D. H. Reigel. 1981. Myelography and cytology in the treatment of medulloblastoma. *International Journal of Radiation Oncology • Biology • Physics* 7 (6):721–5.

Dunbar, S. F., P. D. Barnes, and N. J. Tarbell. 1993. Radiologic determination of the caudal border of the spinal field in cranial spinal irradiation. *International Journal of Radiation Oncology • Biology • Physics* 26 (4):669–73.

Ellenberg, L., Q. Liu, G. Gioia, et al. 2009. Neurocognitive status in long-term survivors of childhood CNS malignancies: A report from the Childhood Cancer Survivor Study. *Neuropsychology* 23 (6):705–17.

Fogliata, A., S. Bergstrom, I. Cafaro, et al. 2011. Cranio-spinal irradiation with volumetric modulated arc therapy: A multi-institutional treatment experience. *Radiotherapy & Oncology* 99 (1):79–85.

Fouladi, M., E. Gilger, M. Kocak, et al. 2005. Intellectual and functional outcome of children 3 years old or younger who have CNS malignancies. *Journal of Clinical Oncology* 23 (28):7152–60.

Freeman, C. R., and J. P. Farmer. 1998. Pediatric brain stem gliomas: A review. *International Journal of Radiation Oncology • Biology • Physics* 40 (2):265–71.

Fukunaga-Johnson, N., H. M. Sandler, R. Marsh, and M. K. Martel. 1998. The use of 3D conformal radiotherapy (3D CRT) to spare the cochlea in patients with medulloblastoma. *International Journal of Radiation Oncology • Biology • Physics* 41 (1):77–82.

Glasgow, G. P., and J. E. Marks. 1983. The dosimetry of a single "hockey stick" portal for treatment of tumors of the cranio-spinal axis. *Medical Physics* 10 (5):672–5.

Gondi, V., B. P. Hermann, M. P. Mehta, and W. A. Tome. 2011. Hippocampal dosimetry predicts neurocognitive function impairment after fractionated stereotactic radiotherapy for benign or low-grade adult brain tumors. *International Journal of Radiation Oncology • Biology • Physics* 83 (4):e487–93.

Gondi, V., R. Tolakanahalli, M. P. Mehta, et al. 2010. Hippocampal-sparing whole-brain radiotherapy: A "how-to" technique using helical TomoTherapy and linear accelerator-based intensity-modulated radiotherapy. *International Journal of Radiation Oncology • Biology • Physics* 78 (4):1244–52.

Gordon, K. B., D. H. Char, and R. H. Sagerman. 1995. Late effects of radiation on the eye and ocular adnexa. *International Journal of Radiation Oncology • Biology • Physics* 31 (5):1123–39.

Gutierrez, A. N., D. C. Westerly, W. A. Tome, et al. 2007. Whole brain radiotherapy with hippocampal avoidance and simultaneously integrated brain metastases boost: A planning study. *International Journal of Radiation Oncology • Biology • Physics* 69 (2):589–97.

Halperin, E. C. 1993. Concerning the inferior portion of the spinal radiotherapy field for malignancies that disseminate via the cerebrospinal fluid. *International Journal of Radiation Oncology • Biology • Physics* 26 (2):357–62.

Happersett, L., M. Hunt, S. Wolden, K. Goodman, L. Hong, and C. Chui. 2004. Evaluation of an intensity modulated radiation therapy treatment technique for patients with medulloblastoma. *International Journal of Radiation Oncology • Biology • Physics* 60 (1):S639–40.

Hartley, K. A., C. Li, F. H. Laningham, M. J. Krasin, X. Xiong, and T. E. Merchant. 2008. Vertebral body growth after craniospinal irradiation. *International Journal of Radiation Oncology • Biology • Physics* 70 (5):1343–9.

Hawkins, R. B. 2001. A simple method of radiation treatment of craniospinal fields with patient supine. *International Journal of Radiation Oncology • Biology • Physics* 49 (1):261–4.

Henk, J. M., R. A. Whitelocke, A. P. Warrington, and E. M. Bessell. 1993. Radiation dose to the lens and cataract formation. *International Journal of Radiation Oncology • Biology • Physics* 25 (5):815–20.

Hirsch, J. F., D. Renier, P. Czernichow, L. Benveniste, and A. Pierre-Kahn. 1979. Medulloblastoma in childhood: Survival and functional results. *Acta Neurochirurgica* 48 (1–2):1–15.

Holupka, E. J., J. L. Humm, N. J. Tarbell, and G. K. Svensson. 1993. Effect of set-up error on the dose across the junction of matching cranial-spinal fields in the treatment of medulloblastoma. *International Journal of Radiation Oncology • Biology • Physics* 27 (2):345–52.

Hood, C., T. Kron, C. Hamilton, et al. 2005. Correlation of 3D-planned and measured dosimetry of photon and electron craniospinal radiation in a pediatric anthropomorphic phantom. *Radiotherapy & Oncology* 77 (1):111–6.

Hua, C., J. K. Bass, R. Khan, L. E. Kun, and T. E. Merchant. 2008. Hearing loss after radiotherapy for pediatric brain tumors: Effect of cochlear dose. *International Journal of Radiation Oncology • Biology • Physics* 72 (3):892–9.

Huang, E., B. S. Teh, D. R. Strother, et al. 2002. Intensity-modulated radiation therapy for pediatric medulloblastoma: Early report on the reduction of ototoxicity. *International Journal of Radiation Oncology • Biology • Physics* 52 (3):599–605.

Jain, N., K. R. Krull, P. Brouwers, et al. 2008. Neuropsychological outcome following intensity-modulated radiation therapy for pediatric medulloblastoma. *Pediatric Blood and Cancer* 51 (2):275–9.

Jakacki, R. I., J. W. Goldwein, R. L. Larsen, G. Barber, and J. H. Silber. 1993. Cardiac dysfunction following spinal irradiation during childhood. *Journal of Clinical Oncology* 11 (6):1033–8.

Jakacki, R. I., C. M. Schramm, B. R. Donahue, F. Haas, and J. C. Allen. 1995. Restrictive lung disease following treatment for malignant brain tumors: A potential late effect of craniospinal irradiation. *Journal of Clinical Oncology* 13 (6):1478–85.

Jereb, B., N. Sundaresan, B. Horten, A. Reid, and J. H. Galicich. 1981. Supratentorial recurrences in medulloblastoma. *Cancer* 47 (4):806–9.

Kalapurakal, J. A., V. Sathiaseelan, T. Bista, and M. H. Marymont. 2000. Adverse impact of multileaf collimator field shaping on lens dose in children with acute leukemia receiving cranial irradiation. *International Journal of Radiation Oncology • Biology • Physics* 48 (4):1227–31.

Karlsson, U., T. Kirby, W. Orrison, and M. Lionberger. 1995. Ocular globe topography in radiotherapy. *International Journal of Radiation Oncology • Biology • Physics* 33 (3):705–12.

Khatua, S., G. Dhall, S. O'Neil, et al. 2010. Treatment of primary CNS germinomatous germ cell tumors with chemotherapy prior to reduced dose whole ventricular and local boost irradiation. *Pediatric Blood and Cancer* 55 (1):42–6.

Kiltie, A. E., J. M. Povall, and R. E. Taylor. 2000. The need for the moving junction in craniospinal irradiation. *British Journal of Radiology* 73 (870):650–4.

Koshy, M., A. C. Paulino, R. B. Marcus, Jr., and J. Ting. 2004. The effect of an extended source-to-skin distance in the treatment of the spinal field in children receiving craniospinal irradiation. *Medical Dosimetry* 29 (1):7–10.

Lau, M. W., H. Guscott, R. Heaton, N. Laperriere, and B. Millar. 2006. Craniospinal irradiation in the supine position—Application of an image-guided radiotherapeutic technique to plan, verify and treat. *International Journal of Radiation Oncology • Biology • Physics* 66 (3):S249–50.

Laughton, S. J., T. E. Merchant, C. A. Sklar, et al. 2008. Endocrine outcomes for children with embryonal brain tumors after risk-adapted craniospinal and conformal primary-site irradiation and high-dose chemotherapy with stem-cell rescue on the SJMB-96 trial. *Journal of Clinical Oncology* 26 (7):1112–8.

Lavey, R. S., A. J. Olch, and J. Finlay. 2006. Treatment of localized intracranial germinoma using chemotherapy followed by whole ventricular radiotherapy with a simultaneous integrated boost. *International Journal of Radiation Oncology • Biology • Physics* 66 (3 Suppl. 1):S110.

Lees, R. E., A. R. Kagan, M. Wollin, S. L. Chan, and J. E. Bellotti. 1988. Pediatric medulloblastoma: A new treatment technique. *Medical Dosimetry* 13 (3):143–7.

Letschert, J. G., D. Gonzalez Gonzalez, J. Oskam, et al. 1991. Results of radiotherapy in patients with stage I orbital non-Hodgkin's lymphoma. *Radiotherapy and Oncology* 22 (1):36–44.

Li, C., R. Muller-Runkel, S. Vijayakumar, L. C. Myrianthopoulos, and F. T. Kuchnir. 1994. Craniospinal axis irradiation: An improved electron technique for irradiation of the spinal axis. *British Journal of Radiology* 67 (794):186–93.

Lin, R., E. B. Hug, R. A. Schaefer, D. W. Miller, J. M. Slater, and J. D. Slater. 2000. Conformal proton radiation therapy of the posterior fossa: A study comparing protons with three-dimensional planned photons in limiting dose to auditory structures. *International Journal of Radiation Oncology • Biology • Physics* 48 (4):1219–26.

Liu, A. K., D. Thornton, J. Backus, W. Dzingle, S. Stoehr, and F. Newman. 2009. Supine craniospinal irradiation setup with two spine fields. *Medical Dosimetry* 34 (3):214–6.

Loeffler, J. S., H. M. Kooy, and N. J. Tarbell. 1999. The emergence of conformal radiotherapy: Special implications for pediatric neuro-oncology. *International Journal of Radiation Oncology • Biology • Physics* 44 (2):237–8.

MacDonald, S. M., and T. I. Yock. 2010. Proton beam therapy following resection for childhood ependymoma. *Child's Nervous System* 26 (3):285–91.

Mah, K., C. E. Danjoux, S. Manship, N. Makhani, M. Cardoso, and K. E. Sixel. 1998. Computed tomographic simulation of craniospinal fields in pediatric patients: Improved treatment accuracy and patient comfort. *International Journal of Radiation Oncology • Biology • Physics* 41 (5):997–1003.

Maor, M. H., R. S. Fields, K. R. Hogstrom, and J. van Eys. 1985. Improving the therapeutic ratio of craniospinal irradiation in medulloblastoma. *International Journal of Radiation Oncology • Biology • Physics* 11 (4):687–97.

Merchant, T. E., R. M. Chitti, C. Li, X. Xiong, R. A. Sanford, and R. B. Khan. 2010. Factors associated with neurological recovery of brainstem function following postoperative conformal radiation therapy for infratentorial ependymoma. *International Journal of Radiation Oncology • Biology • Physics* 76 (2):496–503.

Merchant, T. E., O. Goloubeva, D. L. Pritchard, et al. 2002. Radiation dose-volume effects on growth hormone secretion. *International Journal of Radiation Oncology • Biology • Physics* 52 (5):1264–70.

Merchant, T. E., C. H. Hua, H. Shukla, X. Ying, S. Nill, and U. Oelfke. 2008. Proton versus photon radiotherapy for common pediatric brain tumors: Comparison of models of dose characteristics and their relationship to cognitive function. *Pediatric Blood and Cancer* 51 (1):110–7.

Merchant, T. E., L. E. Kun, M. J. Krasin, et al. 2008. Multi-institution prospective trial of reduced-dose craniospinal irradiation (23.4 Gy) followed by conformal posterior fossa (36 Gy) and primary site irradiation (55.8 Gy) and dose-intensive chemotherapy for average-risk medulloblastoma. *International Journal of Radiation Oncology • Biology • Physics* 70 (3):782–7.

Merchant, T. E., S. R. Rose, C. Bosley, S. Wu, X. Xiong, and R. H. Lustig. 2011. Growth hormone secretion after conformal radiation therapy in pediatric patients with localized brain tumors. *Journal of Clinical Oncology* 29 (36):4776–80.

Merchant, T. E., S. H. Sherwood, R. K. Mulhern, et al. 2000. CNS germinoma: Disease control and long-term functional outcome for 12 children treated with craniospinal irradiation. *International Journal of Radiation Oncology • Biology • Physics* 46 (5):1171–6.

Merchant, T. E., Y. Zhu, S. J. Thompson, M. R. Sontag, R. L. Heideman, and L. E. Kun. 2002. Preliminary results from a Phase II trial of conformal radiation therapy for pediatric patients with localised low-grade astrocytoma and ependymoma. *International Journal of Radiation Oncology • Biology • Physics* 52 (2):325–32.

Michalski, J. M., E. E. Klein, R. Gerber. 2002. Method to plan, administer, and verify supine craniospinal irradiation. *Journal of Applied Clinical Medical Physics* 3 (4):310–6.

Miemietz, M. B., F. Bruns, J. Fruehauf, M. Bremer, and J. H. Karstens. 2007. Radiation of the craniospinal axis in supine position: Long-term experience with a reference-point based method. *International Journal of Radiation Oncology • Biology • Physics* 69 (3 Suppl. 1):S246.

Milano, M. T., L. S. Constine, and P. Okunieff. 2007. Normal tissue tolerance dose metrics for radiation therapy of major organs. *Seminars in Radiation Oncology* 17 (2):131–40.

Miralbell, R., A. Bleher, P. Huguenin, et al. 1997. Pediatric medulloblastoma: Radiation treatment technique and patterns of failure. *International Journal of Radiation Oncology • Biology • Physics* 37 (3):523–9.

Mitchell, J. D., H. Christine, and M. T. Vlachaki. 2007. Role of ovarian transposition based on the dosimetric effects of craniospinal irradiation on the ovaries: A case report. *Medical Dosimetry* 32 (3):204–10.

Monroe, A. T., N. Bhandare, C. G. Morris, and W. M. Mendenhall. 2005. Preventing radiation retinopathy with hyperfractionation. *International Journal of Radiation Oncology • Biology • Physics* 61 (3):856–64.

Mu, X., T. Bjork-Eriksson, S. Nill, et al. 2005. Does electron and proton therapy reduce the risk of radiation induced cancer after spinal irradiation for childhood medulloblastoma? A comparative treatment planning study. *Acta Oncologica* 44 (6):554–62.

Munshi, A., and R. Jalali. 2008. A simple technique of supine craniospinal irradiation. *Medical Dosimetry* 33 (1):1–5.

Narayana, A., S. Jeswani, and A. C. Paulino. 1999. The cranial-spinal junction in medulloblastoma: Does it matter? *International Journal of Radiation Oncology • Biology • Physics* 44 (1):81–4.

Olch, A. J., M. Geurts, B. Thomadsen, R. Famiglietti, and E. L. Chang. 2007. Portal imaging practice patterns of children's oncology group institutions: Dosimetric assessment and recommendations for minimizing unnecessary exposure. *International Journal of Radiation Oncology • Biology • Physics* 67 (2):594–600.

Packer, R. J., A. Gajjar, G. Vezina, et al. 2006. Phase III study of craniospinal radiation therapy followed by adjuvant chemotherapy for newly diagnosed average-risk medulloblastoma. *Journal of Clinical Oncology* 24 (25):4202–8.

Panandiker, A. P., H. Ning, A. Likhacheva, et al. 2007. Craniospinal irradiation with spinal IMRT to improve target homogeneity. *International Journal of Radiation Oncology • Biology • Physics* 68 (5):1402–9.

Parker, W., M. Brodeur, D. Roberge, and C. Freeman. 2010. Standard and nonstandard craniospinal radiotherapy using helical TomoTherapy. *International Journal of Radiation Oncology • Biology • Physics* 77 (3):926–31.

Parker, W., E. Filion, D. Roberge, and C. R. Freeman. 2007. Intensity-modulated radiotherapy for craniospinal irradiation: Target volume considerations, dose constraints, and competing risks. *International Journal of Radiation Oncology • Biology • Physics* 69 (1):251–7.

Parker, W. A., and C. R. Freeman. 2006. A simple technique for craniospinal radiotherapy in the supine position. *Radiotherapy and Oncology* 78 (2):217–22.

Parsons, J. T., F. J. Bova, C. R. Fitzgerald, W. M. Mendenhall, and R. R. Million. 1994a. Radiation optic neuropathy after megavoltage external-beam irradiation: Analysis of time-dose factors. *International Journal of Radiation Oncology • Biology • Physics* 30 (4):755–63.

———. 1994b. Radiation retinopathy after external-beam irradiation: Analysis of time-dose factors. *International Journal of Radiation Oncology • Biology • Physics* 30 (4):765–73.

Paulino, A. C., M. Lobo, B. S. Teh, et al. 2010. Ototoxicity after intensity-modulated radiation therapy and cisplatin-based chemotherapy in children with medulloblastoma. *International Journal of Radiation Oncology • Biology • Physics* 78 (5):1445–50.

Paulino, A. C., and M. Skwarchuk. 2002. Intensity-modulated radiation therapy in the treatment of children. *Medical Dosimetry* 27 (2):115–20.

Penagaricano, J. A., N. Papanikolaou, Y. Yan, and V. Ratanatharathorn. 2004. Application of intensity-modulated radiation therapy for pediatric malignancies. *Medical Dosimetry* 29 (4):247–53.

Penagaricano, J. A., N. Papanikolaou, Y. Yan, E. Youssef, and V. Ratanatharathorn. 2005. Feasibility of cranio-spinal axis radiation with the Hi-Art tomotherapy system. *Radiotherapy and Oncology* 76 (1):72–8.

Phillips, C., D. Willis, J. Cramb, F. Chicas-Angulo, and M. Sexton. 2004. A modified technique for craniospinal irradiation in children designed to reduce acute and late radiation toxicity. *Australasian Radiology* 48 (2):188–94.

Rades, D., R. Baumann, M. Bremer, M. Leuwer, and J. H. Karstens. 2001. Application of a new verification technique allowing craniospinal irradiation in supine position. *Radiotherapy and Oncology* 58 (2):215–7.

Raggi, E., M. A. Mosleh-Shirazi, and F. H. Saran. 2008. An evaluation of conformal and intensity-modulated radiotherapy in whole ventricular radiotherapy for localised primary intracranial germinomas. *Clinical Oncology (Royal College of Radiologists)* 20 (3):253–60.

Rappaport, R., and R. Brauner. 1989. Growth and endocrine disorders secondary to cranial irradiation. *Pediatric Research* 25 (6):561–7.

Rembielak, A., and T. C. S. Woo. 2005. Intensity-modulated radiation therapy for the treatment of pediatric cancer patients. *Nature Clinical Practice Oncology* 2 (4):211–7.

Ries, L. A. G., M. A. Smith, J. G. Gurney, et al. 1999. *Cancer Incidence and Survival among Children and Adolescents: United States SEER Program 1975–1995* (NIH Pub. No. 99-4649). Bethesda, MD: National Cancer Institute.

Roback, D. M., J. M. Johnson, F. M. Khan, G. P. Engeler, and W. A. McGuire. 1997. The use of tertiary collimation for spinal irradiation with extended SSD electron fields. *International Journal of Radiation Oncology • Biology • Physics* 37 (5):1187–92.

Roggli, V. L., R. Estrada, and R. E. Fechner. 1979. Thyroid neoplasia following irradiation for medulloblastoma: Report of two cases. *Cancer* 43 (6):2232–8.

Scharf, C. B., A. C. Paulino, and K. N. Goldberg. 1998. Determination of the inferior border of the thecal sac using magnetic resonance imaging: Implications on radiation therapy treatment planning. *International Journal of Radiation Oncology • Biology • Physics* 41 (3):621–4.

Schroeder, T. M., M. Chintagumpala, M. Fatih Okcu, et al. 2008. Intensity-modulated radiation therapy in childhood ependymoma. *International Journal of Radiation Oncology • Biology • Physics* 71 (4):987–93.

Schultheiss, Timothy E. 2008. The radiation dose-response of the human spinal cord. *International Journal of Radiation Oncology • Biology • Physics* 71 (5):1455–9.

Seppälä, J. K. H., and J. A. J. Kulmala. 2011. Increased beam attenuation and surface dose by different couch inserts of treatment tables used in megavoltage radiotherapy. *Journal of Applied Clinical Medical Physics* 12 (4):3554.

Shibamoto, Y., K. Sasai, N. Oya, and M. Hiraoka. 2001. Intracranial germinoma: Radiation therapy with tumor volume-based dose selection. *Radiology* 218 (2):452–6.

Silber, J. H., J. Radcliffe, V. Peckham, et al. 1992. Whole-brain irradiation and decline in intelligence: The influence of dose and age on IQ score [see comment]. *Journal of Clinical Oncology* 10 (9):1390–6.

Sklar, C. A., and L. S. Constine. 1995. Chronic neuroendocrinological sequelae of radiation therapy. *International Journal of Radiation Oncology • Biology • Physics* 31 (5):1113–21.

Slampa, P., Z. Pavelka, L. Dusek, et al. 2007. Longterm treatment results of childhood medulloblastoma by craniospinal irradiation in supine position. *Neoplasma* 54 (1):62–7.

Smitt, M. C., E. M. McPeak, and S. S. Donaldson. 1998. The advantages of three-dimensional conformal radiotherapy for treatment of childhood cancer. *Radiation Research* 150 (5 Suppl.):S170–7.

Sohn, J. W., M. C. Schell, K. K. Dass, J. H. Suh, and M. Tefft. 1994. Uniform irradiation of the craniospinal axis with a penumbra modifier and an asymmetric collimator. *International Journal of Radiation Oncology • Biology • Physics* 29 (1):187–90.

South, M. S., J. K. Chiu, B. S. Teh, C. Bloch, T. Schroeder, and A. C. Paulino. 2007. Supine craniospinal irradiation: Early results on patterns of failure. *International Journal of Radiation Oncology • Biology • Physics* 69 (3 Suppl. 1):S577.

South, M., J. K. Chiu, B. S. Teh, C. Bloch, T. M. Schroeder, and A. C. Paulino. 2008. Supine craniospinal irradiation using intrafractional junction shifts and field-in-field dose shaping: Early experience at Methodist Hospital. *International Journal of Radiation Oncology • Biology • Physics* 71 (2):477–83.

St. Clair, W. H., J. A. Adams, M. Bues, et al. 2004. Advantage of protons compared to conventional X-ray or IMRT in the treatment of a pediatric patient with medulloblastoma. *International Journal of Radiation Oncology • Biology • Physics* 58 (3):727–34.

Sutton, L. N., J. Radcliffe, J. W. Goldwein, et al. 1999. Quality of life of adult survivors of germinomas treated with craniospinal irradiation. *Neurosurgery* 45 (6):1292–7; discussion 1297–8.

Swanson, E., P. Jursinic, J. Herman, W. Lee, and S. Y. Firat. 2005. A dosimetric comparison of 3-dimensional conformal radiotherapy (3DCRT) and helical TomoTherapy for craniospinal irradiation. *International Journal of Radiation Oncology • Biology • Physics* 63 (2):S265.

Tatcher, M., and A. S. Glicksman. 1989. Field matching considerations in craniospinal irradiation. *International Journal of Radiation Oncology • Biology • Physics* 17 (4):865–9.

Taylor, R. E., C. C. Bailey, K. J. Robinson, et al. 2004. Impact of radiotherapy parameters on outcome in the International Society of Paediatric Oncology/United Kingdom Children's Cancer Study Group PNET-3 study of preradiotherapy chemotherapy for M0-M1 medulloblastoma. *International Journal of Radiation Oncology • Biology • Physics* 58 (4):1184–93.

Thomadsen, B., M. Mehta, S. Howard, and R. Das. 2003. Craniospinal treatment with the patient supine. *Medical Dosimetry* 28 (1):35–8.

Timmermann, B., A. J. Lomax, L. Nobile, et al. 2007. Novel technique of craniospinal axis proton therapy with the spot-scanning system: Avoidance of patching multiple fields and optimized ventral dose distribution. *Strahlentherapie und Onkologie* 183 (12):685–8.

Tinkler, S. D., and H. H. Lucraft. 1995. Are moving junctions in craniospinal irradiation for medulloblastoma really necessary? *British Journal of Radiology* 68 (811):736–9.

Verlooy, J., V. Mosseri, S. Bracard, et al. 2006. Treatment of high risk medulloblastomas in children above the age of 3 years: A SFOP study. *European Journal of Cancer* 42 (17):3004–14.

Wallace, W. H., S. M. Shalet, J. H. Hendry, P. H. Morris-Jones, and H. R. Gattamaneni. 1989. Ovarian failure following abdominal irradiation in childhood: The radiosensitivity of the human oocyte. *British Journal of Radiology* 62 (743):995–8.

Wang, S., Z. Liao, X. Wei, et al. 2006. Analysis of clinical and dosimetric factors associated with treatment-related pneumonitis (TRP) in patients with non-small-cell lung cancer (NSCLC) treated with concurrent chemotherapy and three-dimensional conformal radiotherapy (3D-CRT). *International Journal of Radiation Oncology • Biology • Physics* 66 (5):1399–407.

Wilkinson, J. M., J. Lewis, G. P. Lawrence, H. H. Lucraft, and E. Murphy. 2007. Craniospinal irradiation using a forward planned segmented field technique. *British Journal of Radiology* 80 (951):209–15.

Williams, G. B., L. E. Kun, J. W. Thompson, H. J. Gould, and R. M. S. Stocks. 2005. Hearing loss as a late complication of radiotherapy in children with brain tumors. *Annals of Otology, Rhinology, and Laryngology* 114 (4):328–31.

Woo, S. Y., S. S. Donaldson, R. J. Heck, K. L. Nielson, and C. Shostak. 1989. Minimizing and measuring lens dose when giving cranial irradiation. *Radiotherapy and Oncology* 16 (3):183–8.

Xu, W., A. Janss, R. J. Packer, et al. 2004. Endocrine outcome in children with medulloblastoma treated with 18 Gy of craniospinal radiation therapy. *Neuro-Oncology* 6 (2):113–8.

Yom, S. S., E. K. Frija, A. Mahajan, et al. 2007. Field-in-field technique with intrafractionally modulated junction shifts for craniospinal irradiation. *International Journal of Radiation Oncology • Biology • Physics* 69 (4):1193–8.

Yuh, G. E., L. N. Loredo, L. T. Yonemoto, et al. 2004. Reducing toxicity from craniospinal irradiation: Using proton beams to treat medulloblastoma in young children. *Cancer Journal* 10 (6):386–90.

Zeltzer, P. M., J. M. Boyett, J. L. Finlay, et al. 1999. Metastasis stage, adjuvant treatment, and residual tumor are prognostic factors for medulloblastoma in children: Conclusions from the Children's Cancer Group 921 randomized phase III study. *Journal of Clinical Oncology* 17 (3):832–45.

Chapter 5
Hodgkin Lymphoma

5.1 Clinical Overview

Hodgkin lymphoma (HL) occurs most commonly in adults but is also seen in adolescents and young adults. Over 80% of pediatric patients are at least 10 years of age at diagnosis. HL accounts for 6% of childhood cancers. About 1700 new cases of lymphoma are diagnosed each year with slightly more Hodgkin than non-Hodgkin subtypes. Most patients present with painless, firm adenopathy, usually in the neck. The disease generally starts in a single lymph node and spreads to contiguous lymph node regions (e.g., cervical to mediastinal to paraaortic) and occasionally to lymphatic organs (e.g., paraaortic nodes to spleen, high cervical nodes to tonsil). The most commonly involved lymph nodes are in the cervical, supraclavicular, and mediastinal regions. The staging system for patients with HL is based on the number of involved sites, whether the involved lymph nodes are on one or both sides of the diaphragm, and the presence of extranodal disease or B symptoms (high fevers, drenching night sweats, or unexplained loss of greater than 10% of body weight over the preceding 6 months). About 25% to 35% of patients present with a B symptom that portends a worse prognosis. Approximately 10% of cases present as stage IV disease with metastasis to extralymphatic organs, most commonly the liver, lung, or bone marrow. Children are more likely than adults to present with stage I/II disease and less likely to present with stage IV disease (Hodgson, Hudson, et al. 2007). The event-free survival is 80% to 85% and 10-year overall survival is 85% to 97% for early-stage and 70% to 90% for advanced-stage pediatric HL (Donaldson et al. 2002; Nachman et al. 2002). Survival is significantly better for pediatric and young adult than for older adult HL patients (Cleary et al. 1994).

Staging evaluation includes computed tomographic (CT) scans of the neck, chest, abdomen and pelvis, gallium or positron emission tomography (PET) scan, and, if there are B symptoms or spread to sites on both sides of the diaphragm, bone marrow biopsy.

The malignant cell in the involved lymph nodes is called a Reed-Sternberg cell, which originates from the B-lymphocyte lineage. Hodgkin lymphoma is comprised of two distinct entities: (1) the more common classical HL, which includes subtypes of nodular sclerosis, mixed cellularity, lymphocyte predominant, lymphocyte depleted, and interfollicular, and (2) the non-classical nodular lymphocyte predominant HL.

Treatment for HL has changed dramatically over the last 50 years with reduction in late effects being the primary motivation. Prior to the advent of effective chemotherapy regimens (in the 1960s and 1970s), the primary treatment for HL was radiation therapy to doses of 35 to 44 Gy to the involved and adjoining lymph node regions, termed subtotal or total nodal irradiation (Carmel and Kaplan 1976). Disease control was reasonable, but long-term consequences of radiation, including second malignancies (especially in the breast, thyroid, and lung), skeletal growth deficits, hypothyroidism, and coronary artery disease, have made multiagent chemotherapy, often followed by low dose radiation, the contemporary treatment (Hodgson, Hudson, et al. 2007). Because event-free survival (EFS) is high and 30% to 50% of relapsed patients are cured with additional therapy, the trend has been to attempt to decrease radiation therapy exposure to reduce toxicity. Several cooperative group randomized studies have focused on limiting the dose, volume, and proportion of patients receiving radiation therapy following chemotherapy (Nachman et al. 2002; Dorffel et al. 2003). Response-based, risk-adapted therapy is based on the theory that patients who have a rapid and complete response to chemotherapy have biologically favorable disease that can be cured with less treatment. Commonly, enlarged nodes do not shrink all the way to normal size after treatment due to fibrosis, even though they contain no active residual disease. A decrease in size by >80% or a return to normal size and resolution of PET positivity of all initially involved disease sites is therefore usually considered a complete response. Most patients achieve a complete response to chemotherapy (Figure 5.1). Patients with low-risk disease who have a complete response may be spared radiation therapy, whereas those that have a partial response receive 15 to 25 Gy (recent Children's Oncology Group [COG] protocols such as AHOD0431 specifies 21 Gy). Those with intermediate- or high-risk disease generally undergo low-dose (15 to 25 Gy) radiation. Clinical trials are underway to

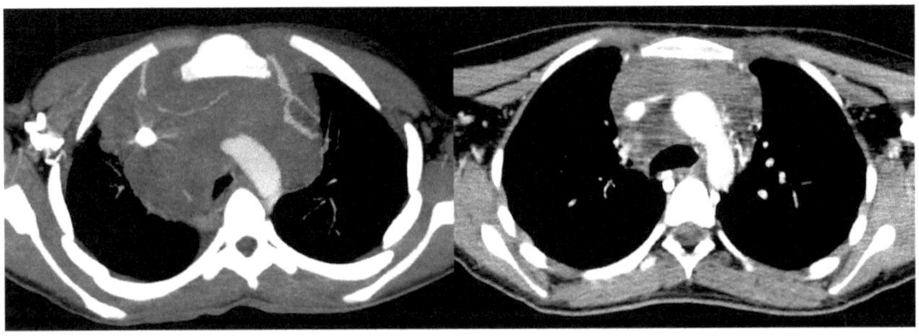

FIGURE 5.1 (Left) Large prechemotherapy mediastinal mass. (Right) Postchemotherapy excellent response with some residual disease.

determine whether certain intermediate risk patients who respond completely to chemotherapy can avoid radiotherapy. Sites containing residual disease after chemotherapy receive involved field irradiation to a dose of 21 Gy in U.S. trials and approximately 20 to 35 Gy in European trials. Involved field irradiation is also often given following chemotherapy to patients who have relapsed after receiving chemotherapy alone.

The classical irradiated volumes for HL often bear names descriptive of the anteroposterior–posteroanterior (AP–PA) field shapes. A "mantle" field encompasses the cervical, supraclavicular, axillary, mediastinal, and hilar lymph node regions. A "mini-mantle" field is similar to the mantle, without coverage of the axillary nodes. A "spade" field covers the spleen and paraaortic lymph nodes. An "inverted-Y" encompasses the paraaortic and iliac nodes. Total nodal irradiation covers all the aforementioned lymph node regions, whereas subtotal nodal irradiation generally omits the iliac nodes. Treatment volumes that include an uninvolved lymph node region adjacent to the grossly involved regions are referred to as extended-field radiotherapy (EFRT). EFRT was the standard when radiation therapy was used as the sole treatment modality. The radiation volume following chemotherapy generally includes the regions initially bearing clinically detectable disease based on examination, CT, or PET scan, but not adjacent node regions, and is referred to as involved-field radiotherapy (IFRT). Guidelines for IFRT field definition are found in Table 5.1 (Hodgson, Hudson, et al. 2007). Studies are being conducted to test whether it is safe to treat only the involved lymph nodes plus a margin (involved-node radiotherapy [INRT]) rather than the entire lymph node region containing the involved nodes. There is limited evidence in the literature that carefully implemented INRT can result in equivalent progression-free survival (PFS) and overall survival (OS) compared to IFRT or EFRT (Campbell et al. 2008).

During the first 10 to 15 years after diagnosis, about half the deaths occurring in HL patients are due to progression or relapse of HL. After this period, the main causes of death and morbidity are cardiovascular disease, secondary cancers, and respiratory diseases (Lee et al. 2000; Aleman et al.

TABLE 5.1 Involved Field Radiation Guidelines

Involved Node(s)	Radiation Field[a]
Unilateral neck	Unilateral neck + ipsilateral supraclavicular
Supraclavicular	Supraclavicular + mid/low neck + infraclavicular[b]
Axilla	Axilla ± infraclavicular/supraclavicular
Mediastinum	Mediastinum + hila + infraclavicular/supraclavicular
Hila	Hila ± mediastinum
Spleen	Spleen ± adjacent paraaortics
Paraortics	Paraaortics ± spleen
Iliac	Iliacs + inguinal/femoral

Source: Hodgson, D. C., M. M. Hudson, and L. S. Constine, *Seminars in Radiation Oncology* 17 (3):230–42, 2007.

[a] Clinical target volume (CTV) encompasses postchemotherapy mediastinal width laterally and prechemotherapy extent in superior–inferior direction.

[b] Upper neck region not treated if supraclavicular involvement is extension of the mediastinal disease.

2003; Gustavsson et al. 2003). A host of other morbidities are also seen from thyroid to gonadal dysfunction. Toxicities of high-dose full volume radiotherapy (RT) are well known but less clear following 15 to 25 Gy IFRT. Since toxicity is dose and volume dependent, the risks of the new treatments should be much less, but it takes more than 10 years for many of the toxicities to be expressed. The cumulative incidence of a grade 3–5 chronic health condition 30 years after diagnosis of HL is about 50% (Oeffinger et al. 2006). Gustavsson et al. (2003) summarized the published literature reporting the impact on survival and late effects of radiation therapy in HL patients in a total of over 27,000 patients.

5.2 General Treatment Planning Guidelines

Imaging technologies such as FDG–PET, MRI, CT, and image fusion are commonly used in the determination of the gross tumor volume (GTV) or clinical target volume (CTV) in HL radiotherapy, whereas bony anatomy and other surrogates for lymph node chains seen on planar images are becoming less frequently used for field definition. FDG–PET has become the standard imaging modality for staging and follow-up of HL, replacing Gallium-67 and has been shown to change the stage (mostly upstaging) in approximately 25% of HL patients staged by other means (Bangerter et al. 1998; Van Den Bossche et al. 2002; Wirth et al. 2002). Esiashvili et al. (2008) found that 70% of patients receiving IFRT had regions either added or excluded due to FDG–PET findings. The most common field adjustments were extending fields for contralateral neck and paraaortic/spleen, while the most common sites excluded from treatment were pleural and pericardial cavities and lung nodules. CT of the neck, chest, abdomen, and pelvis is a routine part of the staging and postchemotherapy follow-up. MRI and bone scans are used less frequently. Image fusion of pre- and postchemotherapy CT along with the corresponding FDG–PET images allows one to target the appropriate nodal and other initially involved tissues (and contiguous nodal chains for IFRT) and avoid areas either never involved with disease (for INRT) or prechemotherapy target volumes that are no longer in the same postchemotherapy anatomical location (Figure 5.1). Note that not all lymph nodes involved with disease will be PET avid so abnormalities seen on CT are considered definitive. Lymph nodes not seen on PET scan but are greater than about 2.5 cm on CT after chemotherapy having been recommended for treatment (recent COG trial).

Patient position typically is supine with arms positioned akimbo (hands on hips with elbows bent). When axillary nodes are involved, having the arms extended laterally maximizes the distance from the axillary nodes to the chest wall, thus enabling a reduction in lung dose. A less attractive alternative is to treat with arms above the head because this makes complete humoral head shielding difficult since the axillary lymph nodes move anterior to the humeral head in this position (Pergolizzi et al. 2000). If axillary lymph nodes do not need treatment, then any of these arm positions that are comfortable and reproducible can be used. The chin is hyperextended to avoid dose to the oral cavity when neck irradiation is employed. A comfortable head and neck cushion, preferably custom shaped for each patient using a vacuum bag device along with an aquaplast mask or other means of reproducing the head position, is always better than a chin strap and is required for safely giving highly conformal treatments where sparing of structures nearby the targeted lymph nodes is desired. Body immobilization also provides meaningful decreases in daily positioning error for patients treated for diseases occurring in the trunk (Bentel et al. 1997; Prosnitz et al. 1997; Fuss et al. 2004; Nevinny-Stickel et al. 2004).

As planning target volumes (PTVs) and field sizes get smaller with the new treatment volume guidelines (INRT vs. IFRT vs. EFRT), it becomes ever more important to use immobilization devices that can help reproduce the patient's shoulders, neck, chest, abdomen, pelvis, and legs. Even for AP–PA treatments, reproducibility of the patient's pose is important to assure proper dose coverage. This is especially true for treatments that include the clavicles, neck, inguinal, and femoral lymph nodes, which are in areas that are prone to large changes in position daily without immobilization. With good immobilization using vacuum bags or alpha cradles, daily variations can generally be kept to less than or equal to 5 mm and in many cases less than or equal to 3 mm. It should be noted though that complete loss of skin sparing on the posterior skin surface is possible with body immobilization systems. For low risk patients, doses are low enough that skin reactions should be minimal, but for patients needing above 30 Gy, their skin should be monitored and skin care prophylaxis employed. Generally, with good immobilization and adequate CTV margins, PTV margins of 0.3 to 0.5 cm can be used safely for AP–PA or more conformal treatments.

CT-based planning has largely replaced two-dimensional planning both because of the ubiquity of CT simulators but also because of the need to carefully define nodal volumes and organs at risk (OARs) for more conformal treatments. Patients should be CT scanned in the treatment position in their immobilization device using CT slice spacing of 2 to 3 mm. As mentioned earlier, image fusion with either diagnostic CT-to-planning CT or PET-CT-to-planning CT is routinely used for target volume definition. The prechemotherapy scan is used to define the superior and inferior extent of the target volume and the postchemotherapy scan is used for the axial dimensions.

A technique for sparing the breasts in female patients was described by Dabaja (2008). Patients lay supine on an inclined board with a 15- to 20-degree tilt from horizontal. The gantry was then rotated 5 to 10 degrees to minimize the divergence of the beam through the lungs, heart, and mandible. A facemask and immobilization bag were used. Dabaja found that both the breasts and the heart moved inferiorly, reducing the percentage of these volumes receiving the prescribed dose.

For supradiaphragmatic fields, blocking of normal structures such as the larynx, mouth, cervical spinal cord, lungs, heart, and humeral heads is often used depending on the total dose and only when involved lymph nodes would not be underdosed. Anterior laryngeal, posterior occipital, and posterior cervical spinal cord blocks are frequently used to spare the larynx, mouth, and cervical spinal cord, respectively. For mantle fields, lung blocking may be enlarged during the course of radiotherapy to limit the lung dose.

Whole heart or whole lung irradiation is warranted when there is pericardial effusion or invasion with tumor or pulmonary nodules, respectively. Doses of 10 to 15 Gy are given, usually with partial transmission blocks such that this dose completes at the same time as the larger field.

For subdiaphragmatic disease with splenic involvement, the left kidney should be blocked carefully to avoid shielding the spleen. Depending on the position of the involved nodes, this can be challenging or even impossible using AP–PA fields. Current protocols suggest limiting the whole kidney dose to 12 Gy and no more than 50% of each kidney should receive more than 15 Gy.

When both supra- and subdiaphragmatic disease is present, these areas are treated sequentially to avoid marrow toxicity. Junctions are often needed due to the large field sizes required. Typically, the junction is constructed to create a dosimetric match at the coronal midplane. Part of the spinal cord will be in the gap but part can be in a three-field overlap region, which becomes larger as the disparity of the sizes of the upper and lower fields increases. Care must be taken to well document the location of the border of the first treatment volume to be able to correctly position the border of the subsequent treatment volume to produce the planned dosimetric match. Methods to reduce the potential hot spots in the spinal cord due to three-field overlap include the use of half-transmission blocks placed on the inferior 1.5 cm of the posterior mantle and superior 1.5 cm of the posterior paraaortic fields (Oh et al. 2000). Shielding of lungs, heart, liver, spleen, kidneys, breasts, and gonads is always the goal, but frequently the location of the targets permit only partial shielding.

For targets and OARs susceptible to motion due to breathing, an assessment should be made to determine the extent to which motion is present. An appropriate internal target volume (ITV) margin for targets and planning-at-risk volume (PRV) margin for OARs should be used, or beam gating should be considered especially when PTV margins are less than 1 cm. For treatment of mediastinal disease, there has been some interest in motion management of the lungs and heart to reduce potential toxicity. Graham et al. (2008) found that treating adults with IFRT during deep-inspiration (DI) breath holding monitored by the Varian RPM system reduced total mean lung dose and V20 by about 20% compared to normal breathing. Stromberg et al. (2000) used active breathing control (ABC) to treat adult HL with IFRT and compared lung dose–volume measures when the PTV was created based on DI or a composite PTV volume using the maximum and minimum phases of the breathing cycle. They also found that DI was the phase with the most significant decrease in lung mass irradiated. This technique reduced the lung mass irradiated by 12% and the mean volume of heart irradiated to 30 Gy decreased from 26% to just 5%. The ABC-at-DI technique can also be helpful in decreasing the overlap of heart and spleen for those who require both mediastinal and splenic/paraaortic fields. Claude et al. (2007) described their experience with ABC-at-DI for children. Using modified mantle fields to 36 Gy that included treatment of the bilateral supraclavicular nodes, mediastinum, and hila, they found that ABC-at-DI reduced the mean lung volume receiving more than 5, 20, and 30 Gy and the mean lung dose by approximately 25%. Mean breast and thyroid dose did not change. A CTV-to-PTV margin of 1.3 cm was used with ABC compared to a 1.8 cm margin for free breathing. The breath-hold period ranged from 15 s to 20 s. ABC was found to be feasible for children over the age of 13. These investigators concluded that ABC-at-DI was reproducible and considerably decreased normal lung dose at the expense of a two- to threefold increase in treatment time.

The concept of IFRT was developed to reduce the toxicities often seen after EFRT while preserving high EFS rates. Although guidelines for IFRT have been published (Yahalom and Mauch 2002; Girinsky et al. 2006, 2007;

Hodgson, Hudson, et al. 2007), consensus does not exist as to the optimal field design. IFRT generally encompasses the initially involved nodal region including the prechemotherapy involved nodes plus the contiguous uninvolved lymph node chain. For example, when any abnormal lymph nodes are found in one side of the neck, that entire side of the neck is treated (but not the bilateral neck). When the mediastinum is involved, the hila are typically treated as well despite the fact that they are anatomically separate. The radial extent of disease in the lateral, anterior, and posterior directions after response to chemotherapy and the prechemotherapy extent in the superior and inferior directions defines the GTV. Table 5.1 shows recently published guidelines for involved field coverage based on lymph node involvement. The CTV can be limited by the postchemotherapy anatomic limits of the nodal region and Campbell et al. (2008) used a maximum of 10 cm along the contiguous lymphatic chain beyond the prechemotherapy-involved nodes as an upper limit to the treated volume. Recent COG protocols modify the IFRT approach to further reduce irradiated volumes by irradiating postchemotherapy volumes in the axial dimensions with bulky involvement at presentation or nonbulky disease with slow response to chemotherapy but include immediately adjacent lymph node compartments (AHOD0831 high risk HL). The superior and inferior extent of the GTV is still determined from the prechemotherapy volumes. An additional boost dose is often prescribed to nodes that have not responded completely to chemotherapy.

Recent studies of relapses in patients treated with chemotherapy alone showed that most recurrences occurred in the initially involved lymph nodes rather than elsewhere in the same or adjacent lymph node chain (Shahidi et al. 2006). This finding has generated great interest in INRT, which treats just the initially involved lymph nodes plus a margin. The CTV is the initial volume of involved lymph nodes before chemotherapy except that normal structures that were displaced by the prechemotherapy lymph nodes are not included in the CTV. For mediastinal disease in complete remission (CR), the CTV can be limited to the lateral boundaries of the normal mediastinum to reduce volume of irradiated lung. In the neck, blood vessels and muscle are not included in the CTV if their position at the time of treatment planning is separate from the CTV even though image fusion shows them to be inside the prechemotherapy lymph node volume. At any site, image fusion of pre- and postchemotherapy images is useful in determining which tissues adjacent to the initial lymph node volume potentially harbor HL at the time of radiation therapy. Similar to IFRT, the length of the CTV is the prechemotherapy lymph node volume. Usually the CTV is expanded by 1 cm axially to make the PTV but is often expanded by up to 2 cm superiorly and inferiorly.

Figures 5.2 to 5.4 show the progression of reducing field sizes for a typical HL case receiving 24 Gy. The target volume includes the mediastinum, and cervical and supraclavicular lymph nodes on the right side of the neck.

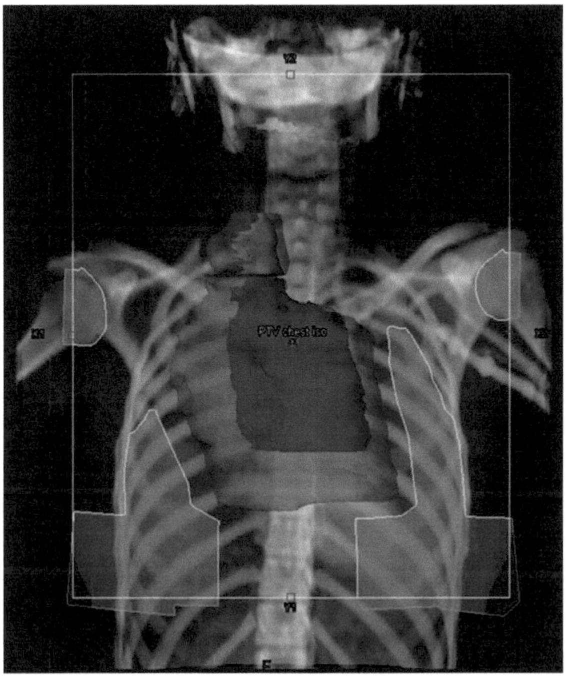

(a)

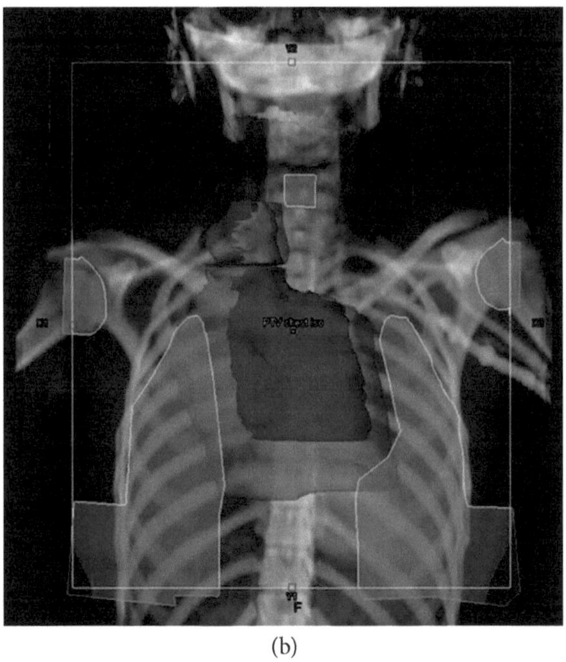

(b)

FIGURE 5.2 (See color insert.) (a) EFRT treating necessary lung. (b) EFRT with lung sparing.

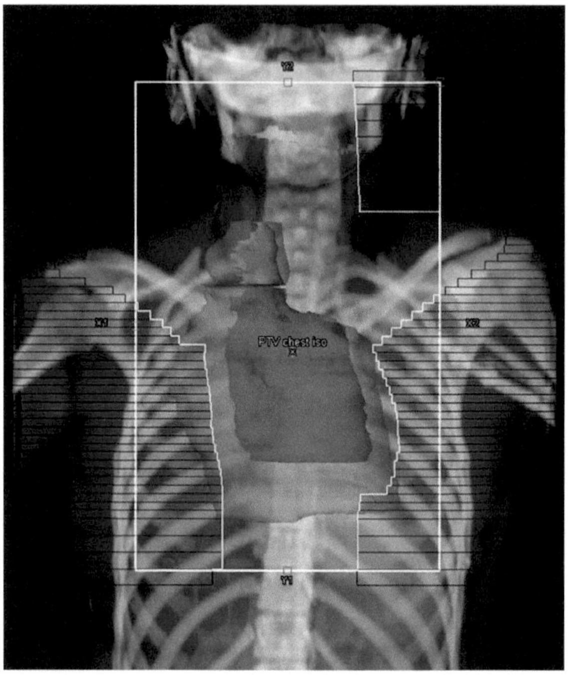

FIGURE 5.3 (See color insert.) IFRT field design.

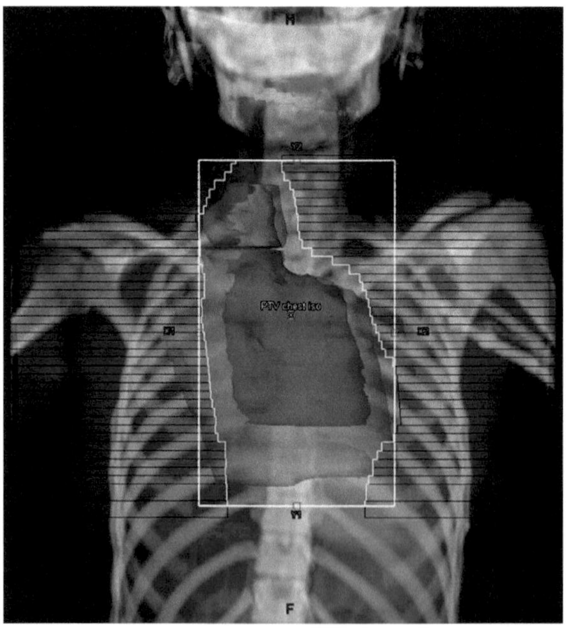

FIGURE 5.4 (See color insert.) INRT field design.

In the figures, the pink and brown volumes are the prechemotherapy mediastinal and neck CTV, and the smaller purple and green volumes are the postchemotherapy residuals. Figures 5.2a and 5.2b show classic AP–PA mantle fields; Figure 5.2a has smaller lung blocks that would be needed to adequately treat the volume of the original disease but would only be used for half the treatment due to the volume of lung treated. Figure 5.2b has larger lung blocks that block some of the initial disease but fully treat the postchemotherapy residual. Figure 5.3 shows an AP–PA IFRT field and Figure 5.4 shows an AP–PA INRT field. Figure 5.5 shows dose–volume histograms (DVHs) for the heart, lungs, and breasts for the EFRT, IFRT, and INRT plans shown in Figures 5.2 to 5.4. Dose coverage of the PTV defined

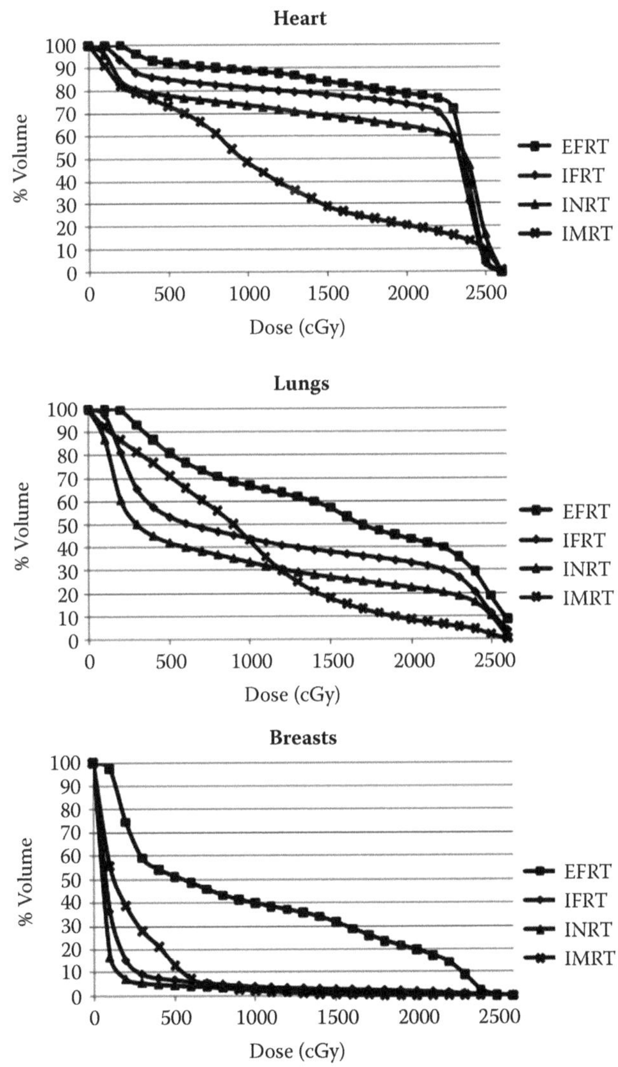

FIGURE 5.5 DVH for heart, lungs, and breasts for female patient treated AP–PA by EFRT, IFRT, INRT, or IMRT to the involved node volume.

in the INRT plan was comparable in the three plans. One can see a progressive decrease in dose to all three normal tissues as the field sizes are reduced, the most impressive being for the breasts followed by the lungs and heart. The clinical significance of these volume reductions will be greater for higher risk cases that require higher doses than 24 Gy.

Although at the time of this writing INRT is considered experimental, there seems to be considerable enthusiasm for testing this approach in an effort to further reduce toxicity for patients who have early stage HL (Girinsky et al. 2006, 2007; Campbell et al. 2008). Examples in the literature include studies by Koh and Hodgson. By using IFRT with the exclusion of the axillary and supraclavicular lymph nodes instead of standard mantle fields, patients with mediastinal disease may benefit by a reduction of the mean dose to the female breast by 64%, the lung by 24%, and the whole heart by 29%. Moving from mantle fields to IFRT was predicted to reduce breast cancer risk by 65% and reducing the prescribed dose by 40% would reduce the cancer risk by a further 40% (Koh et al. 2007). Hodgson, Hudson, et al. (2007) found a factor of 4 reduction of mean dose to the bilateral breasts for IFRT versus mantle fields and a reduction by a factor of 10 for INRT. They also reported dose reduction factors of 2 and 4 for bilateral lung dose as one changes from mantle to IFRT to INRT. Clearly, the magnitude of the reduction in dose to OARs is significant when the field size is reduced but the magnitude of reduction depends on the geometry of each patient's tumor and OARs.

Figure 5.6 shows the AP treatment field for subdiaphragmatic disease. The field encompasses two regions and in cases where field size limits are exceeded, is broken into two junctioned fields. The upper region includes paraaortic lymph nodes and the spleen (if involved) and the lower region includes the iliac and perhaps the inguinal–femoral lymph nodes. Field shaping is achieved partly by the multileaf collimator (MLC) (shown with diagonal lines) and partly with cerrobend blocks (labeled cerro-blk) where MLCs cannot be used. The GTV is shown in gray patches, and the liver, spleen, and left kidney are also shown. Note that parasplenic and paraaortic lymph node involvement forces a significant portion of the left kidney to be irradiated for adequate target coverage. Blocking of the left kidney must be carefully constructed to maximize kidney sparing without jeopardizing target coverage. The right kidney (not shown) is largely excluded from the field. COG protocols have limited the ipsilateral kidney dose to 12 Gy and 50% can receive 15 Gy. No more than 25% of the contralateral kidney should receive 12 Gy. If both kidneys require treatment, they may receive a mean dose not exceeding 12 Gy with no more than 50% of each kidney receiving more than 15 Gy. Usually less than 25% of the liver receives more than the prescribed dose, which is frequently less than 25 Gy, and liver tolerance is not exceeded. Up to 50% of the whole liver can receive more than 15 Gy. If the whole liver is included in the treated volume, the dose is limited to 10.5 Gy (COG AHOD0831).

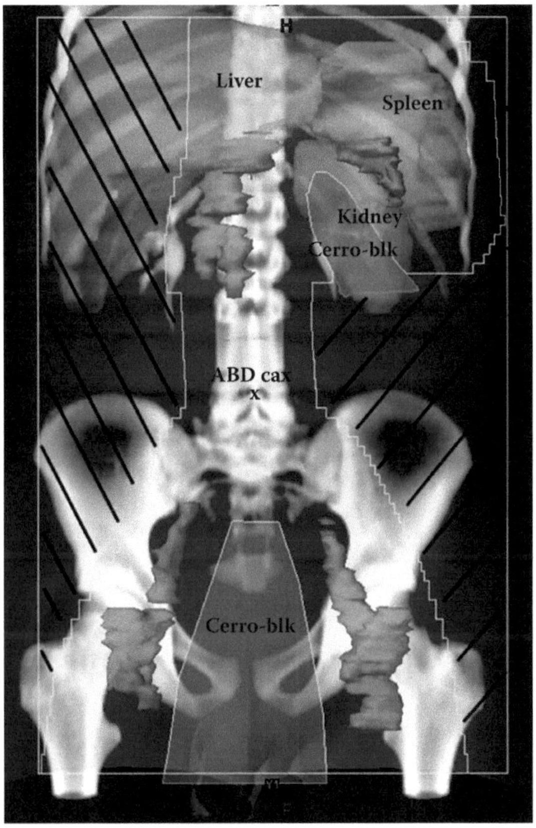

FIGURE 5.6 Paraaortic and inverted-Y field including treatment of the spleen.

5.3 Three-Dimensional Conformal Radiation Therapy (3DCRT) and Intensity-Modulated Radiation Therapy (IMRT) for Hodgkin Lymphoma (HL)

As CT-based treatment planning for HL becomes more common, there is a growing interest in optimizing conventional AP–PA treatment fields or using more complex beam arrangements to better spare normal structures. For PTVs that include both the neck and chest, dose variations of 25% (+15% in the neck and −10% in the inferior medial chest) are common with AP–PA fields due to the varying patient thicknesses across this region as well as decreased dose in narrow openings in the field commonly found in the inferior-medial aspect of the classic mantle and mini-mantle fields in the space between the lung shielding. Verification of treatment planning dose calculation accuracy in narrow off-axis regions of the field and under cerrobend and MLC blocks is highly relevant for HL treatment planning. Historically, source to surface distances (SSDs), patient thicknesses, and off-axis distances have been measured for mantle fields to perform "IRREG" calculations—dose calculations at midplane at locations of the axillae, neck, and superior and inferior mediastinum. Typically, the dose variations were recorded but not compensated (Bentel et al. 1997). Thus, a fairly inhomogeneous dose distribution was likely delivered to the vast majority of patients treated in the past. Similar dose heterogeneity can be seen for patients treated with IFRT techniques. To correct this inhomogeneity, a simple variation on the conventional AP–PA beam arrangement is to add subfields using the MLC. For treatments that target nodes in the neck and chest, subfields that exclude thinner portions of the neck and other subfields, which only irradiate the mediastinum inferior to the central axis can be effective in homogenizing the dose (see Figure 5.7). Modern treatment planning systems provide tools to efficiently plan these fields using an initial calculation to discover where the regions of overdose are and then creating nested subfields to block them, for example, in 5% dose increments. A step-by-step procedure was given by MacDonald et al. (2005) that resulted in doses within ±5% of the prescribed dose throughout large fields typically using just two subfields in addition to the conventional AP–PA fields. More elaborate and thorough dose homogenization can be performed by using the dynamic multileaf collimator (dMLC) technique described by Davis et al. (2006). Bounding AP and PA field were defined and the electronic compensation feature of the treatment planning system (TPS) was used to determine the fluence map that homogenizes the dose within the field. The regions of each field that would normally be blocked by cerrobend, such as the lungs, were painted with a software tool that sets the fluence to zero in these regions. The TPS then created the required MLC leaf motions to approximate the ideal fluence. Two or three carriage splits are generally required. This process can produce doses uniform to within about 2% across a classic mantle field. However, the number of monitor units (MUs) was increased by a factor of 3 but experience with these techniques shows that multiples of up to 6 can be expected. A minor difference between dMLC and conventional fields is that lung dose in the "blocked" region was 19% of the prescribed dose, compared to about 15% if cerrobend blocks had been used.

With multibeam IMRT techniques, one can do much more than homogenize the dose within the field. Especially attractive for supradiaphragmatic disease, IMRT allows sparing of the heart, coronary arteries, esophagus, spinal cord, and thyroid (Girinsky et al. 2006) as well as the lungs and female breasts (Goodman et al. 2005). IMRT is indicated in cases with large treatment volumes where a significant volume of lung and heart would otherwise be treated, in cases with subcarinal disease that receive cardiotoxic

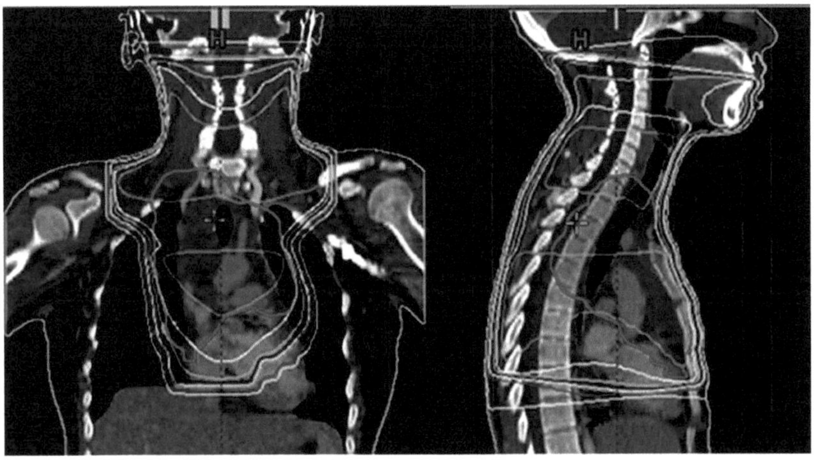

(a)

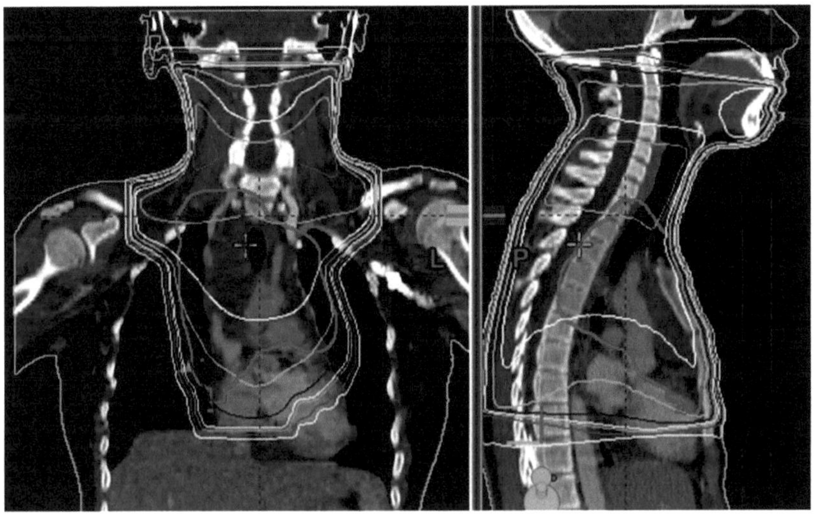

(b)

FIGURE 5.7 (See color insert.) (a) IFRT with one mediastinal subfield (field-in-field). (b) IFRT without mediastinal subfield.

chemotherapy, or for retreatment where spinal cord tolerance could be exceeded (Goodman et al. 2005). Another example is where involved para-aortic or splenic lymph nodes are located anterior to a large volume of the left kidney, making it impossible to both protect the kidney and adequately treat the disease with AP–PA fields.

Again referring to the postchemotherapy PTV in Figures 5.2 to 5.4, an IMRT plan was calculated with nine axial nonopposing 6 MV beams to give 24 Gy to the involved node PTV. The DVHs for heart, lungs, and breasts were calculated and overlaid onto the DVHs for the various AP–PA treatment methods described earlier (Figure 5.5). The heart receives a factor of 2 to 3 less dose for clinically significant dose levels except for doses approaching the prescribed dose. For the lungs, the IMRT (INRT) plan produces a volume reduction compared to the AP–PA INRT plan (which was lower than either of the other AP–PA plans) for doses above about 12 Gy, although the volumes are already quite low for AP–PA INRT. IMRT produces higher volumes of lung receiving doses lower than 12 Gy compared to AP–PA INRT or IFRT methods due to the omnidirectional beam arrangement. This trade-off between the lung volume receiving a low or high dose is probably clinically advantageous. IMRT gave larger volumes of breast tissue doses less than 5 Gy, but these dose levels are probably not clinically significant for the production of second malignancy (Travis et al. 2003).

There have been several dosimetric studies in the literature (in adults) that compared IMRT, 3DCRT, AP–PA, and even proton treatment techniques with varying results for the ability to protect the lungs, heart, and other normal structures. Either IFRT or INRT PTV volumes were studied and in some cases compared. The studies tabulate the percentage reductions as well as the actual percentage of the OAR volumes that receive a particular dose. In many cases, the percentage reduction is large but the absolute percentage volume or the dose level is too small to be clinically relevant. For treatment of bulky mediastinal disease to a mean dose of 34 Gy, IMRT reduced mean lung dose by 12% but increased V20 by 37%, reduced the mean heart dose by 11%, and improved target volume coverage by 18% compared to an AP–PA plan. Using a NTCP model, the chance for a lung complication was reduced from 14% for an AP–PA technique to just 4% with IMRT. IMRT was also useful for simultaneous boost of FDG–PET positive residual nodes (Goodman et al. 2005). Girinsky et al. (2006) and Ghalibafian et al. (2008) used five equally spaced beams with a virtual OAR volume between the PTV and heart to try to improve the sparing of the heart, coronary arteries, lungs, breasts, and esophagus. They were able to reduce the V30 for the heart from 25% with AP–PA to 14% with IMRT as well as achieve a 10% reduction of the mean dose to the origin of the coronary arteries. Girinsky et al. found no improvement in the lung dose metrics using IMRT, and the breast V5 was 26% for IMRT versus 1% for AP–PA, but the breast V20 was similar at about 4%. Esophagus dose was 16% lower while spinal cord dose was 74% lower for IMRT compared to APPA. Cases with subdiaphragmatic disease can also benefit where lymph nodes need to be treated that would jeopardize the kidneys or ovaries.

Chera et al. (2009) compared AP–PA to IMRT with five equally spaced fields and to an AP proton beam for nine patients with HL of the neck or mediastinum but not in the hilar, subcarinal, internal mammary, pericardial, or diaphragmatic nodes. Patients received INRT to 30 Gy. They compared V4 up to V30 for the total body, lungs, breast, and thyroid. For lung, there was no difference between the three treatment delivery techniques for V16 and above either because of lack of statistical significance or lack of clinical significance (all percentage volumes were below about 13%). At V4 and V10, proton values were smallest (11%–13%) and IMRT largest (21%–38%), but these dose levels may be too small to be clinically significant. For breast, no statistically significant differences were found between the three modalities except at V4, where protons and AP–PA were 6%–8% and IMRT was about 25%. It is unlikely that this dose level is clinically relevant. For the thyroid, there was no statistically significant volume difference at any dose level among the three modalities. For the total body, the V4-V16 was significantly smaller (by 100s to 1000s of cc) for protons than either AP–PA or IMRT. For V24 and V30, IMRT provided the smallest volume (a few 100 ccs less than for protons) and AP–PA treated nearly 1000 cc larger volume than did IMRT (Chera et al. 2009).

Weber et al. (2009) compared a 30 Gy prescribed dose by rapid arc (RA) to nine-beam standard IMRT (sIMRT) for 10 early stage female HL patients for PTVs based on either IFRT or INRT volumes. Breast, heart, thyroid, and submandibular gland doses were lowered so significantly using INRT compared to IFRT that no clinically significant difference was found between the two IMRT delivery techniques except perhaps for thyroid, where RA for INRT lowered the V10 to 65% versus 74% with IMRT. MUs for RA were generally about one-third that of sIMRT. Some may find the provided OAR dose–volume constraints useful. In both the Chera and Weber studies, reductions in the PTV by itself largely overshadow the effects of the treatment delivery technology when it comes to reducing OAR dose–volume values. The best treatment technology for a particular patient will greatly depend on the particular geometrical relationships between PTV and OARs and should be determined on a case-by-case basis.

Three to nine equally spaced beams were used for IMRT for HL in the aforementioned papers. However, HL presents a particularly challenging case for IMRT, with targets closely surrounded by OARs. In these situations, more than five beams should be used to maximize the characteristics of IMRT. TomoTherapy techniques have been employed (effectively over 70 beams), which appear to further improve the therapeutic ratio (Mundt and Roeske 2005; Vlachaki and Kumar 2010). Gated TomoTherapy or volumetric modulated arc therapy (VMAT) treatments along with daily image-guided radiation therapy (IGRT) may allow for smaller margins, which may further improve on these dosimetric results. Although IMRT is attractive in certain cases, one should always consider if an AP–PA technique would be appropriate or even superior to IMRT. For example, if just one side of the neck needs treatment, there may be little benefit from IMRT, and there could be some harm in that tissues that do not need any dose will be irradiated.

Current treatment protocols for HL use lower radiation doses and volumes than in the past, which dramatically improves the toxicity profile for treatment of this disease compared to treatment practices of the past. Technological advances in treatment planning and delivery, patient immobilization, and image guidance are key in ensuring that this goal of lower toxicity is not at the expense of target dose coverage and retention of the excellent local control and survival for these patients.

5.4 Organ-at-Risk Doses and Late Effects

5.4.1 Bone and Soft-Tissue Growth

Full dose (35–44 Gy) mantle field irradiation produces bone and soft-tissue growth retardation, which can manifest as spinal and clavicular shortening and underdevelopment of the soft tissues of the neck and chest. These were some of the first recognized side effects of HL therapy given in the 1960s and 1970s (Donaldson and Kaplan 1982). Age at time of treatment, total dose of radiation, fraction size, and volume treated are all significant predictors of ultimate attained height. In a study evaluating the effects on growth of contemporary treatment for pediatric HL, Papadakis et al. (1996) found that patients treated with both radiation and chemotherapy had a significantly smaller final attained height than those treated with either radiation or chemotherapy alone. Younger patients and those treated with higher RT dose suffered the greatest loss of potential height (Papadakis et al. 1996). The height in the sitting position, a measure of upper body length, was more than one standard deviation below the mean in 16 of 23 patients 3 to 12 years old at initial treatment (Mauch et al. 1983). In patients receiving asymmetric irradiation to the clavicles as in lateralized mantle fields, clavicles fully irradiated to doses of just 15 Gy grew 0.5 cm less than partially irradiated clavicles, and this effect was more pronounced for those with a mean age of 10 than those with a mean age of 16 years old (Merchant et al. 2004). There is a fairly steep dose–effect relationship for bone growth effects between about 15 and 30 Gy. Radiation doses greater than 33 Gy to subtotal or total lymphoid volumes produced a mean loss of final stature about 13 cm greater in children irradiated prior to puberty than in older children, resulting in an adult height 2 standard deviations below the U.S. population mean. Reducing the dose or volume reduced the growth deficit (Willman et al. 1994). Reductions in bone growth will still occur even with the 21 to 25 Gy given in modern protocols but should be lessened by the reduction in field sizes now being used.

5.4.2 Thyroid

The thyroid gland is particularly sensitive to radiation. Hypothyroidism has been reported in between 10% and 60% of HL patients who received cervical lymph node irradiation as children (Hudson et al. 2004; Chow et al. 2006; Metzger et al. 2006; Donaldson et al. 2007). Nearly all patients who received full dose mantle irradiation developed some thyroid function abnormalities if not overt hypothyroidism (Morgan et al. 1985). The development of thyroid dysfunction is associated with radiation dose. Thyroid abnormalities developed in 17% of children who received up to 26 Gy compared to in 78% of those who received greater than 26 Gy (Gozdasoglu et al. 1995). In another series, the incidence of hypothyroidism was 39% after a mean dose of 38 Gy compared to 23% after a mean dose of 24 Gy (Koontz 2006). In a study of 105 children, half whom received 15 Gy and the other half 25.5 Gy for supradiaphragmatic HL, 42% developed abnormal thyroid function. Three children developed thyroid nodules (Donaldson et al. 2007). Due to the high sensitivity of the thyroid to radiation dose, both for dysfunction and second malignancy (as discussed later), minimizing the thyroid dose whenever target coverage is not sacrificed is key in reducing the frequency of these complications.

5.4.3 Pulmonary

Multiple acute and chronic pulmonary toxicities have been reported after full dose mantle field radiation therapy alone (Donaldson and Kaplan 1982; Horning et al. 1994; Nysom et al. 1998; Villani et al. 2000). Pulmonary fibrosis is the late phase of radiation-induced lung damage. Diffusion capacity (%DLCO) and (FEV1/FVC) (a measure of lung function) in particular were found to deteriorate but patients were generally asymptomatic (Morgan et al. 1985; Smith et al. 1989; Shapiro et al. 1990; Villani et al. 2000). Symptomatic acute pneumonitis occurred in several children treated with mantle fields to a dose of 44 Gy, one of which was fatal. None of the lower dose patients in the series had this complication (Donaldson and Kaplan 1982). Mantle irradiation with 44 Gy has been shown to be a significant predictor of reduced pulmonary function (Horning et al. 1994). Radiation may also lead to retardation of growth of ribs and sternum leading to pulmonary restriction (Attard-Montalto et al. 1992).

Long-term effects of contemporary treatment with lower radiation doses are being evaluated. One study showed that the incidence of pulmonary toxicity 6 years after mediastinal irradiation was significantly lower in patients given doses less than 20 Gy than in patients given greater than 20 Gy (Bossi et al. 1997). Studies in children receiving chemotherapy followed by low dose mantle irradiation (15–25 Gy) showed the majority had asymptomatic, but significantly reduced %DLCO and pulmonary function up to 2 years after diagnosis (Mefferd et al. 1989; Marina et al. 1995). With 15 to 25 Gy IFRT, abnormal pulmonary function tests were still present in about one-third of children without clinical symptoms (Hunger et al. 1994; Hudson et al. 2004; Chow et al. 2006; Donaldson et al. 2007). While effective chemotherapy permits reduction in the radiation dose and volume, certain chemotherapeutic agents, bleomycin in particular, are toxic to the lungs. Lung toxicity is often seen during treatment in children and adolescents treated with the commonly used bleomycin-containing four-drug ABVD chemotherapy regimen (Hunger et al. 1994; Marina et al. 1995; Oguz et al. 2007). Bleomycin dose is typically reduced or discontinued if pulmonary toxicity is suspected. Alternatives to bleomycin and other drugs toxic to lung tissue are being investigated. We now know that lung toxicity is a complex reaction to radiation dose and volume; chemotherapy agents and dose; timing of therapies; smoking; and comorbidities present before, during, and after therapy.

Limited data are available on the relationship between radiation dosimetric parameters in the treatment of HL and pulmonary toxicity in adults, but data is sparse for children. Most of the dose–response data for lung toxicity comes from adult patients receiving radiation therapy for lung cancer. Children with HL are much younger, either do not smoke or have not smoked for very long, and are less likely to have underlying lung disease. Lung cancer patients have been shown to be more susceptible to radiation pneumonitis than lymphoma patients (Kwa et al. 1998). However, younger pediatric HL survivors have been shown to be more susceptible to radiation-induced lung toxicity than older children (Nysom et al. 1998; Oguz et al. 2007).

Understanding the relationship between radiation dose and volume on lung toxicity will help us design treatment plans which can minimize lung toxicity without compromising the high cure rate. Commonly used dose-volume measures are V13, V20, V30 and mean lung dose (MLD) (Graham et al. 2008). Based largely on lung cancer patient data, a V20 of less than about 35% is considered safe (Marks et al. 2010). There have been at least two dosimetric studies of lung toxicity in adult HL patients. Ng et al. calculated lung doses (uncorrected) using CT planning and measured changes in %DLCO in patients receiving ABVD chemotherapy plus radiotherapy to the mediastinum. They found that patients in which >33% of the lung volume received 20 Gy (23 Gy corrected) and a mean lung dose of >13 Gy (15 Gy corrected) significantly predicted for persistently reduced %DLCO at 6 and 12 months postchemotherapy. A reduction in %DLCO at 1 year was estimated at 1% (0.85% corrected) per Gy of MLD. The majority of these patients were asymptomatic (Ng et al. 2008). It should be noted that the lung doses calculated in this study were uncorrected for the effect of the lower lung density, which is the case for the majority of the lung doses in the literature reported prior to about 2005. Actual lung doses were approximately 15% higher than reported, meaning that the lung tolerance to radiation is somewhat greater than reported. Koh et al. (2006) reported on two RTOG grade 2 radiation pneumonitis (RP) cases out of 64 adult HL patients treated. These patients were CT planned with inhomogeneity corrections applied and the median MLD was 12.2 Gy. These two cases had a V20 value of 41% and 47% and an MLD of 16.4 Gy and 17.6 Gy. A statistically significant correlation was found between RP and V13, V20, V30, and MLD. It should be noted that these two patients also had an autologous stem cell transplant (ASCT), which predisposes for lung damage (Koh et al. 2006). From these data, MLD greater than about 15 Gy or V20 greater than about 40% may be useful limits to aid in planning treatment, but more data is needed with longer follow-up to establish concretely the safe dose–volume values for children.

Dosimetric studies of pediatric HL patients treated with IFRT found an MLD greater than 14 to 16 Gy or a V20 greater than 35% was predictive of grade 2 or higher RP (Brunet et al. 2007; Hua et al. 2008). Patients needing whole lung irradiation as part of their treatment were at higher risk of RP due to their necessarily higher MLD even with a total prescribed boost dose of only 25.5 Gy (Hua et al. 2008). In a study of 99 HL patients, lung DVH data was correlated to lung toxicity. A V20–30Gy equal to 32% was more likely to produce pneumonitis than a value of 22% (Hua et al. 2010). Thus, lung toxicity appears to be reduced by decreasing radiation dose and volume of lung irradiated. Recent COG HL protocols limited the whole lung dose to 12 Gy (corrected) and the whole lung D25 to 21 Gy (corrected) (COG AHOD0831).

5.4.4 Cardiac and Cardiovascular

Cardiac toxicity after full dose mediastinal radiotherapy constitutes the second most common cause of treatment-related death in HL patients; second malignancy is the most common (Lee et al. 2000; Mertens et al. 2008). Mediastinal radiation-associated heart diseases include acute pericarditis, pericardial effusions, pericardial fibrosis, pericardial and myocardial fibrosis, valvular defects (Wethal et al. 2009), conduction defects, and coronary artery disease (Donaldson and Kaplan 1982; Piovaccari et al. 1995; Stewart et al. 1995; Lee et al. 2000; Hull et al. 2003; Adams et al. 2004). Radiation injury to the capillary network is the underlying cause of ischemic myocardial degeneration and heart failure after heart irradiation. The critical target appears to be the endothelial lining of blood vessels, which undergoes a slowly progressive inflammatory response following irradiation. The biochemical and physiological mechanisms of radiation-induced cardiac toxicity was reviewed by van Rijswijk et al. (2008) and Schultz-Hector and Trott (2007). Wethal et al. (2009) reported on the causes and effects of valvular and left ventricular dysfunction due to irradiation.

The 30-year incidence of cardiovascular disease was 34.5% after high-dose mediastinal radiotherapy and the 25-year incidence of congestive heart failure was about 7% after treatment with high-dose mediastinal radiotherapy and anthracycline-containing (e.g., doxorubicin, brand-named Adriamycin) chemotherapy regimens in patients treated before age 41. The relative risk for heart diseases was markedly elevated at 18 for children. The combination of anthracyclines and radiation was significantly more cardiotoxic than radiation alone (Aleman et al. 2007). Doxorubicin-based chemotherapy with 30 to 35 Gy mantle irradiation giving a subcarinal cardiac dose of less than 30 Gy significantly increased the risk of cardiac morbidity in 20-year-olds compared to radiation alone (relative risk [RR] 4.3 versus 2.5) (Myrehaug et al. 2008). Lee et al. (2000) reported that 28% of pediatric HL patients receiving more than 35 Gy EFRT developed cardiovascular toxicity.

Cardiac toxicity data reported for adult patients may not be directly transferrable to children because of the well-known predisposing risk factors present for adults (such as alcohol and tobacco consumption, and cholesterol level) and children's increased sensitivity to therapy (Hancock, Tucker, et al. 1993; Lee et al. 2000; Adams et al. 2004; Aleman et al. 2007). The RR of acute myocardial infarction was highest in patients irradiated before age 20 and decreased with increasing age at treatment. The risk of any cardiovascular event also was significantly higher.

By the late 1970s, it was realized that blocking even part of the heart could dramatically reduce cardiac toxicity. Carmel et al. (1976) found that the incidence of acute pericariditis and pericardial effusion after mediastinal irradiation were reduced from 20% with the whole heart in the mantle fields to 7.5% with partial cardiac shielding and just 2.5% with subcarinal blocking introduced after 25 to 35 Gy.

Several studies have shown decreased cardiotoxicity in children receiving doses less than 30 Gy. Hancock, Donaldson, et al. (1993) reported on cardiac toxicity in 635 children mostly treated with mantle fields combined with chemotherapy. No cardiac deaths were seen if the prescribed dose was less than 30 Gy. There were 12 cardiac deaths (the majority acute myocardial infarction) among those receiving greater than 30 Gy, producing a relative risk for cardiac death of about 30. There were also 106 nonfatal cardiac toxicities, which also occurred mostly in patients receiving greater than 30 Gy. Elevated risk for cardiac toxicity persisted 20 years after treatment (Hancock, Donaldson, et al. 1993). Eighty-three percent of 28 pediatric HL survivors, most of whom received mediastinal doses of less than 30 Gy, had subtle left ventricular and diastolic dysfunction (Iarussi et al. 2005). In a comparison of mantle field versus IFRT, Koh et al. (2007) found that patients receiving mediastinal doses greater than 30 Gy had a significantly higher risk of cardiac death than those receiving lower doses. The recent Childhood Cancer

Survivor Study (CCSS) report on cardiac outcomes in childhood cancer survivors found increased relative risk of congestive heart failure, myocardial infarction, pericardial disease, and valvular abnormalities by twofold to sixfold compared to nonirradiated survivors with doses >15 Gy (Mulrooney et al. 2009).

The risk of cardiac morbidity with IFRT or EFRT to doses less than 25 Gy in children is uncertain due to lack of sufficient follow-up, but there is some data that can be used to guide treatment. An elevated risk for heart disease has been reported among children who received just 13 Gy to a small part of the heart and 3 Gy mean dose to the entire heart (Carr et al. 2005). Five cardiac abnormalities were noted in 78 children given IFRT to 15 to 25 Gy (Hudson et al. 2004). However, another series reported a cardiac abnormality in only 1 out of 43 children, 70% having received just 15 Gy EFRT (Chow et al. 2006). In a group of 110 children with low-risk HL treated with chemotherapy and 15 to 25.5 Gy, only one case of cardiac dysfunction was seen (Donaldson et al. 2007). A mean left ventricular dose ≥17 Gy and a mean central cardiac dose of ≥37 Gy have been reported to be associated with decreased left ventricular ejection fraction but the clinical significance of this finding is unclear (Constine et al. 1997). In two COG studies of which a total of 204 children were treated with IFRT to 25.5 Gy along with chemotherapy, no clinically relevant acute cardiac toxicities were reported (Kung et al. 2006; Tebbi et al. 2006). With modern reduced dose and volume techniques, overall cardiac morbidity has decreased, but the incidence of myocardial infarction has not changed. This may be due to greater radiation sensitivity of the coronary arteries than the cardiac muscle or lack of dose reduction to or shielding of the proximal coronary arteries (Hancock, Tucker, et al. 1993).

Radiation to the neck and mediastinum was also found to be an independent risk factor for cerebrovascular disease, transient ischemic attack (TIA), and stroke, with a 30-year cumulative incidence of about 7% in adults and a doubling of relative risk in patients less than 20 years old at the time of treatment. When the neck is treated, these risks may be reduced if hot spots are reduced in the carotid arteries located in the thinner lateral aspects of the neck (De Bruin et al. 2009).

5.4.5 Reproductive

Both radiotherapy and alkylating agent chemotherapy cause germ cell depletion and gonadal endocrine dysfunction. These effects are of concern for patients receiving abdominal pelvic irradiation (inverted-Y field) for HL. The testes are one of the most radiosensitive tissues, with doses less than 7 Gy causing significant impairment of function (Green et al. 2010). The ovaries are somewhat less sensitive.

Damage may be caused during direct irradiation of the gonads or, more commonly, from scattered radiation during treatment of adjacent tissues. Megavoltage x-ray scatter dose is weakly related to beam energy but about half that of cobalt-60 x-rays. Stovall et al. (2004) reconstructed the gonadal doses from 760 pediatric cancer survivors and found that 194 HL patients received larger gonadal doses than patients with any other disease in the cohort. The range of doses given to the testes was 2.5 to 5 Gy and to the ovaries, 3–45 Gy for tumor doses of 34 to 45 Gy where median dose for the entire cohort for either organ was less than 0.2 Gy. Although the testes are always shielded during HL radiotherapy, ovaries can be in the direct radiation field when pelvic lymph nodes are irradiated. For the testes, full thickness blocking reduces the primary dose by at least 95% but application of leaded clamshell shielding can reduce head scatter and some internal scatter, generally reducing the testes dose to about 1% to 3% of the prescribed dose (Fraass et al. 1985; Nazmy et al. 2007). In seven HL patients treated with abdominal pelvic radiation without chemotherapy whose testes were shielded by clamshells, measurements of testicular dose were made and patients were followed for sperm production. Centola et al. (1994) found that the average testicular dose was about 2 Gy and all seven patients had zero sperm count more than 3 years after therapy.

Recovery of spermatogenesis takes place from surviving stem cells (type A spermatogonia) and is dependent on the dose of radiation. Complete recovery as indicated by a return to preirradiation sperm concentrations and germinal cell numbers takes place within 9 to 18 months following 1 Gy or less, by 30 months with doses of 2 to 3 Gy, and at 5 years or more with doses of 4 Gy and above (Howell and Shalet 1998). For male patients, gonadotoxic chemotherapy is more of a threat to fertility than the scatter dose from pelvic radiation, especially at prescribed doses less than 25 Gy (Byrne et al. 1987; Ortin et al. 1990).

The estimated dose at which half of the oocytes are lost in humans (LD50) is 4 Gy (Wallace et al. 1989). Permanent ovarian failure is predicted after 20 Gy in the younger women versus 6 Gy in the older women (Lushbaugh and Casarett 1976). In another study, the age cutoff for increased radiosensitivity was 25 years old (Thibaud et al. 1998). Chiarelli et al. (1999) studied over 700 female survivors of childhood cancers to assess the risk of menopause and infertility as a function of age at diagnosis, and radiation and alkylating agent dose. They found that doses less than 20 Gy for the entire cohort did not increase the risk of early menopause but risk increased with dose above 20 Gy irrespective of age. Also, patients who received alkylating agents and abdominal pelvic radiation before the age of puberty had a lower risk of this complication (RR = 2.2 versus 3.2) than those who already attained puberty, reflecting the greater number of oocytes present at the earlier age of therapy. The risk of fertility problems for the entire cohort was ninefold compared to patients treated with only surgery (Chiarelli et al. 1999). Where alkylating agents can be withheld and low dose radiotherapy given without direct irradiation of the ovaries, fertility is preserved (Ortin et al. 1990; Hudson et al. 2004).

For the ovaries, repositioning out of the radiation field by oophoropexy can preserve ovarian function but scatter dose must still be considered (Le Floch et al. 1976). Lateral transposition is preferred to medial transposition because medially there can be higher scattered dose and sometimes incomplete shielding of the primary beam (Hadar et al. 1994).

Irradiation of the ovaries and uterus during childhood increases the risk to offspring of stillbirth or neonatal death but not for irradiation of the testes. For women treated before menarche with uterine or ovarian doses between 2.5 and 9.99 Gy, the risk increased sixfold, and for doses greater than 10 Gy, 19-fold (Signorello et al. 2010).

5.4.6 Secondary Malignancy

Second malignancies (SM) are the most common cause of death occurring more than 10 years after treatment among children treated for HL (Lee et al. 2000; Provencio 2008). Survivors have an increased risk of most types of malignancy originating anywhere in the body. HL patients appear to be intrinsically more likely to develop SMs independent of treatment modality, probably due to an immunologic deficiency. Childhood HL survivors have a 7- to 18-fold risk for developing a second malignancy compared to the general population (Lin et al. 2005). For patients diagnosed with HL at age 20, the incidence of a SM by age 50 is 10.5% in men and 24.3% in women, compared with 2.4% and 4.5% in the general population, respectively (Hodgson, Gilbert, et al. 2007). The incidence of any SM in the Late Effects Study Group patients was 10.6% within 20 years and 26.3% within 30 years. At 30 years, the risk of leukemia was 2.1% and of a solid tumor was 23.5% (Bhatia et al. 2003). Among 930 children treated for HL at five American institutions between 1960 and 1990, the 25-year actuarial incidence of SM was 19% (Constine et al. 2008). Within 10 years of developing a SM, 21% of patients were diagnosed with a third malignancy. The second most common SM, after breast cancer in girls, is thyroid cancer, with an incidence of 4.4% within 30 years. On multivariate analysis, the risk for thyroid cancer was greatest in patients treated before age 5 years. Thyroid cancer occurs in childhood HL survivors at a rate 2 to 36 times that of the general population (Lin et al. 2005; Constine et al. 2008), with the risk highest among those given radiation therapy prior to age 10 (Metayer et al. 2000; Swerdlow et al. 2000; Bhatia et al. 2003). Radiation doses as low as 2 Gy increase their probability of developing thyroid cancer (Tucker et al. 1991). The risk for lung cancer was also greatly elevated to 27 times that of the general population (Bhatia et al. 2003).

The relative risk for an SM is higher in children than in adults with HL, perhaps due to an increased susceptibility of developing tissues to mutation. Among patients treated at Harvard-affiliated hospitals, the relative risk for a SM compared to the general population among those diagnosed with HL before age 20 was 10.7, at age 20 to 50 was 4.9, and after age 50 was 2.4 (Ng et al. 2002). The SM rate is significantly higher in girls than boys, mainly due to their increased risk of breast cancer (Sankila et al. 1996; Metayer et al. 2000; Constine et al. 2008).

The SM risk is elevated in children initially treated with chemotherapy alone (Curtis et al. 2006). Alkylating agent-containing (e.g., mechlorethamine) chemotherapy is associated with the development of acute leukemia, typically occurring 2 to 10 years after treatment as well as lung cancer (Travis 2002; Travis et al. 2002). The reported risk of an acute leukemia within 10 years of alkylating agent-containing chemotherapy for pediatric HL patients is 4 to 175 times the general population (Lin et al. 2005). The secondary leukemias are nearly always lethal. When secondary non-Hodgkin lymphomas develop among pediatric HL survivors, they also usually occur during the first decade after treatment and occur at a rate 5 to 20 times that of the general population (Lin et al. 2005).

Radiation therapy is associated with the development of solid tumors. Solid tumors develop starting several years after treatment and continue to accrue at an approximately constant rate of 1% to 1.5% per year for decades. The Late Effects Study Group found that leukemia or non-Hodgkin lymphoma constituted 75% of SMs occurring within 5 years of treatment, but only 20% of SMs occurring thereafter (Bhatia et al. 2003). The SM rate was higher after chemotherapy than radiation therapy during the first 15 years of follow-up (probably due to acute leukemia), while more SMs were diagnosed in irradiated patients beyond 15 years after diagnosis (Hodgson, Gilbert, et al. 2007). Treatment of early stage HL with chemotherapy alone minimizes their risk of SM. However, chemotherapy alone is insufficient to optimize relapse-free survival for advanced-stage patients and early-stage patients who do not have a rapid, complete response to chemotherapy. For patients needing radiation therapy, the combined radiation and chemotherapy approach reducing the radiation dose and volume may minimize the risk of SM in HL patients while providing the best chance for cure. In a meta-analysis of published randomized trials, in which median follow-up periods ranged from 4 to 32 years, the SM risk was significantly lower for combined modality therapy than for radiation therapy alone and even lower with chemotherapy alone (Franklin et al. 2006).

The most common SM in young female survivors of HL treated with radiation therapy is breast cancer (Bhatia et al. 2003; Constine et al. 2008). The risk of breast cancer is highest in girls irradiated prior to age 18, then declines with increasing age (Hancock, Tucker, et al. 1993; Ng et al. 2002;

Travis et al. 2003; Wahner-Roedler et al. 2003). The elevated risk starts within 10 years after treatment and lasts until at least 30 years following irradiation (Ng et al. 2002; Alm El-Din et al. 2008). Breast cancer risk increases with breast radiation dose at least to the 40 Gy level in most but not all series (Tinger et al. 1997; Travis et al. 2003; Guibout et al. 2005; Alm El-Din et al. 2008). A matched case-control study of women under 30 years old at the time of treatment for HL found increasing risk with exposure to the involved portion of the breast above 4 Gy. The highest relative risk for breast cancer was in women who received breast irradiation to greater than 40 Gy, which was 40 times the rate for 0 to 4 Gy. The number of excess breast cancer cases per 1000 young female HL survivors over 25 years was estimated to be 42 after 20 Gy and 83 after 40 Gy breast irradiation (Travis et al. 2003). An analysis of 13 population-based cancer registries in North America and Europe found the relative risk for female breast cancer within 10 years of diagnosis to be 3.4 in women initially treated with chemotherapy alone, 6.8 with radiation therapy alone, and 7.5 with both radiation and chemotherapy (Wolden et al. 2000; Hodgson, Gilbert, et al. 2007). When breast cancer does occur, it is more likely to involve both breasts, either synchronously or metachronously, in irradiated HL patients than the general population (Gervais-Fagnou et al. 1999; Wolden et al. 2000). Among 398 girls less than 19 years old at the time of treatment for HL at five American institutions, the cumulative incidence of breast cancer within 30 years was 24%. Their risk for breast cancer was 37 times that of the general population. Of those women developing breast cancer, 34% had a cancer diagnosed in their other breast within three years. On multivariate analysis, age between 12 and 18 years and early stage HL remained significantly associated with breast cancer risk, whereas radiation dose and volume and use of chemotherapy did not (Basu 2008). The Late Effects Study Group reported on 1380 children in the United States and Europe who were under age 17 at the time of treatment for HL. The most common SM occurring in this group was breast cancer, with 40 cases, 39 of which occurred in females. All the breast cancers developed in women given at least 26 Gy to a mantle field. This cohort had a breast cancer risk 57 times greater than the general population. The cumulative incidence of breast cancer in females was 20.1% by age 45 (Bhatia et al. 2003). In the Harvard experience, the risk for breast cancer among girls diagnosed before age 15 was 112 times and for girls diagnosed between age 15 to 19 was 32 times the general population (Ng et al. 2002). The risk of developing breast cancer is reduced by chemotherapy containing an alkylating agent or irradiation of the ovaries to a dose above 5 Gy, presumably through the mechanism of eliminating hormonal stimulation of the breast tissue (Travis et al. 2003; Constine et al. 2008).

SM risk and volume are associated with radiation dose (Travis 2002; van Leeuwen et al. 2003; Hodgson, Koh, et al. 2007; Constine et al. 2008).

The SM incidence for all sites combined was increased for children treated with mantle field irradiation to 35 Gy or higher compared to those who received chemotherapy alone but not for children who received less than 35 Gy. Eighty-five percent of second solid tumors occurred within the radiation field (Constine 2008). Several studies have found the risk of SM after subtotal or total nodal irradiation to be about double the risk after mantle field irradiation (Biti et al. 1994; Ng et al. 2002). In other studies, radiation volume was not associated with the risk of SM (Franklin et al. 2006; Constine et al. 2008). A meta-analysis of published randomized clinical trials, enrolling mostly adults, found no significant difference in the risk of solid tumors, acute leukemia, or all SMs between patients randomized to extended field versus involved field radiation therapy, although extended field irradiation was associated with a significantly higher risk of breast, but not lung, cancer (Franklin et al. 2006). Although radiation doses were not stated, most of the trials in this study were from the high dose era. Hodgson, Koh, et al. (2007) used a validated model for SM after radiotherapy for HL and predicted that contemporary low dose IFRT should significantly lower the risk of SM compared to full dose mantle radiation therapy. Reduction of the target volume to involved nodes is estimated to significantly further reduce the risk of SM (Ballas et al. 2009).

References

Adams, M. J., S. R. Lipsitz, S. D. Colan, et al. 2004. Cardiovascular status in long-term survivors of Hodgkin's disease treated with chest radiotherapy. *Journal of Clinical Oncology* 22 (15):3139–48.

Aleman, B. M. P., A. W. van den Belt-Dusebout, M. L. De Bruin, et al. 2007. Late cardiotoxicity after treatment for Hodgkin lymphoma. *Blood* 109 (5):1878–86.

Aleman, B. M. P., A. W. van den Belt-Dusebout, W. J. Klokman, M. B. Van't Veer, H. Bartelink, and F. E. van Leeuwen. 2003. Long-term cause-specific mortality of patients treated for Hodgkin's disease. *Journal of Clinical Oncology* 21 (18):3431–9.

Alm El-Din, M. A., S. A. El-Badawy, and A. G. Taghian. 2008. Breast cancer after treatment of Hodgkin's lymphoma: General review. *International Journal of Radiation Oncology • Biology • Physics* 72 (5):1291–7.

Attard-Montalto, S. P., J. E. Kingston, O. B. Eden, and P. N. Plowman. 1992. Late follow-up of lung function after whole lung irradiation for Wilms' tumour. *British Journal of Radiology* 65 (780):1114–8.

Ballas, L., S. L. Wolden, I. Shuryak, et al. 2009. A comparison of involved field vs. involved node radiation therapy in the pediatric Hodgkin lymphoma population: Reduction in second malignancy risk. *International Journal of Radiation Oncology • Biology • Physics* 75 (3, Suppl. 1):S514.

Bangerter, M., F. Moog, I. Buchmann, et al. 1998. Whole-body 2-[18F]-fluoro-2-deoxy-D-glucose positron emission tomography (FDG-PET) for accurate staging of Hodgkin's disease. *Annals of Oncology* 9 (10):1117–22.

Basu, S. K., C. Schwartz, S. G. Fisher, et al. 2008. Unilateral and bilateral breast cancer in women surviving pediatric Hodgkin's disease. *International Journal of Radiation Oncology * Biology * Physics* 72 (1):34–40.

Bentel, G. C., L. B. Marks, R. Krishnamurthy, and L. R. Prosnitz. 1997. Comparison of two repositioning devices used during radiation therapy for Hodgkin's disease. *International Journal of Radiation Oncology • Biology • Physics* 38 (4):791–5.

Bhatia, S., Y. Yasui, L. L. Robison, et al. 2003. High risk of subsequent neoplasms continues with extended follow-up of childhood Hodgkin's disease: Report from the Late Effects Study Group. *Journal of Clinical Oncology* 21 (23):4386–94.

Biti, G., E. Cellai, S. M. Magrini, M. G. Papi, P. Ponticelli, and V. Boddi. 1994. Second solid tumors and leukemia after treatment for Hodgkin's disease: An analysis of 1121 patients from a single institution. *International Journal of Radiation Oncology • Biology • Physics* 29 (1):25–31.

Bossi, G., I. Cerveri, E. Volpini, et al. 1997. Long-term pulmonary sequelae after treatment of childhood Hodgkin's disease. *Annals of Oncology* 8 (Suppl. 1):19–24.

Brunet, B., D. Stewart, M. Eliasziw, and A. Balogh. 2007. Radiation pneumonitis (RP) in Hodgkin's lymphoma (HL) patients who received involved field radiation therapy (IFRT) to the mediastinum between 1999 and 2005. *International Journal of Radiation Oncology • Biology • Physics* 69 (3, Suppl. 1):S541.

Byrne, J., J. J. Mulvihill, M. H. Myers, et al. 1987. Effects of treatment on fertility in long-term survivors of childhood or adolescent cancer. *New England Journal of Medicine* 317 (21):1315–21.

Campbell, B. A., N. Voss, T. Pickles, et al. 2008. Involved-nodal radiation therapy as a component of combination therapy for limited-stage Hodgkin's lymphoma: A question of field size. *Journal of Clinical Oncology* 26 (32):5170–4.

Carmel, R. J., and H. S. Kaplan. 1976. Mantle irradiation in Hodgkin's disease: An analysis of technique, tumor eradication, and complications. *Cancer* 37 (6):2813–25.

Carr, Z. A., C. E. Land, R. A. Kleinerman, et al. 2005. Coronary heart disease after radiotherapy for peptic ulcer disease. *International Journal of Radiation Oncology • Biology • Physics* 61 (3):842–50.

Centola, G. M., J. W. Keller, M. Henzler, and P. Rubin. 1994. Effect of low-dose testicular irradiation on sperm count and fertility in patients with testicular seminoma. *Journal of Andrology* 15 (6):608–13.

Chera, B. S., C. Rodriguez, C. G. Morris, et al. 2009. Dosimetric comparison of three different involved nodal irradiation techniques for stage II Hodgkin's lymphoma patients: Conventional radiotherapy, intensity-modulated radiotherapy, and three-dimensional proton radiotherapy. *International Journal of Radiation Oncology • Biology • Physics* 75 (4):1173–80.

Chiarelli, A. M., L. D. Marrett, and G. Darlington. 1999. Early menopause and infertility in females after treatment for childhood cancer diagnosed in 1964–1988 in Ontario, Canada. *American Journal of Epidemiology* 150 (3):245–54.

Chow, L. M. L., P. C. Nathan, D. C. Hodgson, et al. 2006. Survival and late effects in children with Hodgkin's lymphoma treated with MOPP/ABV and low-dose, extended-field irradiation. *Journal of Clinical Oncology* 24 (36):5735–41.

Claude, L., C. Malet, P. Pommier, P. Thiesse, S. Chabaud, and C. Carrie. 2007. Active breathing control for Hodgkin's disease in childhood and adolescence: Feasibility, advantages, and limits. *International Journal of Radiation Oncology • Biology • Physics* 67 (5):1470-5.

Cleary, S. F., M. P. Link, and S. S. Donaldson. 1994. Hodgkin's disease in the very young. *International Journal of Radiation Oncology • Biology • Physics* 28 (1):77–83.

Constine, L. S., R. G. Schwartz, D. E. Savage, V. King, and A. Muhs. 1997. Cardiac function, perfusion, and morbidity in irradiated long-term survivors of Hodgkin's disease. *International Journal of Radiation Oncology • Biology • Physics* 39 (4):897–906.

Constine, L. S., N. Tarbell, M. M. Hudson, et al. 2008. Subsequent malignancies in children treated for Hodgkin's disease: Associations with gender and radiation dose. *International Journal of Radiation Oncology • Biology • Physics* 72 (1):24–33.

Curtis, R. E., D. M. Freedman, E. Ron, et al., eds. 2006. *New Malignancies Among Cancer Survivors: SEER Cancer Registries, 1973–2000.* NIH Publ. No. 05-5302. Bethesda, MD: National Cancer Institute.

Dabaja, B. 2008. Young female patients with Hodgkin's lymphoma can be safely treated with involved field radiation: New technique to avoid the breasts and decrease the heart dose. *International Journal of Radiation Oncology • Biology • Physics* 72 (1, Suppl. 1):S473.

Davis, Q. G., A. C. Paulino, R. Miller, and J. Y. Ting. 2006. Mantle fields in the era of dynamic multileaf collimation: Field shaping and electronic tissue compensation. *Medical Dosimetry* 31 (3):179–83.

De Bruin, M. L., L. D. A. Dorresteijn, M. B. van't Veer, et al. 2009. Increased risk of stroke and transient ischemic attack in 5-year survivors of Hodgkin lymphoma. *Journal of the National Cancer Institute* 101 (13):928–37.

Donaldson, S. S., and H. S. Kaplan. 1982. Complications of treatment of Hodgkin's disease in children. *Cancer Treatment Reports* 66 (4):977–89.

Donaldson, S. S., M. M. Hudson, K. R. Lamborn, et al. 2002. VAMP and low-dose, involved-field radiation for children and adolescents with favorable, early-stage Hodgkin's disease: Results of a prospective clinical trial. *Journal of Clinical Oncology* 20 (14):3081–7.

Donaldson, S. S., M. P. Link, H. J. Weinstein, et al. 2007. Final results of a prospective clinical trial with VAMP and low-dose involved-field radiation for children with low-risk Hodgkin's disease. *Journal of Clinical Oncology* 25 (3):332–7.

Dorffel, W., H. Luders, U. Ruhl, et al. 2003. Preliminary results of the multicenter trial GPOH-HD 95 for the treatment of Hodgkin's disease in children and adolescents: Analysis and outlook. *Klinische Padiatrie* 215 (3):139–45.

Esiashvili, N., R. Halkar, F. Keller, et al. 2008. Role of FDG-PET in definition of involved-field radiation therapy for pediatric patients with Hodgkin's lymphoma. *International Journal of Radiation Oncology • Biology • Physics* 72 (1, Suppl. 1):S78S.

Fraass, B. A., T. J. Kinsella, F. S. Harrington, and E. Glatstein. 1985. Peripheral dose to the testes: The design and clinical use of a practical and effective gonadal shield. *International Journal of Radiation Oncology • Biology • Physics* 11 (3):609–15.

Franklin, J., A. Pluetschow, M. Paus, et al. 2006. Second malignancy risk associated with treatment of Hodgkin's lymphoma: Meta-analysis of the randomised trials. *Annals of Oncology* 17 (12):1749–60.

Fuss, M., B. J. Salter, P. Rassiah, S. X. Cavanaugh, and T. S. Herman. 2004. Repositioning accuracy of a commercially available double-vacuum whole body immobilization system for stereotactic body radiation therapy. *Technology in Cancer Research & Treatment* 3 (1):59–67.

Gervais-Fagnou, D. D., C. Girouard, N. Laperriere, M. Pintillie, and P. E. Goss. 1999. Breast cancer in women following supradiaphragmatic irradiation for Hodgkin's disease. *Oncology* 57 (3):224–31.

Ghalibafian, M., A. Beaudre, T. Girinsky, M. Ghalibafian, A. Beaudre, and T. Girinsky. 2008. Heart and coronary artery protection in patients with mediastinal Hodgkin lymphoma treated with intensity-modulated radiotherapy: Dose constraints to virtual volumes or to organs at risk? *Radiotherapy & Oncology* 87 (1):82–8.

Girinsky, T., M. Ghalibafian, G. Bonniaud, et al. 2007. Is FDG-PET scan in patients with early stage Hodgkin lymphoma of any value in the implementation of the involved-node radiotherapy concept and dose painting? *Radiotherapy & Oncology* 85 (2):178–86.

Girinsky, T., C. Pichenot, A. Beaudre, M. Ghalibafian, and D. Lefkopoulos. 2006. Is intensity-modulated radiotherapy better than conventional radiation treatment and three-dimensional conformal radiotherapy for mediastinal masses in patients with Hodgkin's disease, and is there a role for beam orientation optimization and dose constraints assigned to virtual volumes? *International Journal of Radiation Oncology • Biology • Physics* 64 (1):218–26.

Goodman, K. A., S. Toner, M. Hunt, E. J. Wu, and J. Yahalom. 2005. Intensity-modulated radiotherapy for lymphoma involving the mediastinum. *International Journal of Radiation Oncology • Biology • Physics* 62 (1):198–206.

Gozdasoglu, S., A. O. Cavdar, E. Babacan, et al. 1995. Late effects of chemoradiotherapy in pediatric Hodgkin's disease. *Journal of Chemotherapy* 7 (5):463–6.

Graham, D. L., B. Brunet, and A. G. Balogh. 2008. Effect of deep inspiration breath hold on lung volumes and radiation dose parameters for involved-field radiation of Hodgkin's and non-Hodgkin's lymphomas of the mediastinum. *International Journal of Radiation Oncology • Biology • Physics* 72 (Suppl. 1):S640.

Green, D. M., T. Kawashima, M. Stovall, et al. 2010. Fertility of male survivors of childhood cancer: A report from the Childhood Cancer Survivor Study. *Journal of Clinical Oncology* 28 (2):332–9.

Guibout, C., E. Adjadj, C. Rubino, et al. 2005. Malignant breast tumors after radiotherapy for a first cancer during childhood. *Journal of Clinical Oncology* 23 (1):197–204.

Gustavsson, A., B. Osterman, and E. Cavallin-Stahl. 2003. A systematic overview of radiation therapy effects in Hodgkin's lymphoma. *Acta Oncologica* 42:589–604.

Hadar, H., D. Loven, P. Herskovitz, O. Bairey, A. Yagoda, and H. Levavi. 1994. An evaluation of lateral and medial transposition of the ovaries out of radiation fields. *Cancer* 74 (2):774–9.

Hancock, S. L., S. S. Donaldson, and R. T. Hoppe. 1993. Cardiac disease following treatment of Hodgkin's disease in children and adolescents. *Journal of Clinical Oncology* 11 (7):1208–15.

Hancock, S. L., M. A. Tucker, and R. T. Hoppe. 1993. Breast cancer after treatment of Hodgkin's disease. *Journal of the National Cancer Institute* 85 (1):25–31.

———. 1993. Factors affecting late mortality from heart disease after treatment of Hodgkin's disease. *JAMA* 270:1949–55.

Hodgson, D. C., E. S. Gilbert, G. M. Dores, et al. 2007. Long-term solid cancer risk among 5-year survivors of Hodgkin's lymphoma. *Journal of Clinical Oncology* 25 (12):1489–97.

Hodgson, D. C., M. M. Hudson, and L. S. Constine. 2007. Pediatric Hodgkin lymphoma: Maximizing efficacy and minimizing toxicity. *Seminars in Radiation Oncology* 17 (3):230–42.

Hodgson, D. C., E.-S. Koh, T. H. Tran, et al. 2007. Individualized estimates of second cancer risks after contemporary radiation therapy for Hodgkin lymphoma. *Cancer* 110 (11):2576–86.

Horning, S. J., A. Adhikari, N. Rizk, R. T. Hoppe, and R. A. Olshen. 1994. Effect of treatment for Hodgkin's disease on pulmonary function: Results of a prospective study. *Journal of Clinical Oncology* 12 (2):297–305.

Howell, S., and S. Shalet. 1998. Gonadal damage from chemotherapy and radiotherapy. *Endocrinology & Metabolism Clinics of North America* 27 (4):927–43.

Hua, C., K. Hoth, L. Kun, and M. Krasin. 2008. Radiation pneumonitis of pediatric patients with sarcoma and Hodgkin's lymphoma: A DVH analysis. *International Journal of Radiation Oncology • Biology • Physics* 72 (1, Suppl.):S494.

Hua, C., K. A. Hoth, S. Wu, et al. 2010. Incidence and correlates of radiation pneumonitis in pediatric patients with partial lung irradiation. *International Journal of Radiation Oncology • Biology • Physics* 78 (1):143–9.

Hudson, M. M., M. Krasin, M. P. Link, et al. 2004. Risk-adapted, combined-modality therapy with VAMP/COP and response-based, involved-field radiation for unfavorable pediatric Hodgkin's disease. *Journal of Clinical Oncology* 22 (22):4541–50.

Hull, M. C., C. G. Morris, C. J. Pepine, and N. P. Mendenhall. 2003. Valvular dysfunction and carotid, subclavian, and coronary artery disease in survivors of Hodgkin lymphoma treated with radiation therapy. *JAMA* 290 (21):2831–7.

Hunger, S. P., M. P. Link, and S. S. Donaldson. 1994. ABVD/MOPP and low-dose involved-field radiotherapy in pediatric Hodgkin's disease: The Stanford experience. *Journal of Clinical Oncology* 12 (10):2160–6.

Iarussi, D., C. Pisacane, P. Indolfi, F. Casale, V. Martino, and M. T. Di Tullio. 2005. Evaluation of left ventricular function in long-term survivors of childhood Hodgkin disease. *Pediatric Blood & Cancer* 45 (5):700–5.

Koh, E.-S., A. Sun, T. H. Tran, et al. 2006. Clinical dose-volume histogram analysis in predicting radiation pneumonitis in Hodgkin's lymphoma. *International Journal of Radiation Oncology • Biology • Physics* 66 (1):223–8.

Koh, E.-S., T. Tran, M. Heydarian, et al. 2007. A comparison of mantle versus involved-field radiotherapy for Hodgkin's lymphoma: Reduction in normal tissue dose and second cancer risk. *Radiation Oncology* 2 (1):13.

Koontz, B. F., J. P. Kirkpatrick, R. W. Clough, et al. 2006. Combined-modality therapy versus radiotherapy alone for treatment of early-stage Hodgkin's disease: Cure balanced against complications [see comment]. *Journal of Clinical Oncology* 24 (4):605–11.

Kung, F. H., C. L. Schwartz, C. R. Ferree, et al. 2006. POG 8625: A randomized trial comparing chemotherapy with chemoradiotherapy for children and adolescents with Stages I, IIA, IIIA1 Hodgkin disease: A report from the Children's Oncology Group. *Journal of Pediatric Hematology/Oncology* 28 (6):362–8.

Kwa, S. L., J. V. Lebesque, J. C. Theuws, et al. 1998. Radiation pneumonitis as a function of mean lung dose: An analysis of pooled data of 540 patients. *International Journal of Radiation Oncology • Biology • Physics* 42 (1):1–9.

Le Floch, O., S. S. Donaldson, and H. S. Kaplan. 1976. Pregnancy following oophoropexy and total nodal irradiation in women with Hodgkin's disease. *Cancer* 38 (6):2263–8.

Lee, C. K. K., D. Aeppli, and M. E. Nierengarten. 2000. The need for long-term surveillance for patients treated with curative radiotherapy for Hodgkin's disease: University of Minnesota experience. *International Journal of Radiation Oncology • Biology • Physics* 48 (1):169–79.

Lin, H. M., M. A. Teitell, H.-M. J. Lin, and M. A. Teitell. 2005. Second malignancy after treatment of pediatric Hodgkin disease. *Journal of Pediatric Hematology/Oncology* 27 (1):28–36.

Lushbaugh, C. C., and G. W. Casarett. 1976. The effects of gonadal irradiation in clinical radiation therapy: A review. *Cancer* 37 (2 Suppl.):1111–25.

MacDonald, S., S. Bernard, A. Balogh, et al. 2005. Electronic compensation using multileaf collimation for involved field radiation to the neck and mediastinum in non-Hodgkin's lymphoma and Hodgkin's lymphoma. *Medical Dosimetry* 30 (2):59–64.

Marina, N. M., C. A. Greenwald, D. L. Fairclough, et al. 1995. Serial pulmonary function studies in children treated for newly diagnosed Hodgkin's disease with mantle radiotherapy plus cycles of cyclophosphamide, vincristine, and procarbazine alternating with cycles of doxorubicin, bleomycin, vinblastine, and dacarbazine. *Cancer* 75 (7):1706–11.

Marks, L. B., S. M. Bentzen, J. O. Deasy, et al. 2010. Radiation dose-volume effects in the lung. *International Journal of Radiation Oncology • Biology • Physics* 76 (3, Suppl):S70–6.

Mauch, P. M., H. Weinstein, L. Botnick, J. Belli, and J. R. Cassady. 1983. An evaluation of long-term survival and treatment complications in children with Hodgkin's disease. *Cancer* 51 (5):925–32.

Mefferd, J. M., S. S. Donaldson, and M. P. Link. 1989. Pediatric Hodgkin's disease: Pulmonary, cardiac, and thyroid function following combined modality therapy. *International Journal of Radiation Oncology • Biology • Physics* 16 (3):679–85.

Merchant, T. E., L. Nguyen, D. Nguyen, S. Wu, M. M. Hudson, and S. C. Kaste. 2004. Differential attenuation of clavicle growth after asymmetric mantle radiotherapy. *International Journal of Radiation Oncology • Biology • Physics* 59 (2):556–61.

Mertens, A. C., Q. Liu, J. P. Neglia, et al. 2008. Cause-specific late mortality among 5-year survivors of childhood cancer: The Childhood Cancer Survivor Study. *Journal of the National Cancer Institute* 100 (19):1368–79.

Metayer, C., C. F. Lynch, E. A. Clarke, et al. 2000. Second cancers among long-term survivors of Hodgkin's disease diagnosed in childhood and adolescence. *Journal of Clinical Oncology* 18 (12):2435–43.

Metzger, M. L., M. M. Hudson, G. W. Somes, et al. 2006. White race as a risk factor for hypothyroidism after treatment for pediatric Hodgkin's lymphoma. *Journal of Clinical Oncology* 24 (10):1516–21.

Morgan, G. W., A. P. Freeman, R. G. McLean, B. H. Jarvie, and R. W. Giles. 1985. Late cardiac, thyroid, and pulmonary sequelae of mantle radiotherapy for Hodgkin's disease. *International Journal of Radiation Oncology • Biology • Physics* 11 (11):1925–31.

Mulrooney, D. A., M. W. Yeazel, T. Kawashima, et al. 2009. Cardiac outcomes in a cohort of adult survivors of childhood and adolescent cancer: Retrospective analysis of the Childhood Cancer Survivor Study cohort. *BMJ* 339:b4606.

Mundt, A. J., and J. C. Roeske. 2005. *Intensity Modulated Radiation Therapy: A Clinical Perspective.* Hamilton, Ontario: BC Decker.

Myrehaug, S., M. Pintilie, R. Tsang, et al. 2008. Cardiac morbidity following modern treatment for Hodgkin lymphoma: Supra-additive cardiotoxicity of doxorubicin and radiation therapy. *Leukemia & Lymphoma* 49 (8):1486–93.

Nachman, J. B., R. Sposto, P. Herzog, et al. 2002. Randomized comparison of low-dose involved-field radiotherapy and no radiotherapy for children with Hodgkin's disease who achieve a complete response to chemotherapy. *Journal of Clinical Oncology* 20 (18):3765–71.

Nazmy, M. S., M. M. El-Taher, E. M. Attalla, et al. 2007. Shielding for scattered radiation to the testis during pelvic radiotherapy: Is it worth? *Journal of Egyptian National Cancer Institute* 19 (2):127–32.

Nevinny-Stickel, M., R. A. Sweeney, R. J. Bale, et al. 2004. Reproducibility of patient positioning for fractionated extracranial stereotactic radiotherapy using a double-vacuum technique. *Strahlentherapie und Onkologie* 180 (2):117–22.

Ng, A. K., M. V. P. Bernardo, E. Weller, et al. 2002. Second malignancy after Hodgkin disease treated with radiation therapy with or without chemotherapy: Long-term risks and risk factors. *Blood* 100 (6):1989–96.

Ng, A. K., S. Li, D. Neuberg, et al. 2008. A prospective study of pulmonary function in Hodgkin's lymphoma patients. *Annals of Oncology* 19 (10):1754–8.

Nysom, K., K. Holm, H. Hertz, and B. Hesse. 1998. Risk factors for reduced pulmonary function after malignant lymphoma in childhood. *Medical and Pediatric Oncology* 30 (4):240–8.

Oeffinger, K. C., A. C. Mertens, C. A. Sklar, et al. 2006. Chronic health conditions in adult survivors of childhood cancer. *New England Journal of Medicine* 355 (15):1572–82.

Oguz, A., T. Tayfun, E. C. Citak, et al. 2007. Long-term pulmonary function in survivors of childhood Hodgkin disease and non-Hodgkin lymphoma. *Pediatric Blood & Cancer* 49 (5):699–703.

Oh, C. E., M. Hunt, J. J. Silva, and J. Yahalom. 2000. Optimal matchline blocking and matchline dosimetry for lymphoma patients. *Medical Dosimetry* 25 (4):231–6.

Ortin, T. T., C. A. Shostak, and S. S. Donaldson. 1990. Gonadal status and reproductive function following treatment for Hodgkin's disease in childhood: The Stanford experience. *International Journal of Radiation Oncology • Biology • Physics* 19 (4):873–80.

Papadakis, V., C. Tan, G. Heller, and C. Sklar. 1996. Growth and final height after treatment for childhood Hodgkin disease. *Journal of Pediatric Hematology/Oncology* 18 (3):272–6.

Pergolizzi, S., N. Settineri, M. Gaeta, et al. 2000. What is the best position of the arms in mantle field for Hodgkin's disease? *International Journal of Radiation Oncology • Biology • Physics* 46 (1):119–22.

Piovaccari, G., R. M. Ferretti, F. Prati, et al. 1995. Cardiac disease after chest irradiation for Hodgkin's disease: Incidence in 108 patients with long follow-up. *International Journal of Cardiology* 49 (1):39–43.

Prosnitz, L. R., D. M. Brizel, and K. L. Light. 1997. Radiation techniques for the treatment of Hodgkin's disease with combined modality therapy or radiation alone. *International Journal of Radiation Oncology • Biology • Physics* 39 (4):885–95.

Provencio, M., I. Millan, P. Espana, et al. 2008. Analysis of competing risks of causes of death and their variation over different time periods in Hodgkin's disease. *Clinical Cancer Research* 14 (16):5300–5.

Sankila, R., S. Garwicz, J. H. Olsen, et al. 1996. Risk of subsequent malignant neoplasms among 1,641 Hodgkin's disease patients diagnosed in childhood and adolescence: A population-based cohort study in the five Nordic countries. Association of the Nordic Cancer Registries and the Nordic Society of Pediatric Hematology and Oncology. *Journal of Clinical Oncology* 14 (5):1442–6.

Schultz-Hector, S., and K.-R. Trott. 2007. Radiation-induced cardiovascular diseases: Is the epidemiologic evidence compatible with the radiobiologic data? *International Journal of Radiation Oncology • Biology • Physics* 67 (1):10–18.

Shahidi, M., N. Kamangari, S. Ashley, et al. 2006. Site of relapse after chemotherapy alone for stage I and II Hodgkin's disease. *Radiotherapy & Oncology* 78 (1):1–5.

Shapiro, S. J., S. D. Shapiro, W. B. Mill, and E. J. Campbell. 1990. Prospective study of long-term pulmonary manifestations of mantle irradiation. *International Journal of Radiation Oncology • Biology • Physics* 19 (3):707–14.

Signorello, L. B., J. J. Mulvihill, D. M. Green, et al. 2010. Stillbirth and neonatal death in relation to radiation exposure before conception: A retrospective cohort study. *Lancet* 376 (9741):624–30.

Smith, L. M., N. P. Mendenhall, M. J. Cicale, E. R. Block, R. L. Carter, and R. R. Million. 1989. Results of a prospective study evaluating the effects of mantle irradiation on pulmonary function. *International Journal of Radiation Oncology • Biology • Physics* 16 (1):79–84.

Stewart, J. R., L. F. Fajardo, S. M. Gillette, and L. S. Constine. 1995. Radiation injury to the heart. *International Journal of Radiation Oncology • Biology • Physics* 31 (5):1205–11.

Stovall, M., S. S. Donaldson, R. E. Weathers, et al. 2004. Genetic effects of radiotherapy for childhood cancer: Gonadal dose reconstruction. *International Journal of Radiation Oncology • Biology • Physics* 60 (2):542–52.

Stromberg, J. S., M. B. Sharpe, L. H. Kim, et al. 2000. Active breathing control (ABC) for Hodgkin's disease: Reduction in normal tissue irradiation with deep inspiration and implications for treatment. *International Journal of Radiation Oncology • Biology • Physics* 48 (3):797–806.

Swerdlow, A. J., J. A. Barber, G. V. Hudson, et al. 2000. Risk of second malignancy after Hodgkin's disease in a collaborative British cohort: The relation to age at treatment. *Journal of Clinical Oncology* 18:498–509.

Tebbi, C. K., N. Mendenhall, W. B. London, et al. 2006. Treatment of stage I, IIA, IIIA1 pediatric Hodgkin disease with doxorubicin, bleomycin, vincristine and etoposide (DBVE) and radiation: A Pediatric Oncology Group (POG) study. *Pediatric Blood & Cancer* 46 (2):198–202.

Thibaud, E., K. Rodriguez-Macias, C. Trivin, H. Esperou, J. Michon, and R. Brauner. 1998. Ovarian function after bone marrow transplantation during childhood. *Bone Marrow Transplantation* 21 (3):287–90.

Tinger, A., T. H. Wasserman, E. E. Klein, et al. 1997. The incidence of breast cancer following mantle field radiation therapy as a function of dose and technique. *International Journal of Radiation Oncology • Biology • Physics* 37 (4):865–70.

Travis, L. B. 2002. Therapy-associated solid tumors. *Acta Oncologica* 41 (4):323–33.

Travis, L. B., M. Gospodarowicz, R. E. Curtis, et al. 2002. Lung cancer following chemotherapy and radiotherapy for Hodgkin's disease. *Journal of the National Cancer Institute* 94:182–92.

Travis, L. B., D. A. Hill, G. M. Dores, et al. 2003. Breast cancer following radiotherapy and chemotherapy among young women with Hodgkin disease. *JAMA* 290 (4):465–75. Erratum appears in 290(10):1318, 2003.

Tucker, M. A., P. H. Jones, J. D. Boice, Jr., et al. 1991. Therapeutic radiation at a young age is linked to secondary thyroid cancer: The Late Effects Study Group. *Cancer Research* 51 (11):2885–8.

Van Den Bossche, B., B. Lambert, F. De Winter, et al. 2002. 18FDG PET versus high-dose 67Ga scintigraphy for restaging and treatment follow-up of lymphoma patients. *Nuclear Medicine Communications* 23 (11):1079–83.

van Leeuwen, F. E., W. J. Klokman, M. Stovall, et al. 2003. Roles of radiation dose, chemotherapy, and hormonal factors in breast cancer following Hodgkin's disease. *Journal of the National Cancer Institute* 95 (13):971–80.

van Rijswijk, S., M. A. Huijbregts, E. Lust, R. J. Strack van Schijndel, M. A. J. M. Huijbregts, and R. J. M. Strack van Schijndel. 2008. Mini-review on cardiac complications after mediastinal irradiation for Hodgkin lymphoma. *Netherlands Journal of Medicine* 66 (6):234–7.

Villani, F. M. D., S. M. D. Viviani, V. M. D. Bonfante, P. M. D. De Maria, F. M. D. Soncini, and A. M. D. Laffranchi. 2000. Late pulmonary effects in favorable stage I and IIA Hodgkin's disease treated with radiotherapy alone. *American Journal of Clinical Oncology* 23 (1):18–21.

Vlachaki, M. T., and S. Kumar. 2010. Helical TomoTherapy in the radiotherapy treatment of Hodgkin's disease: A feasibility study. *Journal of Applied Clinical Medical Physics* 11 (1):3042.

Wahner-Roedler, D. L., D. F. Nelson, I. T. Croghan, et al. 2003. Risk of breast cancer and breast cancer characteristics in women treated with supradiaphragmatic radiation for Hodgkin lymphoma: Mayo Clinic experience. *Mayo Clinic Proceedings* 78 (6):708–15.

Wallace, W. H., S. M. Shalet, J. H. Hendry, P. H. Morris-Jones, and H. R. Gattamaneni. 1989. Ovarian failure following abdominal irradiation in childhood: The radiosensitivity of the human oocyte. *British Journal of Radiology* 62 (743):995–8.

Weber, D. C., N. Peguret, G. Dipasquale, and L. Cozzi. 2009. Involved-node and involved-field volumetric modulated arc vs. fixed beam intensity-modulated radiotherapy for female patients with early-stage supra-diaphragmatic hodgkin lymphoma: A comparative planning study. *International Journal of Radiation Oncology • Biology • Physics* (5):1578–86.

Wethal, T., M. B. Lund, T. Edvardsen, et al. 2009. Valvular dysfunction and left ventricular changes in Hodgkin's lymphoma survivors. A longitudinal study. *British Journal of Cancer* 101 (4):575–81.

Willman, K. Y., R. S. Cox, and S. S. Donaldson. 1994. Radiation induced height impairment in pediatric Hodgkin's disease. *International Journal of Radiation Oncology • Biology • Physics* 28 (1):85–92.

Wirth, A., J. F. Seymour, R. J. Hicks, et al. 2002. Fluorine-18 fluorodeoxyglucose positron emission tomography, gallium-67 scintigraphy, and conventional staging for Hodgkin's disease and non-Hodgkin's lymphoma. *American Journal of Medicine* 112 (4):262–8.

Wolden, S. L., S. L. Hancock, R. W. Carlson, D. R. Goffinet, S. S. Jeffrey, and R. T. Hoppe. 2000. Management of breast cancer after Hodgkin's disease. *Journal of Clinical Oncology* 18 (4):765–72.

Yahalom, J., and P. Mauch. 2002. The involved field is back: Issues in delineating the radiation field in Hodgkin's disease. *Annals of Oncology* 13 (Suppl. 1):79–83.

Chapter 6

Neuroblastoma

6.1 Clinical Overview

Neuroblastoma is the most common tumor in children under 1 year of age and the second most common pediatric solid tumor overall. There are approximately 600 neuroblastoma cases diagnosed annually in the United States. The median age at diagnosis is 18 months. Ninety percent of newly diagnosed patients are under 5 years old and 30% are under 1 year old. Prognosis is especially favorable for infants under 12 to 18 months of age. Neuroblastoma is derived from neural crest cells; about 55% arise in an adrenal gland, and about 35% in the abdominal or thoracic sympathetic ganglia, the two chains of sympathetic nervous system neurons running just lateral to the spinal cord from the upper neck through the coccyx. Approximately 70% of patients present with metastatic disease, most commonly to the bone, bone marrow, lymph nodes, liver, or skin. Neuroblastoma is categorized by the Children's Oncology Group (COG) into low-, intermediate-, and high-risk groups, based on extent of the primary tumor, lymph node involvement, metastatic dissemination, patient age, histologic type, chromosome number, and amplification of the MYCN gene. Survival rates are over 90% for low-risk and only slightly lower for intermediate-risk patients.

Neuroblastoma is highly sensitive to both chemotherapy and radiation therapy. The role of radiation therapy is generally limited to high-risk patients with metastatic disease, who sequentially undergo chemotherapy, surgical resection of the primary disease, myeloablative chemotherapy with stem-cell rescue, radiation therapy, and 13-cis-retinoic acid. This extended treatment regimen produced 3-year event-free survival (EFS) rates of about 50% in two COG randomized trials and 66% in another study, which was significantly higher than the EFS obtained without the addition of myeloablative chemotherapy and 13-cis-retinoic acid (Gatcombe et al. 2009). Radiation therapy was given following recovery from high-dose chemotherapy to the primary tumor bed and sites of metastasis still demonstrating activity on nuclear medicine Iodine-131 metaiodobenzylguanidine (MIBG) scan between the end of induction chemotherapy and the onset of myeloablative chemotherapy. Radiation dose in the current COG trial is 21.6 Gy, with a boost to 36 Gy to any gross residual disease present in the primary tumor site after resection. Patients with metastatic disease, commonly to multiple bone sites, receive 21.6 Gy to each site of metastasis. Radiation therapy is also usually effective for the occasional tumor that does not respond to chemotherapy and for palliation of recurrent disease. Targeted radionuclide therapy using high activity I-131 (18 mCi/Kg) attached to MIBG has been shown to be well tolerated and effective in patients with persistent or recurrent disease after chemotherapy. The role of I-131-MIBG in the initial therapy of high-risk disease is being investigated (Quach et al. 2011).

In the past, total body irradiation (TBI) was commonly used as part of the conditioning regimen prior to bone marrow transplantation in the treatment of stage IV disease. However, TBI has been shown to cause an increase in late effects such as growth delay, second malignancy, pneumonitis, cataracts, and endocrine dysfunction without improving the survival rate compared to pretransplant conditioning with high-dose chemotherapy only, so is no longer used (Flandin et al. 2006).

Because the vast majority of neuroblastoma cases arise in the adrenal gland or abdominal sympathetic ganglion, this chapter will focus on the challenges of treatment planning for tumors in this location. Treatment planning is particularly intricate for the primary tumor, as it is usually in close proximity to the growing spine, liver, and kidneys.

6.2 Target Volume Definition

The target volume is based on the postchemotherapy, presurgical volume. In COG protocols, even if there has been a gross total resection, the tumor volume that existed just before surgery is defined as the gross tumor volume (GTV). International Commission on Radiation Units and Measurements (ICRU) nomenclature would refer to this volume as the clinical target volume (CTV). Occasionally, surgery is performed prior to the start of chemotherapy. In this case, the target volume is still the presurgical volume. In the majority of cases, the surgical resection will result in a gross total resection and 21.6 Gy will be the target dose. However, in 7% to 33% of cases, gross tumor remaining after chemotherapy will be adherent to critical blood vessels or nerves, and will not be able to be safely resected (Simon et al. 2006; Gatcombe et al. 2009). In that case, a 14.4 Gy boost is delivered to the residual disease with a margin. Intra-abdominal tumors can be very large at diagnosis, pushing the kidney and liver out of their natural positions.

Three cases will be presented that illustrate common planning challenges. Case 1 is a patient with a tumor wrapping around the right kidney as seen on the postchemotherapy, presurgical computed tomographic (CT) scan (Figure 6.1). Case 2 is a patient with a large presurgical, postchemotherapy tumor volume with displacement of the left kidney (Figure 6.2). In both cases, a gross total resection was achieved. Although a good partial response to chemotherapy is usually obtained, one can be left with a large target volume to irradiate. Following surgical resection, the anatomy on the planning CT is often nearly normal with the previously shifted organs back in their natural positions. This leaves a challenging situation whereby we must identify all of the surfaces in the planning CT that came into contact with the tumor volume as it existed prior to surgery. There is no perfect method for accomplishing this. Using bony anatomy alone as reference structures can lead to misplacement of the CTV by several centimeters, mostly in the superior–inferior and anterior–posterior directions. This is due to a combination of differences in patient position, breathing state (end inspiration versus free

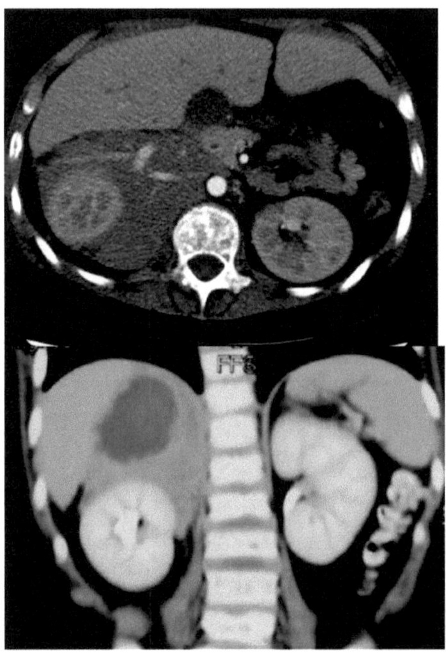

FIGURE 6.1 Case 1: Postchemotherapy, preoperative CT, axial (top) and coronal (bottom) images.

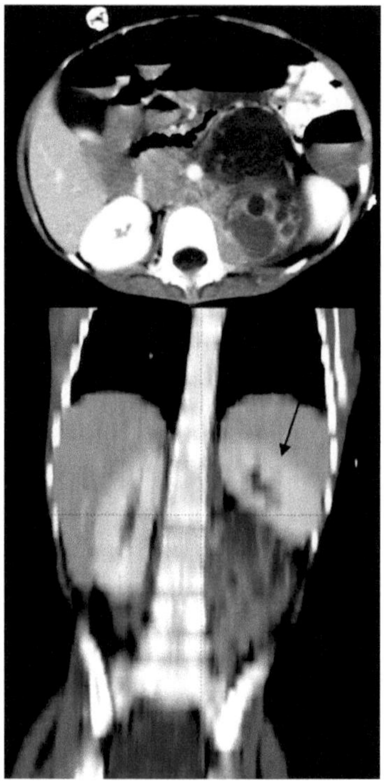

FIGURE 6.2 Case 2: (Upper) Large residual neuroblastoma of left adrenal after chemotherapy. Axial (top) and coronal (lower). (Lower) Left kidney (arrow) pushed superiorly by neuroblastoma.

breathing), and anatomic changes between preoperative CT and treatment planning CT (Chen et al. 2005). Registering the presurgical CT with the planning CT and having a full understanding of the surgical findings can help with this identification process, but often a good deal of approximation is needed. Deformable registration is beneficial but one must carefully check the results in light of current limitations for placing missing tissue in most of these software algorithms. Clips placed at margins of the resection cavity can sometimes be helpful but can also fail to identify target tissue due to folding of surfaces after resection. This volume is then expanded by 1.5 cm to create the CTV. CTV margins are anatomically confined by bone and organs, such as vertebral bodies, kidneys, and liver, that have likely not been invaded by tumor. With good body immobilization and careful matching of portal images with digitally reconstructed radiographs (DRRs) or with cone beam CT, planning target volume (PTV) margins of 0.4 cm can be used. Figure 6.3 shows the PTV for case 1 on the postsurgical planning CT. When a gross total resection is not achieved, then remaining tumor postsurgery is boosted to 36 Gy (case 3). The boost GTV is given a 1.0 cm CTV margin and an appropriate PTV margin.

The current (as of 2012) COG protocol for high-risk neuroblastoma prescribes that consolidative radiation therapy be directed to the primary

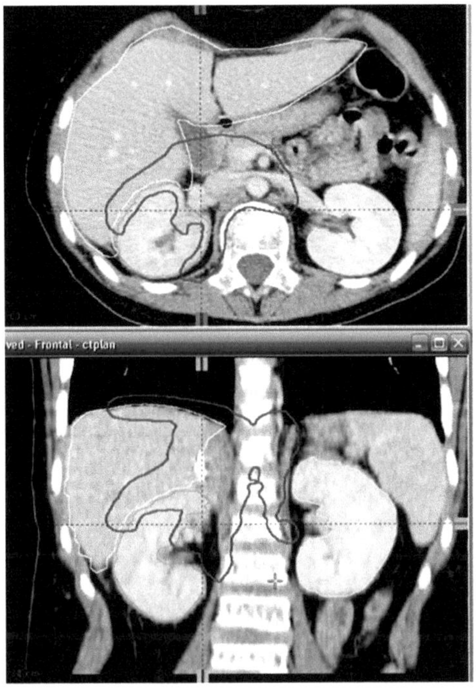

FIGURE 6.3 Case 1: PTV on planning CT, gross tumor seen in Figure 6.1 is gone and right kidney is in normal position.

tumor site regardless of response to chemotherapy or gross total resection of the primary tumor. Regional lymph nodes that were initially involved with the tumor but completely responded to chemotherapy are not included in the GTV. Isolated relapses have been reported to occur just outside the irradiated volume in the nodal groups that were involved at the time of diagnosis. Some institutions therefore routinely include these nodes along with the primary tumor bed in the GTV.

Since most of these patients will be sedated due to their young age, shallow breathing can be expected and little movement of the kidneys or PTV will take place. It is advisable to confirm this, however, either by fluoroscopy when clips in or near the PTV are present or a 4DCT study. In most cases, some or all of the vertebral bodies will need to be assigned as another PTV due to the asymmetrical relationship between the primary target and the vertebral bodies. Several decades ago it was recognized that asymmetrical irradiation of the vertebral bodies (VBs) in children caused asymmetrical bone growth, which resulted in an abnormal curvature of the spine (scoliosis or kyphosis). To decrease the risk of this potentially severe morbidity, VBs at risk for asymmetric irradiation are intentionally included in the uniform high-dose volume (see following section). Ensuring that 100% of these VBs receive at least 18 Gy reasonably assures uniform growth cessation, decreasing the probability of scoliosis and limiting doses above this level better permits reduction of dose to the nearby kidneys. Where the PTV is limited to the region anterior to the VBs and only wraps around the bone by a small, symmetrical amount, then sparing of the VB is possible. It is common to have both VB targets and those that are avoidance structures in the same case.

Immobilization of the chest, abdomen, and pelvis by a custom-formed immobilization bag is important to minimize PTV margins, which in turn reduces the normal tissue irradiated volume. Careful attention to reestablishing the patient's body position each day may take a few extra minutes but can result in achieving reproducibility of 3 mm on a regular basis. Neuroblastoma patients are typically sedated for treatment due to their young age, which aids in achieving this level of reproducibility, because of ease of setup positioning, negligible patient movement, and minimal internal organ motion due to shallow breathing.

6.3 Treatment Planning Techniques and Dosimetry

For abdominal neuroblastoma, the major organs at risk (OARs) are the kidneys, liver, and vertebral bodies. With 3D conformal and especially intensity-modulated radiation therapy (IMRT), treatment can usually be designed to avoid exceeding the tolerance dose of either kidney. Using IMRT with eight coplanar fields (for example, IEC angles 15, 55, 95, 145, 215, 265, 300, 340 degrees), highly conformal treatment can be given to the PTV while limiting both kidneys to well below the tolerance dose. VMAT can achieve even more conformal dose distributions. Since the tumor usually arises from the adrenal gland, the target volume will be associated with one of the kidneys and will therefore generally be left- or right-sided. COG guidelines suggest that the entire ipsilateral and contralateral kidney may receive up to 18 Gy and 14.4 Gy, respectively. When the prescription dose is 21.6 Gy, restricting the liver dose below its tolerance is seldom difficult due to the large volume of liver that is far from the PTV. However, for a prescription dose of 36 Gy, depending on the size and location of the PTV, restricting liver dose can be challenging. The intensive chemotherapy these patients receive prior to radiation therapy significantly lowers their liver tolerance to radiation and puts them at risk for life-threatening hepatic sinusoidal obstruction syndrome, also called veno-occlusive disease. COG guidelines have also suggested that the cumulative radiation dose should be no greater than 23.4 Gy to 50% and 18 Gy to 100% of the liver volume. Prior to the availability of IMRT, the most common treatment technique for abdominal neuroblastoma was opposed anterior and posterior or oblique beams. For example, if the target were on the right, then a left anterior oblique (LAO) and right posterior oblique (RPO) set of fields would be used. The medial

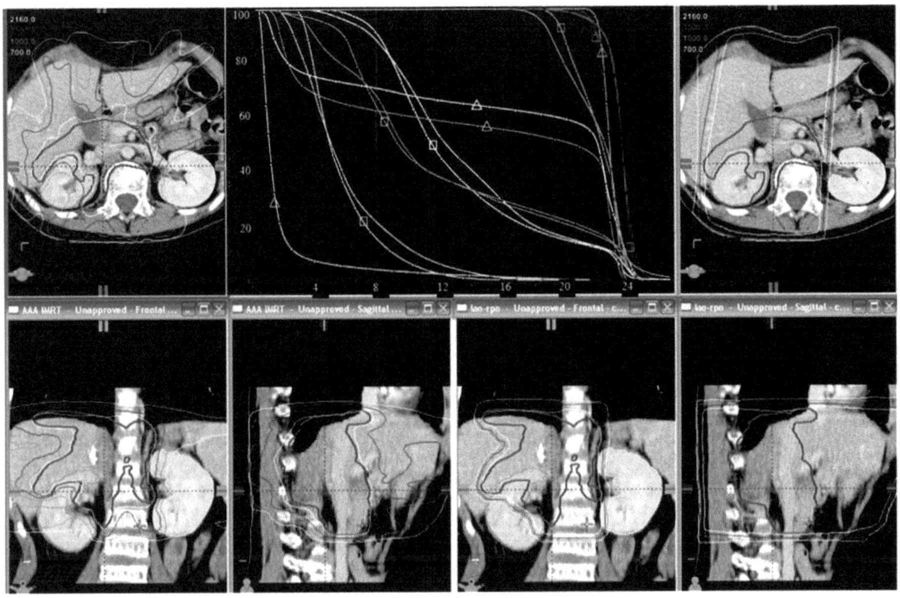

FIGURE 6.4 (See color insert.) Case 1: Eight-beam IMRT (left-side panels, DVH = line) versus opposed obliques (right-side panels, DVH = triangles), VMAT isodoses not shown (DVH = boxes), 21.6 Gy target dose (red). Structures: green is vertebral body OAR, yellow is liver, orange is right kidney, cyan is left kidney. Isodoses: yellow, 21.6 Gy; magenta, 15 Gy; orange, 10 Gy; light blue, 7 Gy.

border adjacent to the spine would be made collinear to maximize sparing of the contralateral kidney. In order to avoid asymmetrical irradiation of the spine, the medial edge of the fields would be extended across midline to fully include the vertebral bodies. The ipsilateral kidney was sacrificed if necessary to completely treat the target volume. The contralateral kidney could be spared with careful positioning of the field edges. Often, large portions of the liver received the prescribed dose. The right side of Figure 6.4 shows this field arrangement. With the advent of 3D conformal techniques, more beams are being used to achieve better normal tissue sparing. Neuroblastoma target volumes tend to be complex shapes close to critical structures so typically more than six beams is warranted to produce the best plan. IMRT will consistently produce better plans than 3D conformal for the challenging cases, but certainly 3D conformal plans can be satisfactory. In the eight coplanar beam IMRT plans that are discussed later, multiple step-and-shoot (MSS) delivery was used with moderate smoothing of the ideal fluence (priority of 40 on a scale of 100). This fluence smoothing reduces the number of monitor units (MUs) without significantly reducing plan quality and is a feature on most planning systems (Anker et al. 2010). It is instructive to compare the dose distributions between the historic treatment and more advanced methods.

Figures 6.4 to 6.6 show a comparison of the dose-volume histograms (DVHs) for the opposed oblique and IMRT plans for cases 1 and 2, without

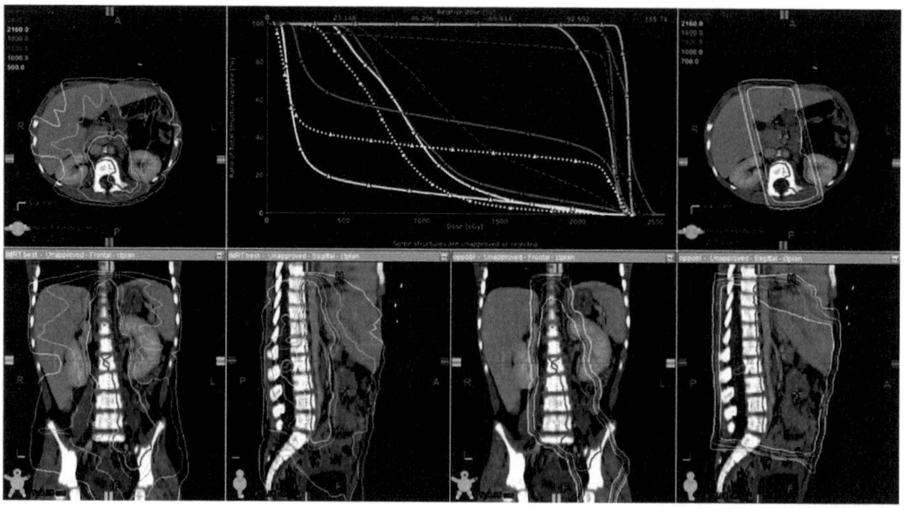

FIGURE 6.5 (See color insert.) Case 2: IMRT (boxes) versus opposed obliques (triangles). Demonstrates longer volume including lymph nodes, VB sparing. Light green, right kidney; magenta, left kidney; yellow dotted, liver; green, vertebral body target; blue dash, vertebral body OAR; red, PTV. Isodoses: yellow, 21.6 Gy; orange, 18 Gy; magenta, 15 Gy; light green, 5 Gy.

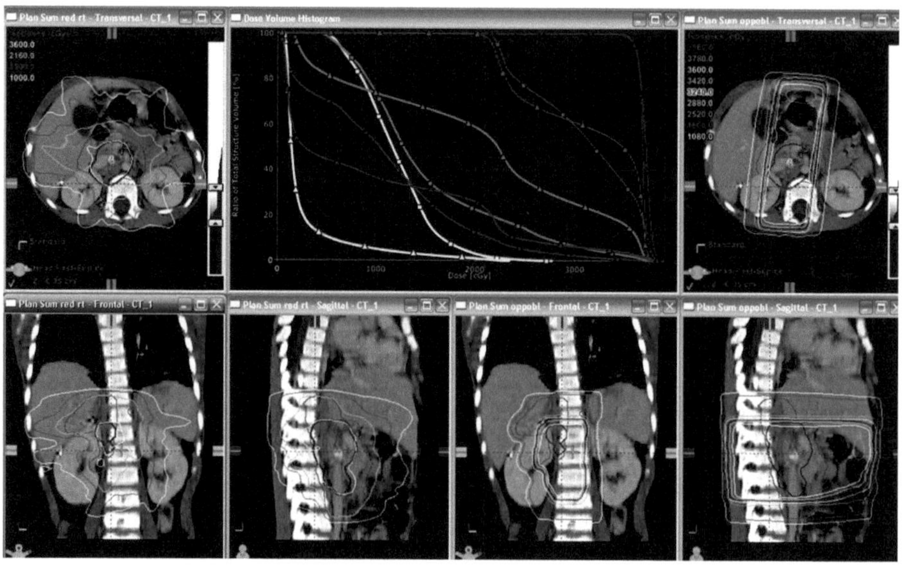

FIGURE 6.6 (See color insert.) Case 3: 21.6 Gy + 14.4 Gy boost, IMRT (boxes) versus opposed obliques (triangles). Isodoses: light green, left kidney; dark green, right kidney; brown, liver; red, 21.6 Gy PTV; blue, 36 Gy PTV. Isodoses: yellow, 36 Gy; green, 21.6 Gy; magenta, 15 Gy; light blue, 10 Gy.

a boost, and case 3, with a boost. Across the three cases, when the opposed oblique beam pairs are used, the V15 of the affected kidney is about 50% without the boost, and 67% with the boost, while the IMRT plans reduce these doses by more than one-half. With IMRT, for case 1, the V15 kidney dose on the affected side is much higher than for case 2, 33% versus 14%, because the PTV wraps around the kidney. The DVH for kidneys is critically dependent on the exact relationship between the PTV and the kidney. In cases where the PTV partially wraps around the kidney, it may not be possible to keep more than 50% of the kidney to less than 15 Gy even using IMRT, especially if the boost PTV is also close to the kidney. At doses well below tolerance (less than 8 Gy), the contralateral kidney receives higher doses from the IMRT plan than the opposed oblique plan because the opposed oblique beams can be shaped to largely miss the contralateral kidney while a multibeam arrangement necessarily exhibits low-dose spreading across midline. The more clinically relevant V15 of the contralateral kidney is usually easily kept below about 10% to 20% for either IMRT or an opposed oblique pair.

For the right-sided target in case 1, the oblique beam plan gives 20 Gy to 60% of the liver compared to only 15% for the IMRT plan. Similar large reductions in volume receiving moderate doses are achieved in case 2 with the IMRT plan than for obliques. When a boost is given as in case 3, in which the target is right-sided but much smaller than in case 1, the liver V20 is below 25% for either technique but IMRT is still more sparing than the opposed oblique pair.

As mentioned earlier, the VBs that are at risk of asymmetric irradiation are included in a separate target volume, whereas VBs that can be spared are contoured as an OAR. Tables 6.1 to 6.3 show the achievable dose–volume parameters for typical cases treated with eight coplanar IMRT beams where either 21.6 Gy is given (after a complete chemotherapy response or gross total resection, right-sided or left-sided PTV, cases 1 and 2) or 21.6 Gy plus a 14.4 Gy boost is given (when there is residual disease after surgery;

TABLE 6.1 DVH for 21.6 Gy to Site of Right Adrenal Primary Using IMRT

Organ or PTV	Dose (Gy)	% Volume
Primary PTV	21.6 (100%)	95
	24.4 (113%)	5
VB PTV	18	100
R kidney	18	27
	15	33
	10	51
L kidney	18	0
	15	2
	10	13
Liver	20	17
	15	39
	10	76
VB OAR	18	22
	15	35

TABLE 6.2 DVH for 21.6 Gy to Site of Left Adrenal Primary Using IMRT

Organ or PTV	Dose (Gy)	% Volume
Primary PTV	21.6 (100%)	95
	23.1 (107%)	5
VB PTV	18	100
R kidney	18	4
	15	10
	10	39
L kidney	18	9
	15	14
	10	42
Liver	20	1
	15	4
	10	22
VB OAR	18	22
	15	33

Note: Note lower liver doses compared to right adrenal case.

TABLE 6.3 DVH for Case 3: 36 Gy to Site of Right Adrenal

Organ or PTV	Dose (Gy)	% Volume
Primary PTV	36 (100%)	95
	38 (106%)	5
R kidney	18	26
	15	35
	10	74
L kidney	18	6
	15	18
	10	65
Liver	20	13
	15	27
	10	52

case 3). Planning the dose distribution for the VBs requires a careful assessment of the potential to keep more than about 70% of the bone to less than 15 Gy with lower doses delivered relatively symmetrically. In cases where the PTV is anterior to the VB or wraps around the VB by a small symmetrical extent and can be limited to inclusion of only 3 mm of the anterior aspect of the VB, a substantial portion can be kept below a dose that will cause bone growth arrest. In cases where this can be accomplished, those

VBs can be treated as OARs for IMRT planning. If not, then they should be designated as targets that should receive at least 18 Gy to the entire VB, a dose that should be enough to impair bone growth. Once bone growth has been impaired, asymmetrical irradiation with doses above 18 Gy to these VBs should not cause spinal curvature. In regions where the kidney is close to the VB, one can indent the VB contour by about 5 mm laterally to help keep dose away from the kidney while still substantially irradiating the VB to the dose necessary to uniformly halt growth. Figure 6.5 illustrates these concepts for case 2 where the vertebral bodies in the superior half of the PTV region are spared because the PTV is positioned anterior to the VB while for the inferior half, where the PTV wraps from anterior to lateral to the VB, the VB is treated as a target.

IMRT plans can also be designed to deliver lower doses to the stomach and intestine in comparison to opposing beams, thereby decreasing acute nausea and vomiting (Paulino et al. 2006; Plowman et al. 2008).

The planning for neuroblastoma is particularly well suited for a study of whether dosimetric improvements can be achieved by type of dynamic delivery, level of fluence smoothing, or number of coplanar beams, delivered either by conventional linear accelerator or by volumetric modulated arc therapy (VMAT). These results generally apply to all treatment sites. Little gain is achieved by sliding window delivery over 10-intensity level step-and-shoot delivery. The OAR DVHs are nearly identical while the PTV DVH is only slightly sharper (about 2% less dose at the D20 level). Using just a 10% fluence smoothing factor compared to about 40% results in somewhat larger differences than does varying the method of IMRT delivery, but still is not clinically significantly different. However, when 10% smoothing is applied, about 25% more MUs are required than with more smoothing (Anker et al. 2010). Finally, comparing 18 or 36 nearly equally spaced beams (all at least 5 degrees from opposing) or VMAT to an eight-beam plan shows surprisingly little improvement in the DVHs. The most significant change is for the left kidney and liver, where 10% to 15% lower volumes receive less than 10 Gy or 14 Gy, respectively (Figure 6.4). Although not shown, the VMAT isodose lines, especially below 20 Gy, are highly conformal to the PTV compared to the eight-beam plan. These results are supported by a theoretical analysis by Bortfeld (2010), who concluded that about 10 beams are sufficient for up to 10 cm diameter regions containing the target and OARs. These findings are likely transferrable to other treatment sites.

There is relatively little in the literature about planning and delivery techniques for neuroblastoma. Paulino et al. (2006) compared anteroposterior–posteroanterior (AP–PA) to IMRT (seven coplanar beams) with or without VB targets. Their kidney dose constraint was V16 should be less than two-thirds of the ipsilateral and one-third of the contralateral kidney. For midline tumors, the mean dose to both kidneys was 19 Gy with IMRT versus

22 Gy with AP–PA technique. For lateralized tumors, the mean dose to the contralateral kidney was 14 Gy with IMRT versus 6 Gy with AP–PA. The entire ipsilateral kidney received almost 100% of the prescribed dose with either technique. He concluded that IMRT was not a good technique for lateralized tumors because of the higher contralateral kidney dose (Paulino et al. 2006). Contrary to Paulino, the dosimetry for the lateralized cases 1 and 2 (Figure 6.5 and Figure 6.6) demonstrates that the ipsilateral kidney can be significantly spared and while the contralateral kidney can receive a higher dose than it would with opposed obliques, the dose is well below tolerance. Plowman et al. (2008) reported two neuroblastoma cases treated with TomoTherapy showing that dose to both kidneys was below tolerance while ensuring dose homogeneity in the lateral direction across the VB. There was a mild dose gradient in the anteroposterior direction, which was considered to produce an acceptable risk of lordosis.

6.3.1 Proton Therapy

Proton therapy has also been used for neuroblastoma. Hillbrand et al. (2008) compared AP–PA, IMRT, scattered protons (3DCPT), and intensity-modulated proton treatment (IMPT) plans. The proton plans consisted of two beam directions, whereas the IMRT plans used seven to nine equally spaced coplanar beams. The prescribed dose was 21 Gy. They used 15 Gy as the tolerance level for the kidney and 12 Gy as tolerance for the liver. The kidney V15 was about 53% for AP–PA, while it was about half that for IMRT, 3DCPT, or IMPT. The liver V12 was about 38% for AP–PA, 22% for IMRT, and 11% for 3DCPT or IMPT. Integral dose at the 10 Gy level was 200 to 300 cc smaller for protons than either photon technique, and at the 2 Gy level, AP–PA and proton techniques were similar but IMRT irradiated about 600 to 700 cc larger volume (Hillbrand et al. 2008).

Hug et al. (2001) described treatment of a 4-year-old with an incompletely resected right adrenal neuroblastoma that wrapped around the vertebral body. Unequally weighted AP and PA fields were used to deliver 25.2 CGE to the initial volume and a left lateral and two posterior oblique patch fields were used to deliver an additional 9 CGE to reach to a cumulative dose of 34.2 CGE to the boost volume. With this technique and target volume geometry, the D50 for the right kidney was 16 CGE and 80% of the liver received less than 10 CGE (Hug et al. 2001). In a comparative study of a multifield 3 mm spot size IMPT plan versus a nine-field IMRT plan, for a 21.6 Gy tumor dose, little clinically relevant dosimetric advantage was found for protons other than the integral dose (Olch et al. 2010). In another study, for an example case of high-risk neuroblastoma where 36 Gy was given to the residual tumor after 21.6 Gy to the initial tumor volume, a single PA IMPT beam generally delivered lower doses to normal organs than a seven-field

IMRT plan with a four-field IMRT boost (Hattangadi et al. 2011). It is unclear whether these dose differences are clinically relevant and IMRT plans with more beams and tighter dose constraints may have further narrowed the differences. As with any target in the abdomen, care must be taken to minimize changes in proton range due to changes in the presence of gas pockets that can move from day to day. To accomplish this, anterior and left lateral beams are generally avoided, with posterior beams being preferred.

6.3.2 Intraoperative Radiotherapy for Neuroblastoma

Intraoperative radiotherapy (IORT) has been used as an alternative to conventional external beam radiotherapy in an attempt to reduce toxicity by excluding critical structures from the irradiated volume. An additional benefit of IORT is the reduced integral dose. Survival and toxicity data for over 200 pediatric neuroblastoma patients treated with IORT are described in the literature. Because the treatment is given immediately following the surgical resection, the surface containing suspected microscopic or gross residual disease can be visually identified and included in the treatment portal, avoiding the target definition uncertainties encountered with image-based postoperative radiotherapy. Critical structures that cannot be moved out of the beam, such as ureters or the renal artery, are protected with lead shielding. Typically, the primary tumor site and involved paraaortic lymph nodes are included in the target volume. Treatment is delivered with one or more adjacent 4 to 16 MeV electron beams giving 8 to 15 Gy prescribed to 80% to 100% of the peak dose in a single fraction using applicators of up to 10 cm in diameter (Haase et al. 1994; Haas-Kogan et al. 2000; Gillis et al. 2007). Treatment is either given inside the linear accelerator vault or by a mobile linear accelerator in the operating room (Gillis et al. 2007). Technical problems with this technique include potential high dose volumes due to overlapping fields, incomplete coverage of tumor either laterally or in depth (Kunieda et al. 2008) movement of tumor edges outside the field by respiratory movement (Kuroda et al. 2003), and difficulty securing the treatment applicator inside the surgical cavity due to the beam angle required (Sugito et al. 2007). IORT is sufficient without external beam therapy if a gross total resection has been achieved (Haas-Kogan et al. 2000). Conventional external beam radiotherapy is given after IORT if the resection was incomplete (Oertel et al. 2006). Although most studies indicate few recurrences in the IORT field (Leavey et al. 1997; Kuroda et al. 2003; Sugito et al. 2007), they have been reported (Gillis et al. 2007; Kunieda et al. 2008). IORT is generally reported to be well tolerated with few long-term complications (Haase et al. 1994; Leavey et al. 1997). Late effects that have been reported have been associated with stenosis of the aorta or renal artery, vascular occlusion, and hypertension (Haas-Kogan et al. 2000; Zachariou et al. 2002; Gillis et al. 2007).

6.4 Organ-at-Risk Doses and Late Effects

6.4.1 Kidney

Excessive radiation to the kidney causes degeneration and sclerosis of the renal arterioles. The resultant diminished blood flow leads to degeneration of renal glomeruli and tubules and fibrosis. Renal dysfunction produces hypertension, edema, and uremia, and can be fatal. The entire kidney in children appears to tolerate up to 15 Gy given in 1.8 Gy daily fractions and the total kidney V20 and V30 has been recommended to be less than 17% and 10%, respectively (Bolling et al. 2011). Between these extremes, the dose–response data is variable. The incidence of symptomatic radiation nephropathy has been reported to be 20% after 20 Gy and steadily increases with increasing dose to 100% at 40 Gy to the entire kidney (Cassady 1995). The time interval between fractions may be critical for kidney sparing, as a high frequency of chronic renal dysfunction has been reported in children with neuroblastoma given just 12 to 14 Gy TBI in twice daily fractions (Tarbell et al. 1990).

6.4.2 Liver

Both radiation and chemotherapy can cause fibrosis and occlusion of the central veins of the hepatic lobules, resulting in necrosis of the surrounding hepatocytes and collapse of the lobules. This condition is called sinusoidal obstructive syndrome (SOS) or veno-occlusive disease and can be fatal. The myeloablative chemotherapy regimens used to treat high-risk neuroblastoma result in death due to SOS in a small percent of patients within weeks of treatment. Symptoms of less severe radiation hepatitis include hepatomegaly, ascites, jaundice, and abdominal pain. Pretreatment with intensive chemotherapy sensitizes the liver of neuroblastoma patients to radiation. Half of children treated with 12 to 25 Gy to the liver in conjunction with chemotherapy developed abnormal liver enzyme levels (Tefft et al. 1970).

6.4.3 Vertebral Bodies and Musculoskeletal

Neuroblastoma usually arises in close proximity to lumbar or thoracic vertebral bodies. These vertebrae are at least partially included in the PTV of the primary tumor bed. Irradiation decreases the growth of the vertebrae and other bones (Figure 6.7.). Paulino et al. (2005) found that the probability of scoliosis in neuroblastoma patients was not significantly increased with up to 17.5 Gy, but was 49% after 17.5 to 23 Gy and 56% after 23 to 36 Gy. Scoliosis and, less commonly, kyphosis occurred in some patients despite uniform irradiation of the vertebrae (Paulino and Fowler 2005a). Others

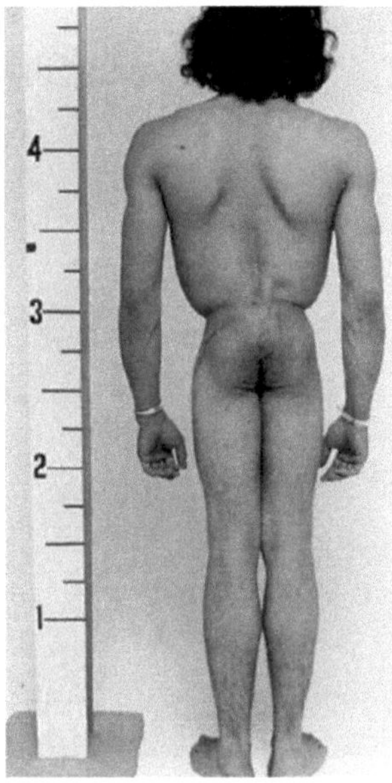

FIGURE 6.7 A 21-year-old man who was treated for neuroblastoma at 1 year of age with 36 Gy to the pelvis. (From Kroll, S. S., et al., *Annals of Surgical Oncology* 1 (6):473–9, 1994. With permission.)

have reported some decrease in vertebral body growth after 20 to 27 Gy (Probert and Parker 1975; Shalet et al. 1987) and significantly more growth restriction after 36 Gy (Hartley et al. 2008). Hogeboom et al. (2001) showed that patients whose spine was treated for the entire length of the flank for Wilms' tumor, generally longer than the length of treated volume for neuroblastoma, with a median dose of 35 Gy did not cause any more height deficit measured at age 18 than for 20 Gy. In children whose abdomen was irradiated when they were less than 12 months of age, the estimated loss of height at age 17 to 18 years was 2 cm after 10 Gy compared to 8 cm after more than 10 Gy (Evans et al. 1991). Inhomogeneity of radiation dose across the vertebral body is highly discouraged, as it results in differential growth across the width of the bone (causing scoliosis, or lateral deviation of the spine) or from the anterior to posterior surfaces (causing kyphosis, or hunchback).

Neuroblastoma patients tend to be very young, even presenting as neonates or infants less than 1 year old. These very young patients are not usually treated with radiation but there is a report on late effects of 20 such patients treated with doses between 12 to 30 Gy. The most common late effect was musculoskeletal toxicity, occurring with a 15-year rate of 47.3% for those receiving radiotherapy and 3.3% for those not receiving radiotherapy ($p = 0.02$). Five of six infants less than 6 months of age developed musculoskeletal toxicity but only one of seven older than 6 months. Infants with International Staging System (INSS) stage 4 (metastatic) neuroblastoma and n-myc amplification have a 3-year event-free survival of just 14% and may therefore receive a survival benefit from locoregional radiation therapy as part of their treatment but with a very high risk of late effects (Paulino et al. 2002).

6.4.4 Gonadal Dose

At least one study has focused on gonadal dose from radiotherapy for abdominal neuroblastoma. For 5- to 15-year-old patients, the distance from the ovaries or testes to the field edge was 6 to 13 cm and 14 to 23 cm, respectively. The ovarian dose was 1.5% of the tumor dose (45 cGy) and the testicular dose was 0.5% (15 cGy) for a 30 Gy prescribed dose. The gonadal doses for 6 MV x-rays were 1.3 times as high as for 18 MV x-rays. Gonadal doses were up to 1.5 times higher with lead blocks than with an MLC. The risk of gonadal damage to patients or development of hereditary disorders in their offspring was deemed insignificant compared to normal incidence (Mazonakis et al. 2007). However, direct irradiation of the gonads is a significant risk factor for gonadal failure (Laverdiere et al. 2009).

6.4.5 Secondary Malignancies

Neuroblastoma patients treated with both chemotherapy and radiation therapy have a significantly higher probability of developing a second malignancy than patients treated with chemotherapy alone or the general population. Those treated with chemotherapy alone do not have an increased second malignancy risk compared with the general population (Rubino et al. 2003). A variety of types of malignancy have been reported among irradiated patients (Iwata et al. 2001; Rubino et al. 2003; Paulino and Fowler 2005b) with the risk for cancer developing in an organ increasing with the radiation dose given to that organ (Rubino et al. 2003). The cumulative incidence of second malignant neoplasms was 3.5% at 25 years and 7.0% at 30 years after diagnosis (Laverdiere et al. 2009).

References

Anker, C. J., B. Wang, M. Tobler, et al. 2010. Evaluation of fluence-smoothing feature for three IMRT planning systems. *Journal of Applied Clinical Medical Physics* 11 (2):3035.

Bolling, T., I. Ernst, H. Pape, et al. 2011. Dose-volume analysis of radiation nephropathy in children: Preliminary report of the risk consortium. *International Journal of Radiation Oncology • Biology • Physics* 80 (3):840-4.

Bortfeld, T. 2010. The number of beams in IMRT—Theoretical investigations and implications for single-arc IMRT. *Physics in Medicine & Biology* 55 (1):83–97.

Cassady, J. R. 1995. Clinical radiation nephropathy. *International Journal of Radiation Oncology • Biology • Physics* 31 (5):1249–56.

Chen, A. B., J. Killoran, H. Kim, and H. Mamon. 2005. Treatment planning for resected abdominal tumors: Differences in organ position between diagnostic and radiation-planning computed tomography scans. *International Journal of Radiation Oncology • Biology • Physics* 63 (5):1613–20.

Evans, A. E., P. Norkool, I. Evans, N. Breslow, and G. J. D'Angio. 1991. Late effects of treatment for Wilms' tumor. A report from the National Wilms' Tumor Study Group. *Cancer* 67 (2):331–6.

Flandin, I., O. Hartmann, J. Michon, et al. 2006. Impact of TBI on late effects in children treated by megatherapy for Stage IV neuroblastoma: A study of the French Society of Pediatric oncology. *International Journal of Radiation Oncology • Biology • Physics* 64 (5):1424–31.

Gatcombe, H. G., R. B. Marcus, Jr., H. M. Katzenstein, M. Tighiouart, and N. Esiashvili. 2009. Excellent local control from radiation therapy for high-risk neuroblastoma. *International Journal of Radiation Oncology • Biology • Physics* 74 (5):1549–54.

Gillis, A. M., E. Sutton, K. D. Dewitt, et al. 2007. Long-term outcome and toxicities of intraoperative radiotherapy for high-risk neuroblastoma. *International Journal of Radiation Oncology • Biology • Physics* 69 (3):858–64.

Haas-Kogan, D. A., B. M. Fisch, W. M. Wara, et al. 2000. Intraoperative radiation therapy for high-risk pediatric neuroblastoma. *International Journal of Radiation Oncology • Biology • Physics* 47 (4):985–92.

Haase, G. M., D. P. Meagher, Jr., L. K. McNeely, et al. 1994. Electron beam intraoperative radiation therapy for pediatric neoplasms. *Cancer* 74 (2):740–7.

Hartley, K. A., C. Li, F. H. Laningham, M. J. Krasin, X. Xiong, and T. E. Merchant. 2008. Vertebral body growth after craniospinal irradiation. *International Journal of Radiation Oncology • Biology • Physics* 70 (5):1343–9.

Hattangadi, J. A., B. Rombi, T. I. Yock, et al. 2011. Proton radiotherapy for high-risk pediatric neuroblastoma: Early outcomes and dose comparison. *International Journal of Radiation Oncology • Biology • Physics* 83 (3):1015–22.

Hillbrand, M., D. Georg, H. Gadner, R. Potter, and K. Dieckmann. 2008. Abdominal cancer during early childhood: A dosimetric comparison of proton beams to standard and advanced photon radiotherapy. *Radiotherapy and Oncology* 89 (2):141–9.

Hogeboom, C. J., S. C. Grosser, K. A. Guthrie, P. R. Thomas, G. J. D'Angio, and N. E. Breslow. 2001. Stature loss following treatment for Wilms' tumor. *Medical & Pediatric Oncology* 36 (2):295–304.

Hug, E. B., M. Nevinny-Stickel, M. Fuss, D. W. Miller, R. A. Schaefer, and J. D. Slater. 2001. Conformal proton radiation treatment for retroperitoneal neuroblastoma: Introduction of a novel technique. *Medical & Pediatric Oncology* 37 (1):36–41.

Iwata, A., T. Hirota, K. Konno, et al. 2001. Osteosarcoma as a second malignancy after treatment for neuroblastoma. *Pediatric Hematology & Oncology* 18 (7):465–9.

Kroll, S. S., S. Y. Woo, A. Santin, et al. 1994. Long-term effects of radiotherapy administered in childhood for the treatment of malignant diseases. *Annals of Surgical Oncology* 1 (6):473–9.

Kunieda, E., S. Hirobe, T. Kaneko, T. Nagaoka, S. Kamagata, and G. Nishimura. 2008. Patterns of local recurrence after intraoperative radiotherapy for advanced neuroblastoma. *Japanese Journal of Clinical Oncology* 38 (8):562–6.

Kuroda, T., M. Saeki, T. Honna, H. Masaki, and Y. Tsunematsu. 2003. Clinical significance of intensive surgery with intraoperative radiation for advanced neuroblastoma: Does it really make sense? *Journal of Pediatric Surgery* 38 (12):1735–8.

Laverdiere, C., Q. Liu, Y. Yasui, et al. 2009. Long-term outcomes in survivors of neuroblastoma: A report from the Childhood Cancer Survivor Study. *Journal of the National Cancer Institute* 101 (16):1131–40.

Leavey, P. J., L. F. Odom, M. Poole, L. McNeely, W. R. Tyson, and G. M. Haase. 1997. Intra-operative radiation therapy in pediatric neuroblastoma. *Medical and Pediatric Oncology* 28 (6):424–8.

Mazonakis, M., F. Zacharopoulou, S. Kachris, C. Varveris, J. Damilakis, and N. Gourtsoyiannis. 2007. Scattered dose to gonads and associated risks from radiotherapy for common pediatric malignancies: A phantom study. *Strahlentherapie und Onkologie* 183 (6):332–7.

Oertel, S., A. G. Niethammer, R. Krempien, et al. 2006. Combination of external-beam radiotherapy with intraoperative electron-beam therapy is effective in incompletely resected pediatric malignancies. *International Journal of Radiation Oncology • Biology • Physics* 64 (1):235–41.

Olch, A. J., S. Lorentini, M. Schwarz, B. Rombi, and K. K. T. Wong. 2010. 3 mm spot size IMPT vs. IMRT for 3 complex pediatric cases. *International Journal of Radiation Oncology • Biology • Physics* 78 (3):S801.

Paulino, A. C., M. S. Ferenci, K.-Yueh Chiang, A. W. Nowlan, and R. B. Marcus, Jr. 2006. Comparison of conventional to intensity modulated radiation therapy for abdominal neuroblastoma. *Pediatric Blood & Cancer* 46 (7):739–44.

Paulino, A. C., and B. Z. Fowler. 2005a. Risk factors for scoliosis in children with neuroblastoma. *International Journal of Radiation Oncology • Biology • Physics* 61 (3):865–9.

———. 2005b. Secondary neoplasms after radiotherapy for a childhood solid tumor. *Pediatric Hematology & Oncology* 22 (2):89–101.

Paulino, A. C., N. A. Mayr, J. H. Simon, and J. M. Buatti. 2002. Locoregional control in infants with neuroblastoma: Role of radiation therapy and late toxicity. *International Journal of Radiation Oncology • Biology • Physics* 52 (4):1025-1031.

Plowman, P. N., K. Cooke, and N. Walsh. 2008. Indications for tomotherapy/intensity-modulated radiation therapy in paediatric radiotherapy: Extracranial disease. *British Journal of Radiology* 81 (971):872–80.

Probert, J. C., and B. R. Parker. 1975. The effects of radiation therapy on bone growth. *Radiology* 114 (1):155–62.

Quach, A., L. Ji, V. Mishra, et al. 2011. Thyroid and hepatic function after high-dose 131 I-metaiodobenzylguanidine (131 I-MIBG) therapy for neuroblastoma. *Pediatric Blood & Cancer* 56 (2):191–201.

Rubino, C., E. Adjadj, S. Guerin, et al. 2003. Long-term risk of second malignant neoplasms after neuroblastoma in childhood: Role of treatment. *International Journal of Cancer* 107 (5):791–6.

Shalet, S. M., B. Gibson, R. Swindell, and D. Pearson. 1987. Effect of spinal irradiation on growth. *Arch Dis Child* 62 (5):461–4.

Simon, T., B. Hero, R. Bongartz, et al. 2006. Intensified external-beam radiation therapy improves the outcome of stage 4 neuroblastoma in children >1 year with residual local disease. *Strahlentherapie und Onkologie* 182 (7):389–94.

Sugito, K., T. Kusafuka, M. Hoshino, et al. 2007. Intraoperative radiation therapy for advanced neuroblastoma: The problem of securing the IORT field. *Pediatric Surgery International* 23 (12):1203–7.

Tarbell, N. J., E. C. Guinan, L. Chin, P. Mauch, and H. J. Weinstein. 1990. Renal insufficiency after total body irradiation for pediatric bone marrow transplantation. *Radiotherapy and Oncology* 18 (Suppl. 1):139–42.

Tefft, M., A. Mitus, L. Das, G. F. Vawter, and R. M. Filler. 1970. Irradiation of the liver in children: Review of experience in the acute and chronic phases, and in the intact normal and partially resected. *American Journal of Roentgenology* 108 (2):365–85.

Zachariou, Z., H. Sieverts, M. J. Eble, S. Gfrorer, and A. Zavitzanakis. 2002. IORT (intraoperative radiotherapy) in neuroblastoma: Experience and first results. *European Journal of Pediatric Surgery* 12 (4):251–4.

Chapter 7
Wilms' Tumor

7.1 Clinical Overview

Wilms' tumor was first categorized by the German surgeon Max Wilms in 1899. Over 90% of all primary tumors of the kidney in children are of the Wilms' tumor type (also called nephroblastoma). The other types are rhabdoid tumor and clear cell sarcoma of the kidney. About 500 Wilms' tumors are diagnosed annually in the United States. Eighty percent of cases present with involvement of a single location in one kidney (Figure 7.1), 14% with multifocal involvement of one kidney, and 6% with tumors in both kidneys (Figure 7.2). The mean age at diagnosis is 31 months for bilateral disease and 44 months for unilateral disease. Patients usually present with asymptomatic swelling of the abdomen noticed by a parent. As reported in a survey of the last 40 years of treatment of Wilms' tumor, overall survival is above 90%, up from about 45% in 1950 (D'Angio 2007; Kalapurakal et al. 2010). Wilms' tumor is highly sensitive to both radiation and chemotherapy. The initial treatment of unilateral disease varies by continent. In North America, surgical resection is advocated to obtain maximal staging, pathology, and biology information from the resected tissue. In Europe, preoperative chemotherapy is used to make the resection easier, decrease the probability of intraoperative spillage of malignant cells into the abdomen, and assess response to therapy. Both continents give preoperative chemotherapy for bilateral disease to shrink the tumors and permit maximum preservation of

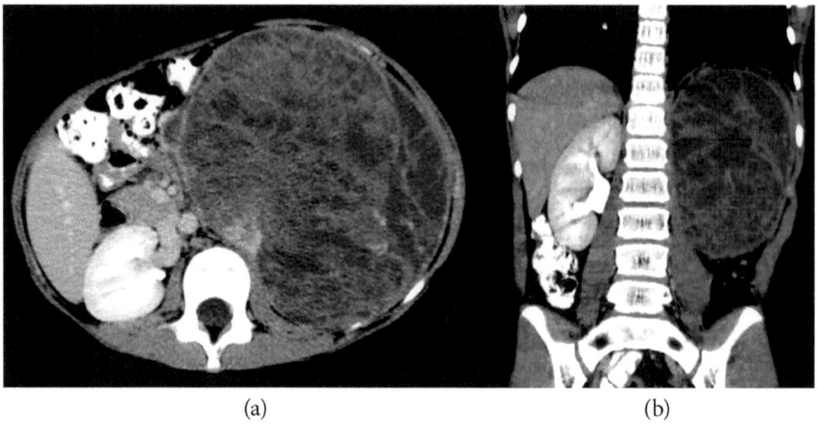

FIGURE 7.1 (a) Left kidney mass, axial CT. (b) Left kidney mass coronal CT.

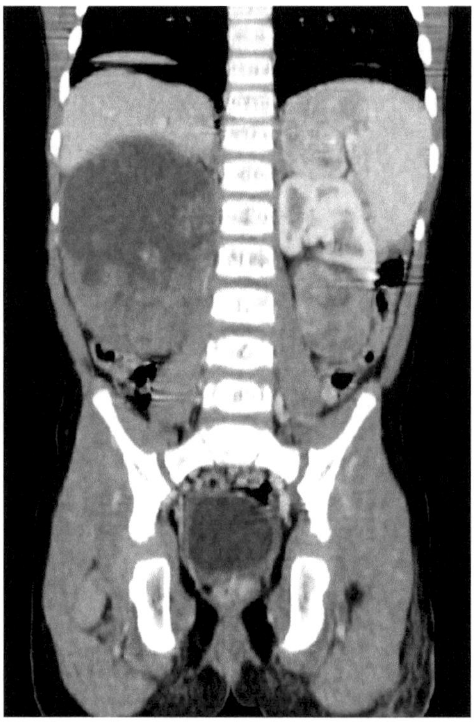

FIGURE 7.2 Bilateral Wilms' affecting the right kidney and half of the left kidney.

kidney tissue after resection. Pathologic tumor specimens are classified as having a favorable or, if anaplasia is present, unfavorable histology. The category "high-risk" renal cancer is comprised of unfavorable histology Wilms' tumor, rhabdoid tumor, and clear cell sarcoma of the kidney. These three entities have the same staging criteria and receive the same treatment.

Favorable histology Wilms' tumors receive radiation therapy when there is gross residual tumor at the primary site following resection, positive surgical margins, peritoneal tumor nodules, regional lymph node involvement, or spillage of tumor cells through a ruptured capsule into the flank or peritoneal cavity (stage III disease) or distant metastasis (stage IV disease). All high-risk renal cancers regardless of stage or completeness of resection receive irradiation to at least the preoperative tumor volume. Lymph node involvement or a localized spill of tumor cells is treated with irradiation of the entire ipsilateral flank. If there has been diffuse dissemination of tumor cells throughout the peritoneum, either by seeding or tumor rupture, the whole abdomen is irradiated (WAI). A dose of 10.5 to 10.8 Gy is given for microscopic residual tumor in the abdomen or tumor spillage for either newly diagnosed or recurrent Wilms' tumor. Gross residual tumor at the primary site, in the peritoneum or liver receives 21.6 to 30.6 Gy. In the unusual circumstance of either stage III diffuse anaplasia or a patient older than 16 years of age, the dose for microscopic residual is raised to 19.8 Gy.

For bilateral Wilms', chemotherapy is given prior to surgical resection in order to permit maximum preservation of the kidney bearing the lesser burden of tumor. Following tumor response to chemotherapy, the more affected kidney is removed and as much tumor is resected from the less affected kidney as feasible consistent with preserving function of that kidney. Often, small areas of gross tumor are left on the unresected kidney. These areas are treated with 21.6 Gy.

Common sites of metastases are the lungs (80% of all metastases in Wilms' tumor patients), lymph nodes, and liver. If any lung metastasis are evident on chest x-ray at diagnosis, the entire bilateral lungs are treated to 12 Gy if the patient is over 12 months of age or 10.5 Gy if under 12 months of age. Other sites of distant metastasis are treated with 19 to 25 Gy. If the liver is diffusely involved, then the entire liver is treated to 19.8 Gy.

It is recommended that radiation therapy begin within 10 days of surgery, which often is quite difficult, as staging and classifying the resected tumor histologically generally takes several days postoperatively and consent and deep sedation by an anesthesiologist during simulation and treatment must be arranged. Analysis of extensive Children's Oncology Group (COG) clinical trial data, however, has demonstrated that there is no decrease in the high rate of local control when radiation therapy is started up to 3 weeks postoperatively (Kalapurakal et al. 2003). Therefore, one should feel comfortable with starting radiation up to several days beyond the mandated postoperative day 10.

The reader can refer to any of the several good reviews of the management of Wilms' tumor for more information on clinical management (Kalapurakal et al. 2004; Gommersall et al. 2005; Spreafico and Bellani 2006; Tongaonkar et al. 2007).

7.2 Field Design

Typically, the treatment volume for Wilms' tumor is the preoperative Anteroposterior-posteroanterior (AP_PA) tumor volume, and may include the flank, entire paraaortic lymph node chain, whole abdomen, and whole lungs. Anteroposterior–posteroanterior (AP–PA) fields are commonly used for these treatments (see Figures 7.3 to 7.6), however, there is a role for more conformal techniques when partial kidney or other normal tissue sparing is required.

7.2.1 Flank Irradiation

The treatment field includes the entire involved kidney and associated tumor with a 1 cm margin as seen on CT/MR scan performed prior to any surgery or chemotherapy. At any levels where the medial border would cut across a vertebral body, it is extended to 1 cm beyond the vertebral body to limit the development of scoliosis. If the tumor extended to the contralateral side far enough to cause the projected treatment field to include portions of the remaining kidney, the field is scaled back at that level to extend to only 1 cm beyond the vertebral body. This is done to preserve function of the remaining kidney.

If the Wilms' has spread to a paraaortic lymph node, the treatment field is expanded beyond the flank to encompass the entire paraaortic lymph node chain, with the medial border 1 cm beyond the vertebral bodies from the crus of the diaphragm superiorly to the bottom of L5 inferiorly. Figure 7.3 shows simulator images of these fields.

7.2.2 Whole Abdomen Irradiation

The whole abdomen fields (AP–PA) include the entire peritoneal cavity, extending bilaterally 1 cm beyond the skin surface from 1 cm above the dome of the diaphragm superiorly to the bottom of the obturator foramina inferiorly. The femoral heads and portions of the heart and lungs are blocked to minimize chronic toxicity. When the prescribed dose is above 15 Gy, a kidney block must be used to limit the whole kidney dose to less than 14.4 Gy. This can be accomplished by using a cerrobend island block in the PA field for the appropriate fraction of the treatment (Figure 7.4a).

7.2.3 Whole Lung Irradiation

The whole lungs can be treated either alone, if there is no indication for abdominal irradiation, or in conjunction with the whole abdomen or flank. When both lung and abdominal treatment is indicated, both are encompassed in single large AP–PA fields at a dose of 1.5 Gy daily. In patients over 12 months of age, the field is reduced to encompass just the whole lung when

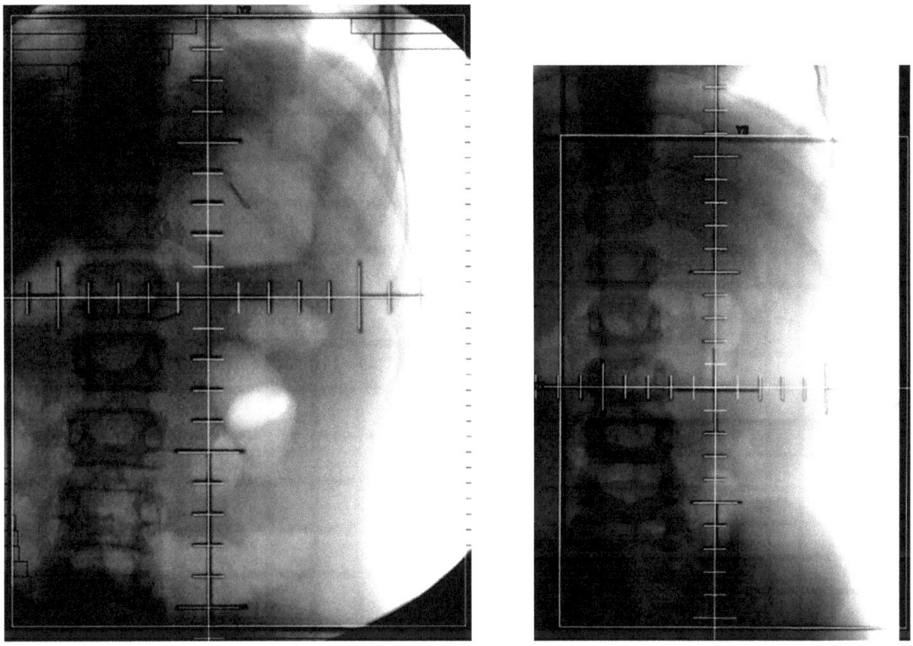

(a) (b)

FIGURE 7.3 **(a) Left flank Simulator image. Because of local spillage, fields cover the diaphragm. (b) Left flank Simulator image. Stage II Wilms'.**

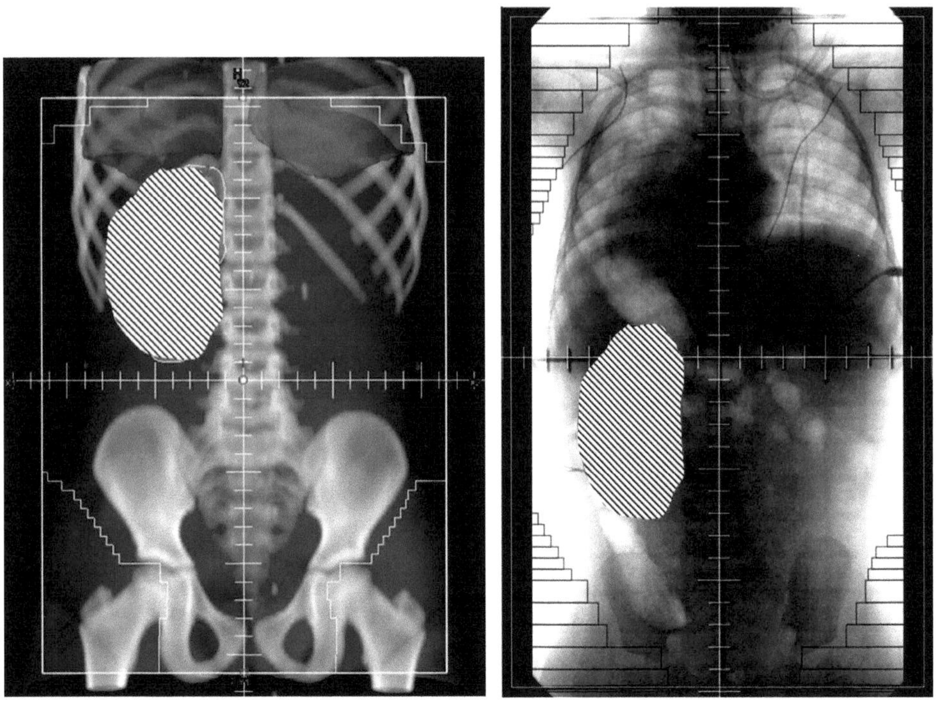

(a) (b)

FIGURE 7.4 **(a) Whole abdomen PA field with kidney block. (b) Whole chest and abdomen PA field with kidney block.**

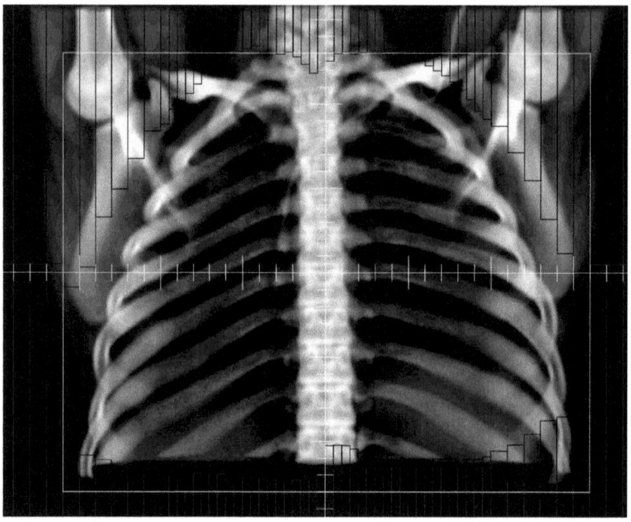

FIGURE 7.5 AP whole lung field.

the dose reaches 10.5 Gy. The whole lung dose may be different due to differences in separation between the midabdomen and lung. This reduced field is treated as the last fraction and the dose necessary to bring the total lung dose to 12 Gy is given. The field reduction can be accomplished by bringing in the inferior jaw independently without changing the isocenter or superior and lateral borders. Figure 7.5 shows the whole lung field (AP). Figure 7.6a shows a coronal CT plane for a case with lung metastases and a left kidney tumor. Figure 7.6b shows the field (AP) used for treating the lung metastases together with the left flank.

Historically, the whole lung treatment was calculated without corrections for lung density. Future protocols will likely require the correction. With the correction, the dose to the prescription point (middle of AP separation on sagittal midline) is reduced by about 4% due to lack of scattering to this point due to the presence of the low-density lung volume. This means that the monitor units (MUs) will be about 4% greater with the correction than without, or conversely, the prescription point dose will be 4% lower if the correction is not applied. This decrease in MUs calculated without the correction serves to lower the magnitude of the increase in lung dose that actually occurs. With the correction, the mean lung dose is 13 to 14 Gy (about a 15% higher dose than the 12 Gy prescribed) and the D35 is 13.5 to 14.5 Gy (Figure 7.7a,b). Without applying inhomogeneity correction, the mean lung dose will be reported to be slightly less than 12 Gy (Figure 7.7c) due to the thicker contours inferior to the central axis. In reality, because of the location of the prescription point (the sagittal midline in solid tissue), the actual lung dose is nearly the same whether or not the correction

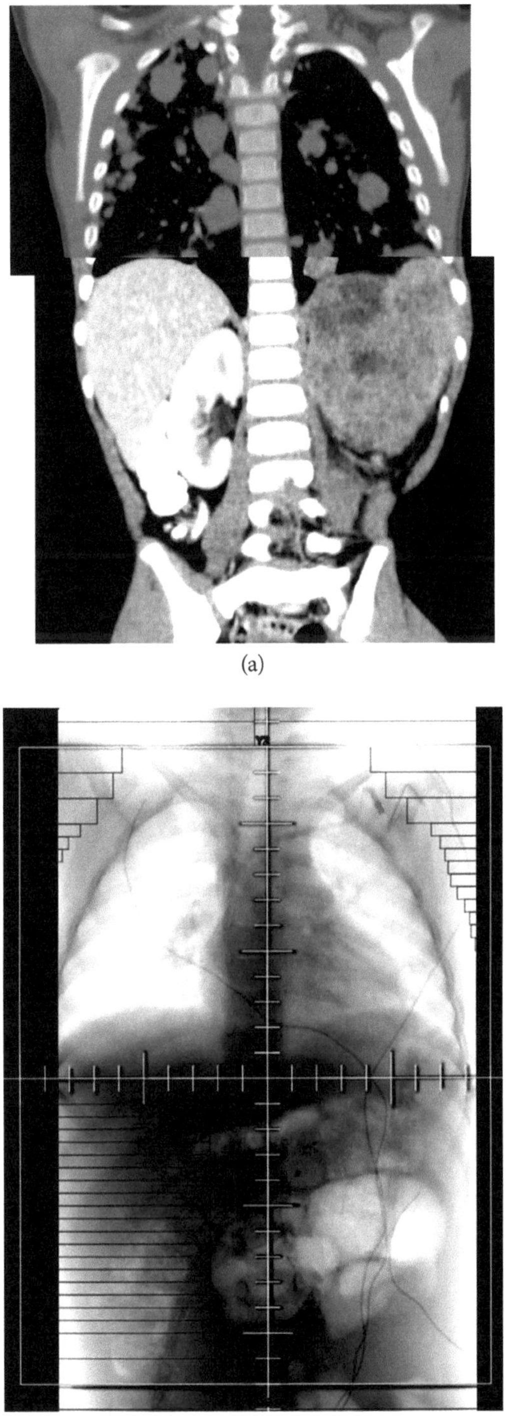

(a)

(b)

FIGURE 7.6 (a) Coronal CT showing lung metastases and left kidney tumor. (b) AP field used to treat the disease in panel (a).

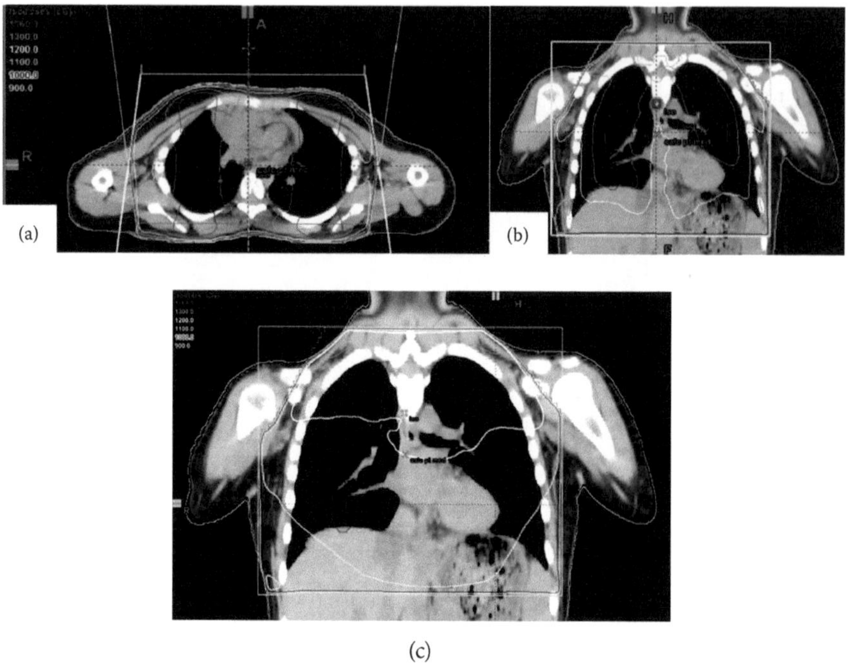

FIGURE 7.7 (See color insert.) (a) Axial and (b) coronal AP–PA whole lung with prescription point in the mediastinum, inhomogeneity correction is turned on. (c) Same as in panel b without inhomogeneity correction.

is applied (the difference is only due to the 4% different MUs), but with the correction, one sees the true lung dose. The doses vary within about 1 Gy as the size of the patient's AP separation varies between 12 and 20 cm. In the future, in addition to requiring inhomogeneity corrections, the protocol prescription may change to refer to the mean lung dose rather than the dose at a point in solid tissue on the midline.

7.2.4 Bilateral Wilms' Tumor

For bilateral Wilms' tumor, the radiation therapy guidelines for unilateral Wilms' apply to each tumor. Radiation therapy is given postoperatively if there is residual disease or a high-risk tumor. Treatment guidelines are modified, if necessary, to assure that the dose to the remaining kidney does not exceed 14.4 Gy. Here the gross tumor volume (GTV) is the surgical bed if margins were involved or a gross-residual tumor based on the postchemotherapy and postoperative CT or MRI. The clinical target volume (CTV) is created using an anatomically confined margin of 1 cm added to the GTV. A margin that accounts for both daily setup error and internal motion is added to the CTV to create the planning target volume (PTV). Although an AP–PA conformal technique can be used for polar tumors (upper or lower pole), treatment with multifield 3D conformal or intensity-modulated radiation therapy (IMRT) may provide the best chance of preserving function of the remaining kidney, especially for renal hilar tumors. Intraoperative radiotherapy (IORT) has also been used to limit the volume of remaining kidney irradiated, giving a single dose of about 15 Gy to just the surgical margin or involved portion of the kidney (Halberg et al. 1991; Nag et al. 2003).

7.3 Intensity-Modulated Radiation Therapy (IMRT) Treatment Planning

Where more conformal techniques are used, the CTV margin is 1 cm for primary and 0.5 cm for boost irradiation of residual disease, and the PTV margin is 0.5 to 1 cm (if respiratory motion is considered). Motion is important to assess and incorporate into planning when appropriate especially when conformal normal tissue sparing techniques are used. The sedated patient typically demonstrates minimal organ motion.

Figure 7.2 shows a coronal plane from a CT scan of a bilateral Wilms' tumor case where the involved right kidney containing unfavorable histology Wilms' tumor with diffuse anaplasia was entirely removed but only a partial nephrectomy was performed to preserve renal function in the left kidney that contained favorable histology Wilms' tumor in its lower pole. The surgical margin of the partial nephrectomy was microscopically positive. Guidelines direct that 19.8 Gy be given to the right flank because of the diffuse anaplasia and 10.8 Gy be given to the left side due to the microscopic residual disease. Because only a portion of one kidney remained, the left-sided CTV was reduced to spare the remaining left kidney that was not close to the involved surgical margin. The entire right flank volume and partial left flank volume were treated to 10.8 Gy with a nine-field IMRT plan. The length of left flank containing the kidney with 1 cm margin was treated, sparing the superior half of the kidney. The interior of the liver was also spared during IMRT so that less than 25% of the liver would receive more than 20 Gy. The right flank was boosted with AP–PA fields to a total of 19.8 Gy. Figure 7.8 shows the dosimetry for this bilateral Wilms' tumor case. The brown color wash is 19.8 Gy and green is 10.8 Gy. The axial plane is through the center of the superior half of the left kidney and the coronal plane is through the center of the left kidney. Each demonstrates the dose reduction to the superior half of the left kidney and the coronal shows the sparing of the interior of the liver.

Occasionally, the liver is diffusely involved and is treated with 19.8 Gy with local boosts up to 30.6 Gy for discrete lesions involving a small portion of the liver. Treatment may be given using AP–PA fields with a posterior kidney block, if necessary, to limit the dose to the remaining kidney to 14.4 Gy to less than 67% of its volume. The block will prevent portions of

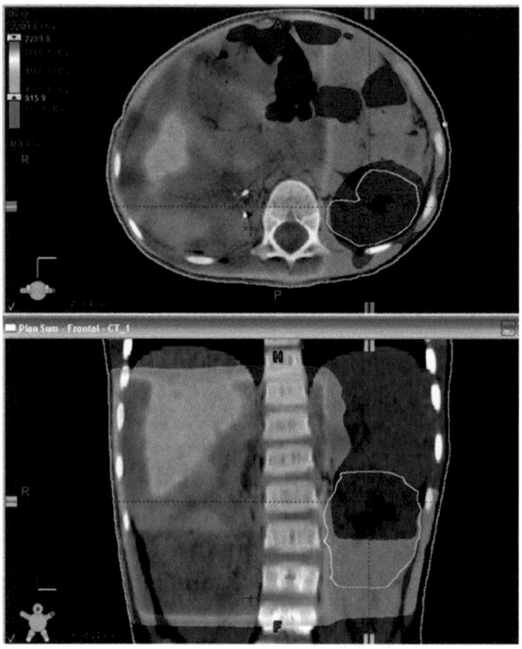

FIGURE 7.8 (See color insert.) Bilateral Wilms', IMRT to right flank and partial left flank for 10.8 Gy used to spare partial left kidney. AP–PA fields boosted the right flank to 19.8 Gy. The central part of liver is spared such that ≥25% of the whole liver receives 20 Gy, green color wash = 10.8 Gy, brown = 19.8 Gy.

the liver from receiving the full prescribed dose, especially when the right kidney is being blocked. Alternatively, IMRT may be used to treat the entire liver while limiting the dose to the kidney below its tolerance. Figure 7.9 shows the dosimetry using an eight-field IMRT plan, with the whole liver receiving 19.8 Gy, whereas the majority of the left kidney is kept below 5 Gy. Fogliata et al. (2007) showed the dose distributions from seven different IMRT planning systems for a similar case, showing sparing of the right kidney was feasible.

An application of IMRT to whole lung irradiation for cardiac sparing has been reported by Kalapurakal et al. (2012). A dosimetric study was performed for treatment of the whole lungs to 12 Gy using a nine-beam IMRT plan calculated using inhomogeneity corrections. Compared to standard AP–PA whole lung treatment, IMRT was able to reduce the cardiac and ventricular V8-V12 by 16% to 77%. Further sparing of the heart could be possible if the vertebral bodies were omitted from the target volume. Although 12 Gy to the whole heart is not generally thought to be toxic, the National Wilms Tumor Study Group (NWTSG) has shown that heart failure is the second most common cause of death (other than by the disease itself) in survivors of Wilms' tumor.

Although Hong et al. (2002) demonstrated that whole abdominal irradiation using IMRT with just five gantry angles was feasible with somewhat

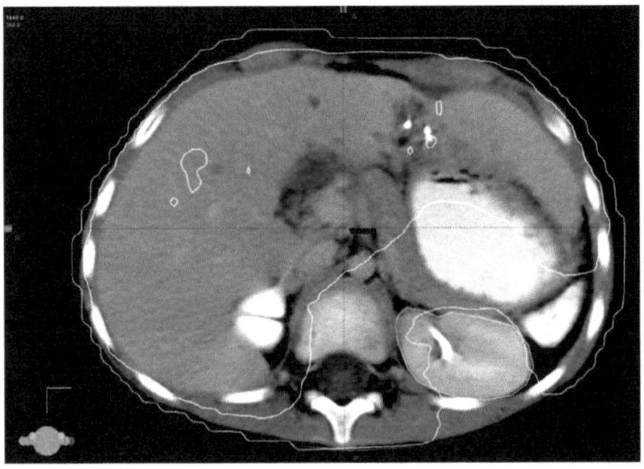

FIGURE 7.9 IMRT to treat whole liver and spare remaining contralateral kidney. Isodose line encompassing liver is 19.8 Gy, isodose line carving around left kidney is 5 Gy.

improved PTV dose coverage, the added complexity may not be warranted for the low doses employed.

7.4 Proton Therapy

Several groups have reported proton therapy for Wilms'. In an early report, heavy ion beams were compared to 15 MV AP–PA photon beams for abdominal or flank irradiation. Sparing of the large bowel and vertebral bodies was shown (Gademann and Wannenmacher 1992). A more recent paper showed that proton therapy for Wilms' can reduce the dose to the remaining kidney and to the liver over photon AP–PA or IMRT treatments (Hillbrand et al. 2008). Kidney sparing during whole liver treatment for liver metastases was compared for protons, volumetric modulated arc therapy (VMAT), and helical TomoTherapy (HT) (Figure 7.10). Clinically relevant doses were comparable for the three modalities (Fogliata et al. 2009). However, whenever proton beams traverse the anterior abdomen, care must be taken to consider the effects of changing densities within the bowel during treatment, which can significantly alter proton dose distributions.

7.5 Organ-at-Risk Doses and Late Effects

Despite the relatively low doses given for Wilms' tumor treatment, late effects were present in 36% to 69% of Wilms' tumor survivors (Paulino et al. 2000; Han et al. 2009; Sasso et al. 2010; van Dijk et al. 2010). Affected patients had fewer and milder late effects than did patients with other pediatric malignancies (Han et al. 2009). Several late effects studies have been published by the large Wilms' tumor treatment cooperative groups such as

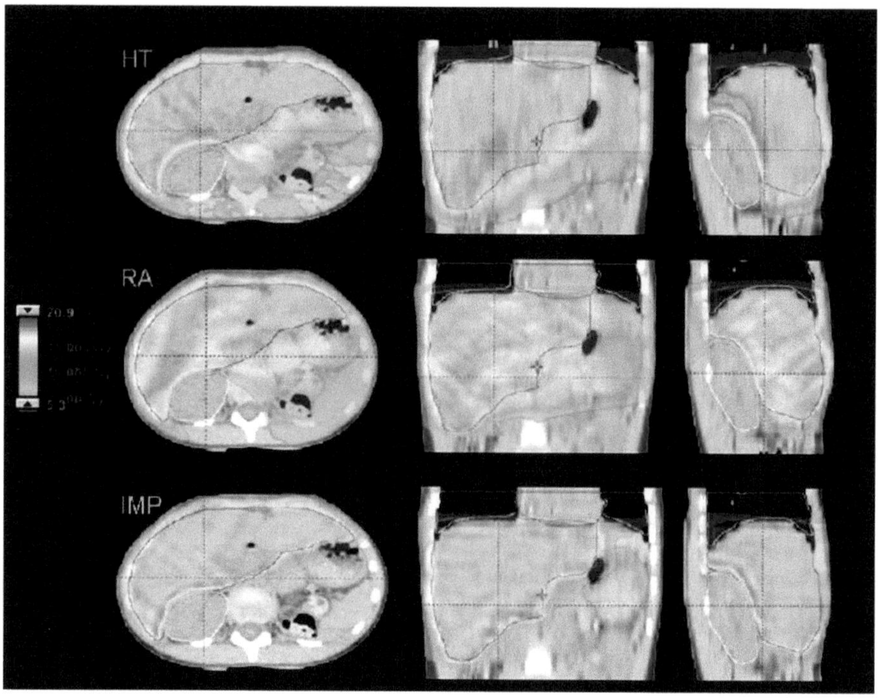

FIGURE 7.10 (See color insert.) Protons versus RA versus HT for Wilms' case where the whole liver is treated with sparing of the ipsilateral kidney. (After Fogliata, A., et al., *Radiation Oncology* 4:2, 2009.)

the National Wilms Tumor Study Group, and the UK Children's Cancer Study Group, as well as the Childhood Cancer Survivor Study group and other large single institutions.

7.5.1 Renal Function

Preservation of the function of the remaining kidney in unilateral Wilms' (or portion of the remaining kidney in bilateral Wilms') is a major concern. After nephrectomy in an otherwise normal individual, the remaining kidney expands to compensate. This compensation is sufficient in most normal people but may not be in a patient treated with nephrotoxic chemo and radiation therapy (Bolling et al. 2011). Bilateral Wilms' presents a major challenge in preserving renal function, since one kidney is surgically removed leaving the other diseased kidney to be salvaged. Nephron-sparing surgery can be performed if at least one-third of the kidney can remain (Giel et al. 2007). This kidney remnant is irradiated with the hope that the tumor will be eradicated without making the patient anephric, requiring lifelong dialysis.

Evidence of renal dysfunction in Wilms' tumor patients is around 30%. The NWTSG report on the worst outcome, end-stage renal disease, found the 20-year cumulative incidence to be low, 1% and 12% for unilateral and bilateral Wilms', respectively (Breslow et al. 2005). Children receiving chemotherapy with radiation doses to the remaining kidney of less than 12 to 15 Gy do not generally suffer renal impairment (Kirkbride and Plowman 1992; Levitt et al. 1992; Daw et al. 2009; Bolling et al. 2011). In an earlier report from NWTSG, 6 of 15 children (3 unilateral and 3 bilateral) who received between 12 and 20 Gy to the remaining kidney developed radiation nephritis. However, one bilateral Wilms' study failed to show a dose–response relationship (Smith et al. 1998).

7.5.2 Growth and Stature

Loss of stature is another concern for patients receiving whole abdomen or flank irradiation. The field design for flank irradiation requires the medial border of the AP–PA fields to include the spine with a 1 cm margin. These fields encompass a large fraction of the child's entire length of spine. Scoliosis has been reported in 10% to 43% of patients but few require orthopedic intervention despite care being taken to include the entire width of the vertebral body (Wallace et al. 1990; Paulino et al. 2000; Trobs et al. 2001; Sasso et al. 2010). The incidence of scoliosis was a function of dose with a rate of 8%, 46%, and 63% for 10 to 12 Gy, 12.1 to 23.9 Gy, and 24 to 40 Gy, respectively (Paulino et al. 2000). Irradiation at a younger age was found to cause a greater rate of decline of stature but height deficits were not clinically significant for doses less than 20 Gy (Wallace et al. 1990; Hogeboom et al. 2001). For patients receiving 12 to 18 Gy whole lung irradiation, underdevelopment of the chest and female breasts has been reported (Attard-Montalto et al. 1992; Trobs et al. 2001).

7.5.3 Cardiac Function

Cardiac toxicity is a concern for patients receiving whole lung or left flank irradiation especially in the context of treatment with cardiotoxic anthracycline chemotherapy given for Wilms' tumor. Pericarditis is the most frequently diagnosed complication but congestive heart failure is also seen. Cumulative doxorubicin dose was the most important risk factor for the occurrence of congestive heart failure in the NTWSG study. Current NTWSG studies use lower anthracycline doses than earlier studies resulting in a decrease in the rate of cardiac toxicity (Green et al. 2001). The entire heart is exposed to 12 Gy for whole lung treatment while a portion of the heart including the left ventricle is irradiated by up to 19.8 Gy during irradiation of the left flank or whole abdomen. In the NTWSG study, the cumulative rate of congestive heart failure was 4.4% at 20 years after diagnosis for all patients treated with doxorubicin. When using radiation dose as a continuous variable, they found a statistically significant relative risk of 1.6 to 1.8 per 10 Gy whole lung or left abdomen irradiation but using dose as a categorical variable, neither the 10 to 20 Gy or greater than 20 Gy lung irradiation groups had a statistically significant relative risk compared to no lung irradiation (Green et al. 2001). In a large French–British cohort study of pediatric patients treated for a variety of diseases, those who received 5 to 15 Gy average heart dose had a 12-fold higher long-term risk of death from cardiac disease (Tukenova et al. 2010). Another large Wilms' tumor study found a much smaller radiation-related relative risk for cardiac toxicity risk of 1.05/Gy (van Dijk et al. 2010). Sorensen et al. (1995) were unable to demonstrate a dose–response relationship in their analysis of cardiac function in anthracycline-exposed Wilms' tumor patients.

7.5.4 Second Malignancy

Although doses for Wilms' tumor tend to be less than 22 Gy now, in the past much higher doses were regularly used. Two recent studies with thousands of Wilms' tumor patients showed that the cumulative incidence of a SM by age 40 was 6.8% (Taylor et al. 2008; Breslow et al. 2010). The smaller study also projected a 12% SM rate at the attained age of 50. Other findings of interest were that the SMs tended to occur in the high-dose volume and that since 1990 when more patients were being treated with more intense chemotherapy to avoid radiation, an increase in the risk of leukemia was seen.

References

Attard-Montalto, S. P., J. E. Kingston, O. B. Eden, and P. N. Plowman. 1992. Late follow-up of lung function after whole lung irradiation for Wilms' tumour. *British Journal of Radiology* 65 (780):1114–8.

Bolling, T., I. Ernst, H. Pape, et al. 2011. Dose-volume analysis of radiation nephropathy in children: Preliminary report of the risk consortium. *International Journal of Radiation Oncology • Biology • Physics* 80 (3):840–4.

Breslow, N. E., A. J. Collins, M. L. Ritchey, Y. A. Grigoriev, S. M. Peterson, and D. M. Green. 2005. End stage renal disease in patients with Wilms tumor: Results from the National Wilms Tumor Study Group and the United States Renal Data System. *Journal of Urology* 174 (5):1972–5.

Breslow, N. E., J. M. Lange, D. L. Friedman, et al. 2010. Secondary malignant neoplasms after Wilms tumor: An international collaborative study. *International Journal of Cancer* 127 (3):657–66.

D'Angio, G. J. 2007. The National Wilms Tumor Study: A 40 year perspective. *Lifetime Data Analysis* 13 (4):463–70.

Daw, N. C., D. Gregornik, J. Rodman, et al. 2009. Renal function after ifosfamide, carboplatin and etoposide (ICE) chemotherapy, nephrectomy and radiotherapy in children with Wilms tumour. *European Journal of Cancer* 45 (1):99–106.

Fogliata, A., G. Nicolini, M. Alber, et al. 2007. On the performances of different IMRT treatment planning systems for selected paediatric cases. *Radiation Oncology* 2 (1):7.

Ferrari, A., and M. Casanova. 2005. New concepts for the treatment of pediatric nonrhabdomyosarcoma soft tissue sarcomas. *Expert Review of Anticancer Therapy* 5 (2):307–18.

Fogliata, A., S. Yartsev, G. Nicolini, et al. 2009. On the performances of Intensity Modulated Protons, RapidArc and Helical TomoTherapy for selected paediatric cases. *Radiation Oncology* 4:2.

Gademann, G., and M. Wannenmacher. 1992. Charged particle therapy to pediatric tumors of the retroperitoneal region: A possible indication. *International Journal of Radiation Oncology • Biology • Physics* 22 (2):375–81.

Giel, D. W., M. A. Williams, D. P. Jones, A. M. Davidoff, and J. S. Dome. 2007. Renal function outcomes in patients treated with nephron sparing surgery for bilateral Wilms tumor. *Journal of Urology* 178 (4, Pt. 2):1786–9; discussion 1789–90.

Gommersall, L. M., M. Arya, I. Mushtaq, and P. Duffy. 2005. Current challenges in Wilms' tumor management. *Nature Clinical Practice Oncology* 2 (6):298–304; quiz 1 p following 324.

Green, D. M., Y. A. Grigoriev, B. Nan, et al. 2001. Congestive heart failure after treatment for Wilms' tumor: A report from the National Wilms' Tumor Study group. *Journal of Clinical Oncology* 19 (7):1926–34.

Halberg, F. E., M. R. Harrison, O. Salvatierra, Jr., M. T. Longaker, W. M. Wara, and T. L. Phillips. 1991. Intraoperative radiation therapy for Wilms' tumor *in situ* or ex vivo. *Cancer* 67 (11):2839–43.

Han, J. W., S. Y. Kwon, S. C. Won, Y. J. Shin, J. H. Ko, and C. J. Lyu. 2009. Comprehensive clinical follow-up of late effects in childhood cancer survivors shows the need for early and well-timed intervention. *Annals of Oncology* 20 (7):1170–7.

Hillbrand, M., D. Georg, H. Gadner, R. Potter, and K. Dieckmann. 2008. Abdominal cancer during early childhood: A dosimetric comparison of proton beams to standard and advanced photon radiotherapy. *Radiotherapy and Oncology* 89 (2):141–9.

Hogeboom, C. J., S. C. Grosser, K. A. Guthrie, P. R. Thomas, G. J. D'Angio, and N. E. Breslow. 2001. Stature loss following treatment for Wilms tumor. *Medical & Pediatric Oncology* 36 (2):295–304.

Hong, L., K. Alektiar, C. Chui, et al. 2002. IMRT of large fields: Whole-abdomen irradiation. *International Journal of Radiation Oncology • Biology • Physics* 54 (1):278–89.

Kalapurakal, J. A., J. S. Dome, E. J. Perlman, et al. 2004. Management of Wilms' tumour: Current practice and future goals. *Lancet Oncology* 5 (1):37–46.

Kalapurakal, J. A., D. M. Green, G. Haase, J. R. Anderson, J. S. Dome, and P. E. Grundy. 2010. Outcomes of children with favorable histology wilms tumor and peritoneal implants treated in National Wilms Tumor Studies-4 and -5. *International Journal of Radiation Oncology • Biology • Physics* 77 (2):554–8.

Kalapurakal, J. A., S. M. Li, N. E. Breslow, et al. 2003. Influence of radiation therapy delay on abdominal tumor recurrence in patients with favorable histology Wilms' tumor treated on NWTS-3 and NWTS-4: A report from the National Wilms' Tumor Study Group. *International Journal of Radiation Oncology • Biology • Physics* 57 (2):495–9.

Kalapurakal, J. A., Y. Zhang, A. Kepka, et al. 2012. Cardiac-sparing whole lung IMRT in children with lung metastasis. *International Journal of Radiation Oncology • Biology • Physics* [epub ahead of print].

Kirkbride, P., and P. N. Plowman. 1992. Radiotherapy to the surviving kidney after unilateral nephrectomy in bilateral Wilms' tumour. *British Journal of Radiology* 65 (774):510–6.

Levitt, G. A., E. Yeomans, C. Dicks Mireaux, F. Breatnach, J. Kingston, and J. Pritchard. 1992. Renal size and function after cure of Wilms' tumour. *British Journal of Cancer* 66 (5):877–82.

Nag, S., D. Tippin, S. Smith, C. Bauer, and F. B. Ruymann. 2003. Intraoperative electron beam treatment for pediatric malignancies: The Ohio State University experience. *Medical & Pediatric Oncology* 40 (6):360–6.

Paulino, A. C., B. C. Wen, C. K. Brown, et al. 2000. Late effects in children treated with radiation therapy for Wilms' tumor. *International Journal of Radiation Oncology • Biology • Physics* 46 (5):1239–46.

Sasso, G., N. Greco, P. Murino, and F. S. Sasso. 2010. Late toxicity in Wilms tumor patients treated with radiotherapy at 15 years of median follow-up. *Journal of Pediatric Hematology/Oncology* 32 (7):e264–7.

Smith, G. R., P. R. Thomas, M. Ritchey, and P. Norkool. 1998. Long-term renal function in patients with irradiated bilateral Wilms tumor. National Wilms' Tumor Study Group. *American Journal of Clinical Oncology* 21 (1):58–63.

Sorensen, K., G. Levitt, D. Sebag-Montefiore, C. Bull, and I. Sullivan. 1995. Cardiac function in Wilms' tumor survivors. *Journal of Clinical Oncology* 13 (7):1546–56.

Spreafico, F., and F. F. Bellani. 2006. Wilms' tumor: Past, present and (possibly) future. *Expert Review of Anticancer Therapy* 6 (2):249–58.

Taylor, A. J., D. L. Winter, K. Pritchard-Jones, et al. 2008. Second primary neoplasms in survivors of Wilms' tumour—A population-based cohort study from the British Childhood Cancer Survivor Study. *International Journal of Cancer* 122 (9):2085–93.

Tongaonkar, H. B., S. S. Qureshi, P. A. Kurkure, M. A. Muckaden, B. Arora, and T. B. Yuvaraja. 2007. Wilms' tumor: An update. *Indian Journal of Urology* 23 (4):458–66.

Trobs, R. B., M. Hansel, T. Friedrich, and J. Bennek. 2001. A 23-year experience with malignant renal tumors in infancy and childhood. *European Journal of Pediatric Surgery* 11 (2):92–8.

Tukenova, M., C. Guibout, O. Oberlin, et al. 2010. Role of cancer treatment in long-term overall and cardiovascular mortality after childhood cancer. *Journal of Clinical Oncology* 28 (8):1308–15.

van Dijk, I. W., F. Oldenburger, M. C. Cardous-Ubbink, et al. 2010. Evaluation of late adverse events in long-term Wilms' tumor survivors. *International Journal of Radiation Oncology • Biology • Physics* 78 (2):370–8.

Wallace, W. H., S. M. Shalet, P. H. Morris-Jones, R. Swindell, and H. R. Gattamaneni. 1990. Effect of abdominal irradiation on growth in boys treated for a Wilms' tumor. *Medical & Pediatric Oncology* 18 (6):441–6.

Chapter 8

Soft-Tissue Tumors

(Rhabdomyosarcoma and Other Soft-Tissue Sarcomas)

8.1 Clinical Overview

Approximately 850 to 900 (1.2/100,000) soft-tissue sarcomas (STSs) are diagnosed in children younger than 20 years of age in the United States annually, comprising 7% of all childhood cancer cases. Rhabdomyosarcoma (RMS) constitutes 40% of all STSs and the majority of STSs diagnosed in the first decade of life (Ries et al. 1999). A STS is categorized as RMS if normal striated skeletal muscle differentiation is seen in the tumor. RMS cells are of the "small round blue cell"-type, which is the same cell type as childhood tumors like neuroblastoma, Ewing's sarcoma, and lymphoma. About two-thirds of RMSs are of the more favorable embryonal histology, and one-third are the unfavorable alveolar histologic variant. RMS is the third most common extracranial solid tumor of childhood after Wilms' tumor and neuroblastoma. STSs can occur anywhere in the body. They often present as a painless mass. Other signs and symptoms are associated with their site of origin. Approximately 40% occur in the head and neck, 20% in genitourinary sites, 20% in extremity sites, and the rest in other locations. Of the head and neck sites, 25% (10% of all RMSs) occur in the orbit and 40% (15% of all RMSs) occur in parameningeal sites. Current treatment protocols assign patients to low-, intermediate-, or high-risk groups, which determine their treatment and assumed prognosis. This system is based on the Intergroup Rhabdomyosarcoma (IRS) clinical group and stage, as well as histology and age. Stages 1 through 4 are based on the location, invasion, size of tumor, nodal spread, and presence of distant metastases. Favorable sites include the orbit, nonparameningeal head and neck, paratesticular (<10 years old), gynecological, and biliary tract tumors, whereas unfavorable sites include the extremity, bladder, prostate, trunk, retroperitoneal, and parameningeal tumors. Parameningeal RMSs have a poor prognosis due to their tendency for bony erosion and intracranial spread (Donaldson and Anderson 1997; Crist et al. 2001; Meazza et al. 2006). The IRS grouping system, groups 1 to 4, is based on the degree of surgical resection performed. Development of effective risk-based multimodal therapy in randomized trials have increased long-term survival in RMS from 25% in 1970 to more than 85% in current studies, with stage-specific rates increasing over time (Crist et al. 2001; Punyko, Mertens, Baker, et al. 2004). Comprehensive clinical reviews of RMSs were published by Paulino and Okcu (2007), Gillespie et al. (2006), and Walterhouse and Watson (2007).

There is a large variety of other STSs that are collectively categorized as nonrhabdomyosarcoma (NRMS). The NRMSs are named according to the normal mesenchymal tissue component of the tumor, for example, fat cells in liposarcoma and cartilage in chondrosarcoma. Approximately 75% of STSs in the 10- to 19-year age group are NRMS. Depending on tumor grade and size, NRMS may be treated with a postoperative dose of 55.8 Gy or 45 Gy preoperatively with a 10.8 to 19.8 Gy postoperative boost for positive margins. A current Intergroup Rhabdomyosarcoma Study Group (IRSG) trial is assessing whether selected tumors will be as well controlled with lower radiation doses. Survival rates for NRMS are generally somewhat higher than for RMS (Ries 1999). Spunt et al. (2008) and Ferrari (2005) gave good overviews of NRMS.

All patients receive chemotherapy. The primary tumor is resected if the resultant functional and cosmetic effects are acceptable. Radiation therapy is employed for unresectable or postoperative residual disease. RMS tends to be more sensitive to both radiation and chemotherapy than NRMS. Traditionally, doses for RMS have been 41.4 Gy for microscopic disease and 50.4 Gy for gross disease. Orbital RMS typically receives 45 Gy, and with a status of microscopic residual and node negative or alveolar histology, 36 Gy. Metastases are present at diagnosis in 20% of RMS, usually in the lung. These patients are treated with whole lung irradiation to 15 Gy in 10 fractions (12 Gy for patients 6 years of age or younger) in addition to chemotherapy, resulting in a 25% survival rate. Radiation doses are being tailored to the completeness of surgical resection of the primary tumor and the presence of involved regional lymph nodes. A recent report on local control with reduced-dose radiotherapy for low-risk RMS on Children's Oncology Group (COG) D9602 validated this approach (Breneman et al. 2012). Hyperfractionation versus conventional fractionation was tested in an IRS trial but not found to be beneficial (Donaldson et al. 2001). Raney et al. (2001) has summarized the IRSG protocols I to IV and describes the next protocol, IRS-V.

8.2 External Beam Treatment Planning

8.2.1 Target Volume Definition

If the tumor is unresectable or if radiotherapy (RT) will be given before surgery but after chemotherapy, the gross tumor volume (GTV) is the visible or palpable disease postchemotherapy. Otherwise, the GTV is the visible or palpable disease at diagnosis, and any residual tumor after induction chemotherapy. After total resection, the "GTV" should also include the margin of the resection cavity and any surfaces that were originally in contact with the tumor prior to surgery. These surfaces may have moved after surgery as organs and tissues displaced by the tumor return to their normal positions. The GTV should not include the original spatial location of tumor volume but should track the contaminated surfaces at their location at the time of radiotherapy. It also includes the entire draining lymph node chain only if involved with the tumor. The clinical target volume (CTV) is the GTV plus 1 cm (plus 1.5 to 2 cm for NRMS) but can be modified to account for anatomical barriers to tumor spread. The planning target volume (PTV) is typically the CTV plus 0.5 to 1.0 cm, depending on site and immobilization techniques. A cone down is sometimes performed for noninvasive tumors where there is a large rapid decrease in tumor size after 36 Gy (or 45 Gy for NRMS) and has been shown to produce equivalent local control and survival as treating the large volume for the entire treatment (Chen 2003). The volume for the final 14.4 Gy or (up to 19.8 Gy for NRMS) may be a smaller GTV2, CTV2, and PTV2 representing tumor volume following response to chemotherapy. Table 8.1 summarizes the volumes, margins, and prescribed doses for each disease.

Using CT to MRI image fusion-based planning resulted in excellent local control using intensity-modulated radiation therapy (IMRT) with 1.0 cm CTV and 0.5 cm PTV margins for head and neck RMS (Wolden 2005; McDonald et al. 2008). Two centimeter anatomically constrained CTV margins plus no more than 1 cm PTV margin have been used for NRMS of the extremities without finding marginal failures instead of the 5 cm margin often recommended in adult sarcoma trials (Krasin et al. 2010).

Site-specific considerations for STS have been developed to optimize the response to therapy and reduce normal tissue toxicity. For bladder, prostate, perineum, pelvis, biliary, and abdominal tumors, surgical debulking is recommended before radiation therapy. For extremities, it is advisable to avoid circumferential irradiation of extremity lymphatics and treating across a joint. Somewhat larger CTV margins are also used in this situation. For

TABLE 8.1 Clinical and Planning Target Margins and Prescribed Dose for Ewing, Rhabdomyosarcoma, and Nonrhabdomyosarcoma

Diagnosis	Role of RT	Target Margin	Prescribed Dose
Ewing family tumors	Definitive EBRT	GTV = postchemotherapy tumor volume	Small primary (tumor <80 mm):
		CTV = GTV+10 mm	45 Gy to PTV1
		PTV1 = CTV+3–7 mm	55.8 Gy to PTV2
		PTV2 = GTV+3–7 mm	Large primary (tumor ≥80 mm): 45 Gy to PTV1 64.8 Gy to PTV2
	Postoperative EBRT	GTV = postoperative bed	50.4 Gy to PTV
		CTV = GTV+10 mm	
		PTV = CTV+3–7 mm	
	Postoperative brachytherapy	GTV = postoperative bed	40 Gy low dose rate over 5 days to CTVb
		CTVb = GTVb +10 mm lateral margin and 20 mm proximal/distal margins	
Rhabdomyosarcoma	Definitive EBRT	GTV = postchemotherapy tumor volume	36.0–41.4 Gy to PTV 1
		CTV = GTV+15 mm	45.0–50.4 Gy to PTV2
		PTV1 = CTV+3–7 mm	
		PTV2 = GTV+3–7 mm	
	Postoperative EBRT	GTV = postoperative bed	36–41.4 Gy to PTV
		CTV = GTV+15 mm	
		PTV = CTV+3–7 mm	

Nonrhabdomyosarcoma soft-tissue sarcoma	Definitive or preoperative EBRT	GTV = postchemotherapy tumor volume	Definitive: 66–70 Gy to PTV
		CTV = GTV+20 mm	Preoperative: 45.0–50.4 Gy to PTV
		PTV = CTV+3–7 mm	
	Postoperative EBRT	GTV = postoperative bed	63.0–66.6 Gy to PTV
		CTV = GTV+20 mm	
		PTV = CTV+3–7 mm	
	Adjuvant brachytherapy and EBRT	GTV = postoperative bed	15–20 Gy from brachytherapy to CTVb
		CTVb = GTV+10 mm lateral margin and 20 mm proximal/distal margins	45–50.4 Gy to PTV1 (from pre- and postoperative EBRT)
	Postoperative brachytherapy alone	GTV = postoperative bed	40–45 Gy low dose rate over 5 days to CTVb
		CTVb = GTV+10 mm lateral margin and 20 mm proximal/distal margins	

Source: Hua, C., et al., *International Journal of Radiation Oncology • Biology • Physics* 70 (5):1598–1606, 2008.

Notes: CTV, clinical target volume; CTV1, primary-phase clinical target volume; CTVb, clinical target volume for brachytherapy; EBRT, external-beam radiotherapy; GTV, gross tumor volume; PTV, planning target volume; PTV1, primary-phase planning target volume; PTV2, boost-phase planning target volume; RT, radiotherapy.

orbital tumors, only the tumor volume with margin needs irradiation while sparing the lens, cornea, lacrimal gland, and optic chiasm. The CTV should not extend outside the bony orbit unless involved with the tumor. For intrathoracic tumors that displaced a large amount of lung parenchyma that returned to normal position after surgery, the CTV should not include normal lung that was not in contact with the presurgical tumor and migrated back into normal anatomic location in the region that was debulked.

Immobilization and motion management are important considerations in the treatment of STS, and especially challenging because these tumors can occur practically anywhere in the body. With knowledge of and, better still, control of the target motion, the smaller the margin on the CTV, the smaller each field can be, and the smaller the volume of normal tissue needlessly irradiated. RMS can be found in and around the chest and in the subdiaphragmatic region where motion due to breathing must be controlled. 4DCT methods should be employed to at least determine the internal target volume (ITV). In the absence of 4DCT capability, fluoroscopy can be used to watch and approximately measure the movement of the bony anatomy near the tumor volume. Techniques already used in adult cases can be adapted for pediatric use.

In most cases, the supine patient position is preferred. For cases in the head and neck, head fixation with either a mouthpiece-based system (preferred) or mask can provide daily reproducibility of 1 to 3 mm if carefully utilized. If the target volume extends to the low neck, shoulder fixation may be necessary, which can be managed by either a head and shoulder thermoplastic mask or facemask with upper body fixation by vacuum or foam bag. Body immobilization by vacuum or foam bag is also advantageous for treatments in the trunk and extremities. Daily reproducibility of 3–4 mm in any one direction can be achieved for sites below the neck with careful application of the immobilization devices. Frequent kilovoltage (kV) or cone beam imaging (if available) should be employed to ensure at least this level of positional accuracy.

8.2.2 Treatment Planning

Treatment planning for STSs can be a significant challenge due to the concern for late effects for normal tissues near or inside the target volume combined with the relatively high radiation doses required, especially for NRMS. In the head and neck region, xerostomia, dental abnormalities, facial growth retardation, neuroendocrine dysfunction, and visual and hearing abnormalities can occur. In the pelvis, musculoskeletal growth delay, bowel obstruction, and gonadal dysfunction can occur. In the extremity, fractures, growth inhibition, fibrosis, atrophy, peripheral nerve damage, and loss of range of motion can occur. For all of the aforementioned, the combined effects of radiation, surgery, and chemotherapy come into play. Note that many of these side effects are not of significant concern with the adult patient. With such a comprehensive list of potential side effects, one has to ask how can we dare irradiate the child and obtain good quality of life years later should the disease be cured. Modern radiotherapy techniques will make a substantial reduction in side effects compared to those children treated with earlier technology, but wise use of the technology can at best mitigate the late side effects that are inevitable with the high dose and typically large volume irradiation of normal growing tissues in children. Merchant (1997) gave an overview of the use of conformal techniques for pediatric sarcomas. More recently, IMRT (including volumetric modulated arc therapy [VMAT]) has been used extensively for these tumors, and reports indicate excellent local control can be achieved while minimizing late effects (Chan et al. 2003; Wolden 2005; Puri et al. 2006; Combs et al. 2007; Hua et al. 2008; McDonald et al. 2008; Sterzing et al. 2009; Matuszak et al. 2010). VMAT results in very similar dose–volume histograms (DVHs) as standard IMRT but with much shorter treatment time and about one-third fewer monitor units (MUs) (Matuszak et al. 2010).

The planning process starts with CT treatment planning (with MRI fusion when needed) with the patient in the custom immobilization device and no more than 3 mm CT slice spacing. In general, treatment planning objectives should include coverage of at least 95% of the PTV with at least 95% of the prescribed dose for the primary volume and even better coverage for the boost when given. Rarely should the volume receiving greater than 110% of the dose be greater than 10% of the PTV (Hua et al. 2008). When the PTV is adjacent to a parallel-type normal structure (such as the kidney, liver, and brain), depending on the anatomy, the adjacent aspect of the PTV may have to be underdosed by about 5% to 10% to stay below the dose–volume tolerance. When the adjacency is with the spinal cord or optic chiasm, a larger indentation of the PTV dose may be needed, especially for NRMS cases. In 31% of patients in the study by Hua et al. (2008), due to the proximity of critical structures to the target, the primary phase CTV was not completely covered by 95% of the prescribed dose. It is unclear as to the consequence of this small volume underdose. Organs at risk (OARs) may include many structures not normally contoured in adult sarcomas, such as various bones and soft tissues. Hua et al. (2008) gave an extensive list of normal structures contoured and their dose–volume statistics for various STS sites in 72 patients. When a boost is planned, the composite dose distribution of the primary and boost should be summed and evaluated before giving the first day's treatment. This avoids the situation where some or the entire primary volume dose has been given and then the addition of the boost dose reveals that tolerance doses will be exceeded. It should be noted that the dose to the primary volume will be greater than planned and prescribed once the boost dose is summed due to the contribution of the dose from the boost. Simultaneous integrated boosts do not have this problem but choosing appropriate daily doses can be challenging depending on the disparity of the primary and boost doses.

Although we do not see the typical adult head and neck, breast, and prostate tumors in children, STS in these sites is the pediatric version of these tumors and they can be planned with similar strategies as in the adult cases. In the head and neck, the most common location for primary RMS, the tumor can invade the brain (parameningeal RMS) or the orbit (orbital RMS), along with the rest of the regional structures. Because of the anatomical constraints to a complete surgical excision of the tumor in these locations, a large tumor volume that is intimate with many normal structures is often what is presented to the treatment planning team. Michalski et al. (2004) reviewed the patients treated on IRS II–IV and found no benefit for whole brain radiation, even for patients with intracranial extension, and the current IRSG study eliminated the use of whole brain radiotherapy. It was also found that a dose of at least 47.5 Gy seemed to be associated with lower rates of local failure, especially for the larger tumors. Figure 8.1 shows a parameningeal RMS that infiltrates the brain, sinus, oropharynx, and includes or abuts the optic nerves, chiasm, and retina. A nine-beam noncoplanar IMRT plan was used to treat this (Figure 8.2). A head fixation system with vacuum-assisted mouthpiece was used. Although most head and neck tumors are

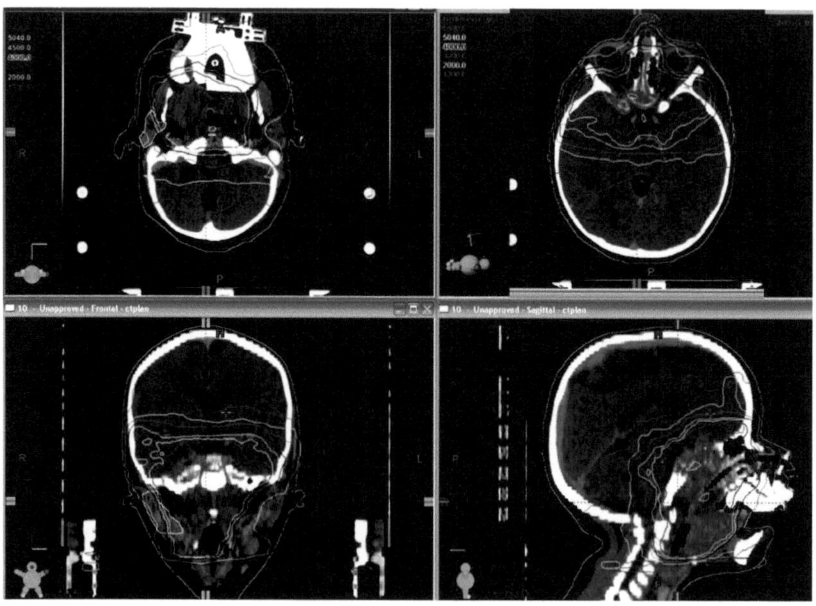

FIGURE 8.1 Parameningeal RMS dose distribution.

Beam Name	IEC Gantry Angle (deg)	IEC Couch Angle (deg)
RSO	270	40
RPO	265	0
RAO	300	0
RASO	340	20
LASO	25	320
LAO	55	0
LPO	95	0
LSO	90	320
ANTSAG	310	90

FIGURE 8.2 Parameningeal RMS plan: nine noncoplanar beam angles.

treated with coplanar beams, if the PTV is not too large nor extends too far inferiorly, noncoplanar beams can provide a superior dose distribution to coplanar beams without irradiating unnecessarily large volumes of normal tissue inferior to the target. Organs at risk drawn for this case were the optic nerves, optic chiasm, retina, lenses, pituitary, cochlea, parotids, spinal cord, and brain. Although the prescribed dose, 50.4 Gy, is well below the tolerance dose of the optic nerves and chiasm (54 Gy), one must be careful to avoid an 8% or greater hot spot occurring in those structures, which would cause their tolerance dose to be exceeded. The whole retina can tolerate 45 Gy and there are data that suggest that higher doses to less than 60% of the retina are tolerated without clinical manifestations of visual damage (Takeda et al. 1999). In this case, the PTV was adjacent to both retinae with a maximum dose of 50 Gy being allowed. Due to the proximity of the lenses to the PTV along with the prescribed dose of 50.4 Gy, the lenses receive greater than 10 Gy in this plan. In similar cases, it may be possible to restrict the lens dose to less than 10 Gy, but with significant compromise of target dose homogeneity. Proton therapy may be better able to avoid cataracts in a case like this. Because the PTV either wholly or partially included the pituitary gland and both cochlea, meaningful sparing of these structures was not attempted. As seen in Figure 8.1, the PTV overlaps the right parotid but is medial to the left parotid. The mean right parotid dose (35 Gy) exceeded the goal of 26 Gy (adult parotid dose tolerance guidelines; we do not have good pediatric parotid dose tolerance data), but the mean left parotid dose was about 17 Gy. This child should retain left parotid salivary production. The PTV abutted the vertebral bodies providing the necessary distance to keep the spinal cord well below 45 Gy. Because this RMS invades the brain, one of the treatment planning goals should be to minimize the volume of brain receiving more than 20 Gy to reduce the risk of long-term cognitive deficits. The entire brain can be contoured, but that volume is huge compared to the volume of brain actually adjacent to the PTV, resulting in less sensitive control of the DVH during optimization. One strategy that works well is to expand the PTV by 4 cm, then use Boolean operations to create a new volume that is the intersection of the expanded PTV and the brain, then crop this reduced brain volume from the actual PTV by about 1 cm. This gives the brain volume most likely to incur the dose spread but leaves an unconstrained region next to the PTV to avoid indentation of the prescribed dose. In this case, just 15% of the whole brain received more than 20 Gy. Combs et al. (2007) and Hua et al. (2008) described their treatment techniques and clinical experience with IMRT for head and neck RMS patients.

Periorbital RMS originates in the orbital region but can also invade adjacent brain and soft tissues. Figure 8.3a shows an extensive tumor that pushes the right eye forward while also invading the tissues along the facial muscles of the right side of the face and inferiorly through the floor of the right orbit to the sinus and below that to the upper palate and soft tissues surrounding the mandible with involvement of the right parotid gland. Many of the same normal structures as the parameningeal case were at risk. Note

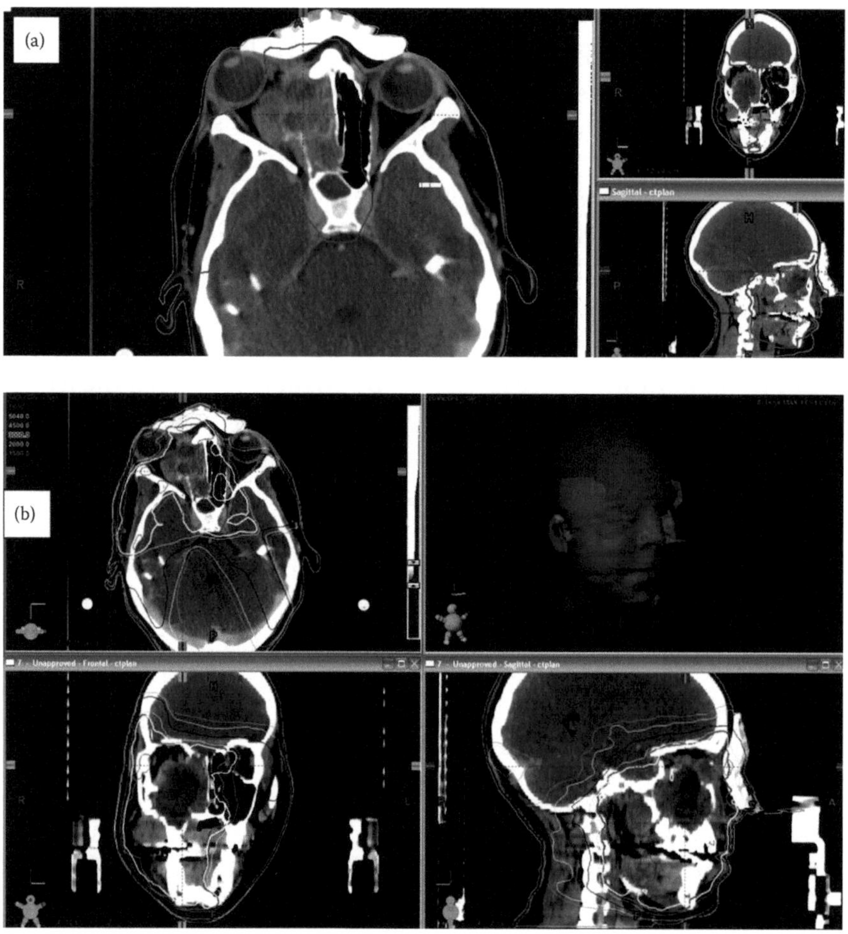

FIGURE 8.3 (See color insert.) Periorbital RMS. (a) PTV in three plane views and 3D view. (b) Isodoses in three views.

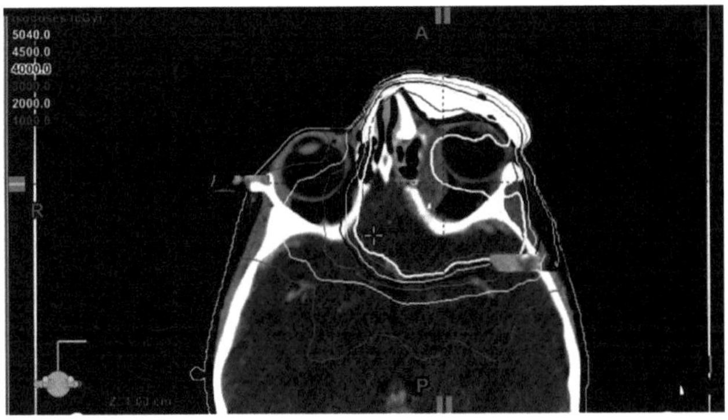

FIGURE 8.4 (See color insert.) Reducing the retinal dose to below 50 Gy using a nine-beam noncoplanar IMRT beam arrangement with lower dose constraint for the retina.

that the mouthpiece was not able to be used in this case because of tumor extension to the upper palate. A plastic nasofrontal fixation plate was customized to be conformal to the shape of the patient with the use of dental impression material (Figure 8.3). Figure 8.3b shows the isodoses distribution in the axial, coronal, and sagittal planes. Figure 8.4 shows a case where the PTV envelopes the left eye and invades the brain but is otherwise much more localized than in the case of Figure 8.3. The grossly visible tumor is seen adjacent and medial to the left eye. Custom bolus made with dental impression material is seen nicely conforming to the contours of the orbital and nasal region and is separate from the head fixation system, which uses a vacuum-assisted mouthpiece. An orbital RMS that invades the brain is treated like a parameningeal tumor. For this patient, the physician wanted to limit the retinal dose to 45 Gy. Figure 8.4 shows the yellow 50.4 Gy isodoses line indenting in around the left eye to produce this relative sparing (less than 10% of the retina received 45 Gy, 1% received 50 Gy). For these tumors, extension to or near the skin surface is common, warranting the addition of bolus as seen in Figure 8.4.

RMS of the breast is a rare occurrence and of bilateral breasts even more so. In either case, adult patient planning techniques can be applied. One difference may be that for RMS, tumor cells may be suspected to have infiltrated the skin, necessitating the application of bolus over the entire intact breast, as seen in Figure 8.5. In this case, the child had bilateral RMS with just the right breast having gross disease remaining after chemotherapy. The right breast was prescribed 50.4 Gy and the left breast, 41.4 Gy. IMRT can be used as a two-dimensional wedge with opposing tangential beams to provide a homogeneous dose in three dimensions. Figure 8.5a shows an 18-field (pseudovolumetric arc) plan that avoids the potential hot spot at the medial match of bilateral tangential beams at the expense of a large volume of lung, 55%, receiving 10 Gy even though only 7% of the total lung gets

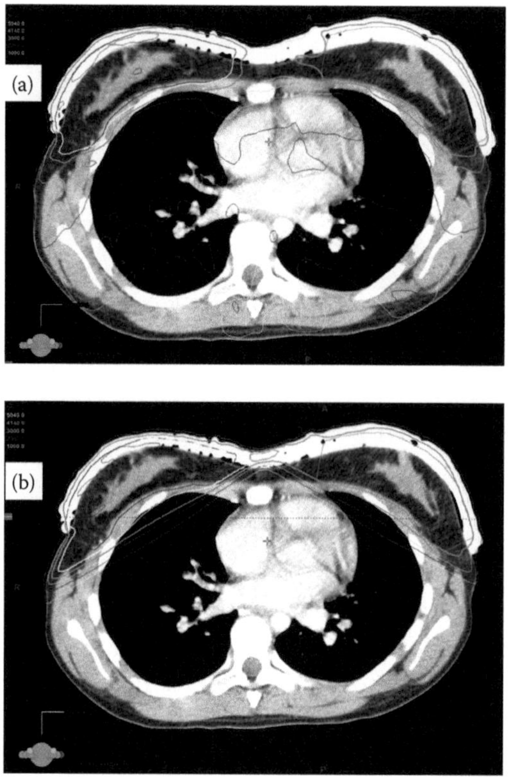

FIGURE 8.5 RMS of bilateral breasts, 50.4 Gy to right breast, 41.4 Gy to left breast. (a) 18-beam IMRT. Total lung 20 Gy to 7%, 10 Gy to 55%. (b) Bilateral tangential IMRT fields.

20 Gy. Figure 8.5b shows bilateral tangential IMRT fields that match at the sagittal midline without creating a hot spot at that junction. In most cases, the tangential beam plan would be preferred unless there were other associated target volumes that could not be covered with tangents.

RMS arising in the prostate or bladder is more frequently seen than in the breast. Figure 8.6 shows a coronal and axial CT of a large mass in the prostate, a completely different appearance than in adults. Here, a much larger volume is involved although a much smaller dose is needed than in adult prostate cancer. Nodal disease adjacent to the primary can create an irregular PTV that IMRT techniques can accommodate. As in adult prostate cancer, the bladder, rectum, and femoral heads are organs at risk, although the relatively lower dose of 50.4 Gy makes this an easier planning exercise. Figure 8.7 shows the dosimetry for two prostate RMS cases: the first is for the case shown in Figure 8.6 for which this large volume was treated to 45 Gy. Protection of the bony pelvis and femoral heads was the primary objective. The second case was treated to 50.4 Gy and the rectum, bladder, and left femoral head were organs at risk. Both were treated with nine-beam coplanar IMRT plans. In the case of Figure 8.7b, nodal disease was also

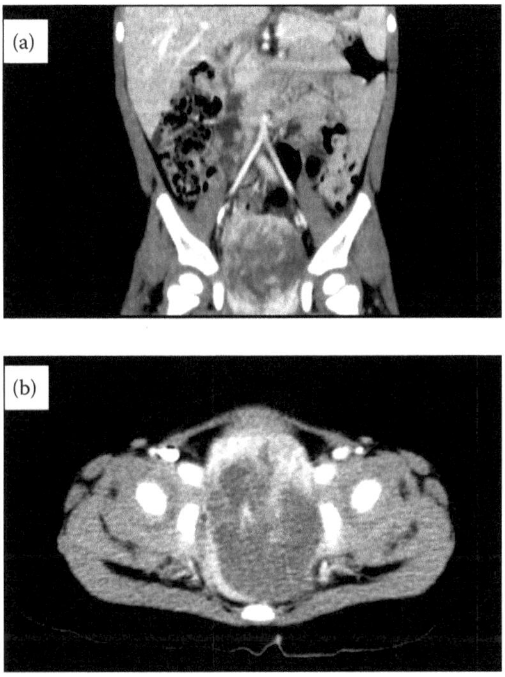

FIGURE 8.6 Prostate RMS. (a) Coronal CT. (b) Axial CT.

targeted along with the primary tumor. Due to uncertainty in the location of the CTV relative to the bladder, the CTV and PTV were drawn to include a large volume of the bladder. During the planning process, dose was carved out around the three critical structures while adequately covering the PTV.

Due to the fact that RMS arises in soft-tissue compartments that can occupy a large space and invade adjacent spaces, rather extensive tumors can be seen. Figure 8.8 shows a RMS of the abdomen and pelvis, wrapping around the right lateral aspect of the bladder and adjacent to both kidneys. The tumor also approaches the right femoral head and the testes. The length of this PTV is over 50 cm long, necessitating two isocenters and a matched dose at the junction. The superior and inferior portions of the PTV were separate overlapping structures. A nine-field coplanar plan was calculated for the superior half of the total PTV. For the inferior half, the same nine-field arrangement was used and the dose grid for the superior half was used as a starting point by the optimization program. This effectively created a feathered match in the overlapping region with reasonable dose uniformity. A large percentage of the surface of the bladder was in contact with the PTV, but with a 50.4 Gy tumor dose, well below bladder tolerance, acceptable late toxicity can be expected. Only about 30% of each kidney received more than 15 Gy, 15% of the bladder more than 40 Gy, and only a few percent of the right femoral head received more than 45 Gy. Organ-at-risk dose recommendations are given in Table 8.2.

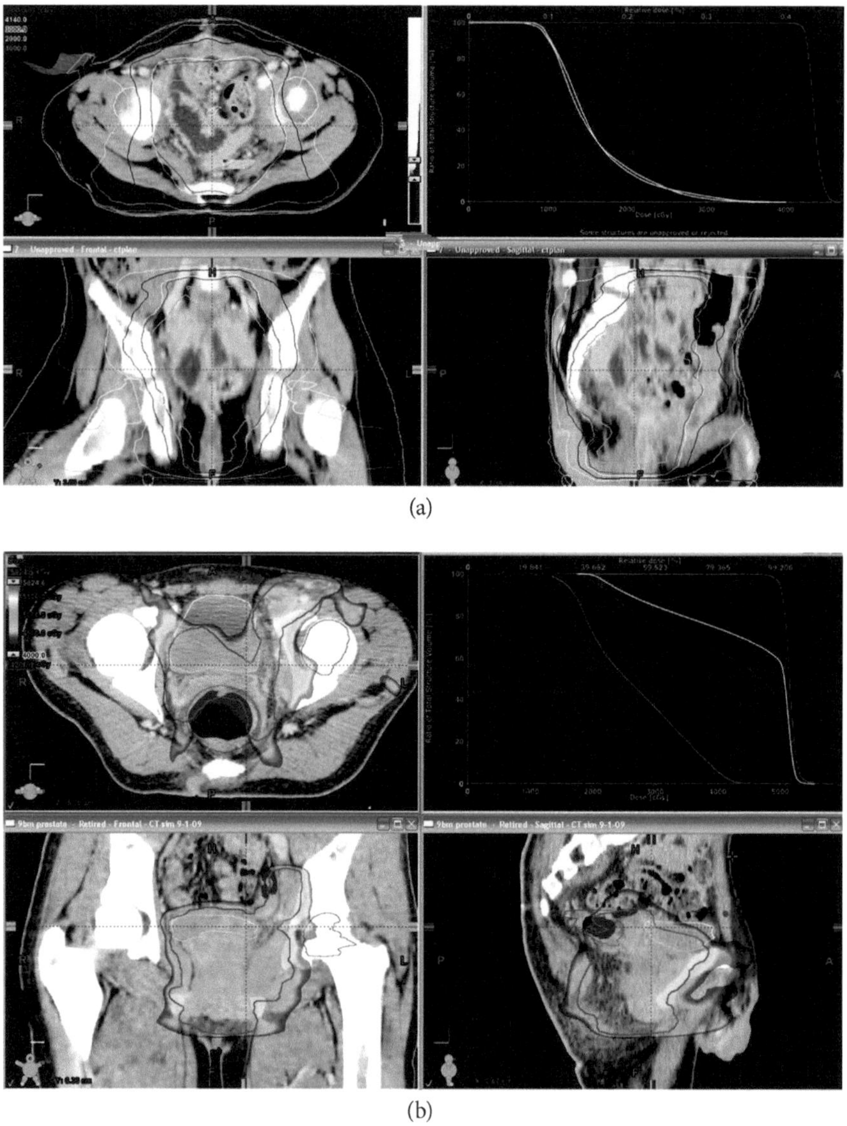

FIGURE 8.7 (See color insert.) (a) Prostate RMS dosimetry. 45 Gy prescribed dose. DVH: red, PTV; yellow, femoral heads. (b) Prostate RMS adjacent to bladder. 50.4 Gy prescribed dose. DVH: green, L femur; purple, rectum; yellow, bladder; red, PTV.

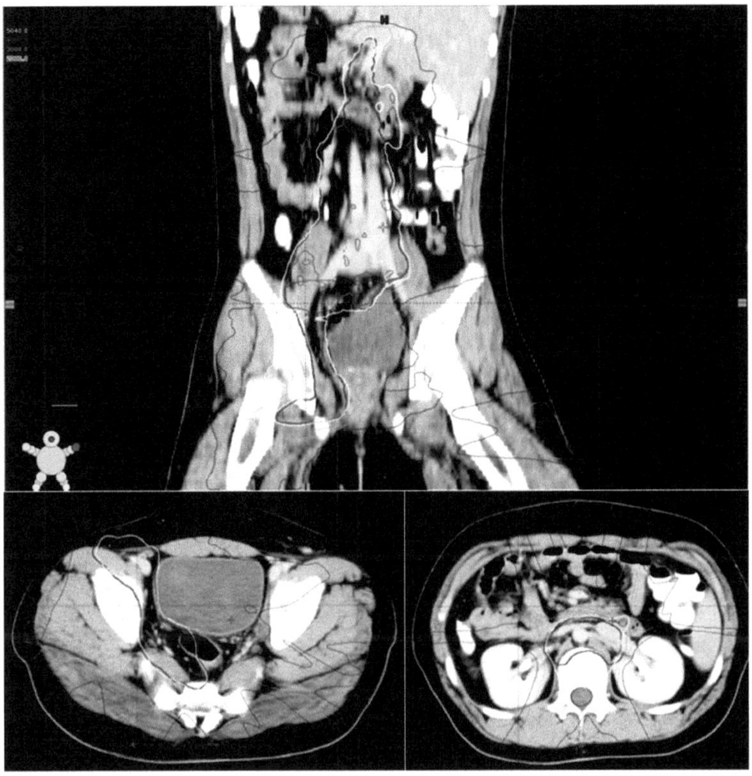

FIGURE 8.8 (See color insert.) Extensive abdominal and pelvic RMS, coronal CT shows extent while axials show target (red) wrapping around bladder and adjacent to both kidneys. Isodoses: red, 55 Gy; yellow, 50.4 Gy (prescribed); orange, 40 Gy; green, 30 Gy; blue, 20 Gy.

TABLE 8.2 Organ-at-Risk Dose–Volume Recommendations

Organ	Volume (%)	Dose (cGy)
Bladder	100	4500
Heart	100	3000
Liver	100 50	2340 3000
Rectum	100	4500
Optic chiasm/nerves	100	5400
Small bowel	50	4500
Spinal cord	Max	5000
Bilateral kidney	50 100	2400 1440
Bilateral lung	20 100	2000 1500
Lens	100	1000
Lacrimal gland	100	4140

8.3 Brachytherapy and Intraoperative Electron Beam (IORT)

Although most STS patients are treated with external beam radiotherapy, more localized forms of radiotherapy have been utilized to both treat the tumor and protect the many intervening normal tissues. Because of the infiltration of RMS around critical structures often with limited ability to completely resect the tumor, brachytherapy and IORT procedures have been performed during or after the limited surgical procedure as either definitive treatment or as an adjunct to external beam radiation therapy. Brachytherapy, both high- and low-dose rate, has been used for vaginal, vulvar, uterine, extremity, head, and neck; and IORT treatments have been used in many sites in the chest, abdomen, extremities, and pelvis for both primary tumors and lymphadenopathy. Also, IORT and brachytherapy methods can be useful for treatment of very young children to delay the start of external beam radiotherapy until a later age. Fairly large tumors (up to about 20 cm diameter) can be treated with these intraoperative methods. COG protocols for RMS and NRMS allow these types of treatments. Table 8.3 compares IORT and brachytherapy techniques.

IORT is frequently, but not always, delivered as a boost to external beam radiotherapy with electron beams of between 6 to 20 MeV and dose of 10 to 20 Gy (Calvo et al. 1989; Haase et al. 1994; Schomberg et al. 1997; Nag, Tippin, Smith, et al. 2003; Oertel et al. 2006). It is used at the time of resection, whether or not there is a gross total resection. Frequently, IORT is used for retreatment when the volume irradiated by external beam treatment would be too morbid. A single dose of 15 Gy is radiobiologically equivalent to about 31 Gy for the tumor, much less than the 50.4 Gy usually needed, but the late reacting normal tissue equivalent is about 55 Gy in 2 Gy fractions. The ability to shield or retract normal structures from the beam provides the needed sparing. Local control is generally good but late morbidity is seen in all studies.

Both low- and high-dose-rate brachytherapy has been used for pediatric STS. The Institut Gustave Roussy published a large series of 45 pediatric RMSs treated with Ir-192 implants predominantly in the pelvis, perineum, and head and neck. Either interstitial or intracavitary applicators were used to deliver 60 to 65 Gy over 5 to 7 days. Most patients had only a limited surgical resection and nearly all had chemotherapy prior to brachytherapy. External beam was only used for nodal disease. The prechemotherapy volume was implanted unless it was very large, in which case just the residual tumor volume was treated. With 5 years mean follow-up, 78% of the patients were alive without disease. The late complication rate was 18%, all severe, occurring in early patients who received over 65 Gy or over 60 Gy to a very large volume. They suggest that limiting brachytherapy to just the

TABLE 8.3 Comparisons of IORT and Brachytherapy Techniques

Parameter	IOHDR	IOERT	LDR	Fractionated HDR	Permanent I-125
Treatment time	Min	Min	2–7 days	6–8 days	Months
Personnel radiation hazard	Nil	Nil	High	Nil	Low
Typical doses used	10–15 Gy	10–15 Gy	45 Gy	36 Gy	140–160 Gy
Sites	Any	Limited	Any	Any	Deep
Dosimetry[a]					
Surface dose	200%	80%	N/A	N/A	N/A
2 cm depth dose	30%	90%	N/A	N/A	N/A
Radiobiology					
Reoxygenation	None	None	Yes	Yes	Yes
Redistribution	None	None	Yes	Yes	Yes
Repair of SLD	None	None	Yes	Yes	Yes
Rx of micro disease	With EBRT	With EBRT	Alone or with EBRT	Alone or with EBRT	Alone or with EBRT
Rx of gross disease	Limited	Limited	Alone or with EBRT	Alone or with EBRT	Alone or with EBRT
Rx in previous RT field	Limited	Limited	Yes	Possible	Yes

Source: Nag, S., et al., *Medical & Pediatric Oncology* 40 (6):360–6, 2003.

Notes: IOHDR, intraoperative high-dose-rate brachytherapy; IOERT, intraoperative electron beam radiotherapy; LDR, low-dose-rate brachytherapy; fractionated HDR, fractionated high-dose-rate brachytherapy; EBRT, external beam radiotherapy; N/A, not applicable; SLD, sublethal damage; Rx, treatment.

[a] Dosimetry profile for a 6 MeV electron beam prescribed to 90% isodose at 0.5 cm depth. The dose prescription for brachytherapy techniques is 100% at 0.5 cm depth.

residual disease would be preferable to lessen late effects (Gerbaulet et al. 1985). Mold-based brachytherapy has been used for primary or recurrent head and neck and orbital tumors (Blank et al. 2009, 2010). After chemotherapy and radical tumor resection, a layered mold was constructed to fit into the surgical defect. Flexible catheters were fitted between the layers for afterloading Ir-192 seeds. Low-dose-rate or pulsed-dose-rate remote afterloading was used to deliver 40 to 50 Gy. This is followed by reconstruction of the surgical deficit. Overall survival was similar to that reported for external beam cases but without the severe fibrosis and asymmetric growth problems. Low-dose-rate brachytherapy has also been performed for prostate or bladder neck RMS in boys (Martelli et al. 2009). Sixty Gy in 6 days was delivered after a conservative surgical procedure that preserved the bladder, neck and urethra, and avoids total cystectomy or prostatectomy and external beam radiotherapy. The median age of these children was just 23 months. Twenty-four of 26 boys were alive after 4 years follow-up and only 1 patient subsequently had bladder dysfunction requiring total cystectomy. The rest of the patients had normal or at least acceptable degrees of continence. Merchant described brachytherapy treatment of 31 children with NRMS in extremities, pelvis, axilla, shoulder, and other sites. They used brachytherapy alone (50 Gy) for high-grade tumors that are completely resected and in combination with external beam radiotherapy (25 Gy brachytherapy plus 45 Gy external beam) for high- and intermediate-risk incompletely resected tumors. Catheters for Ir-192 seeds were placed at the time of the surgical procedure with extension 2 cm beyond the tumor bed and 10 Gy/day was delivered at 0.5 cm from the planes(s) of the implant. The most common side effects were wound problems, fibrosis, and cellulitis. Altogether, there were two local and two regional failures and the median survival for the 25 surviving patients was about 3 years (Merchant et al. 2000).

Where intraoperative electrons cannot access certain sites, such as narrow cavities and the base of the skull, sidewalls of the pelvis, abdomen and thorax, intraoperative high-dose-rate (IOHDR) brachytherapy has also been used for pediatric sarcomas (Merchant et al. 1998; Nag et al. 2001; Goodman et al. 2003). IOHDR has been given in doses of 10 to 12 Gy for microscopic disease and 12 to 15 Gy for gross disease followed by 27 to 30 Gy external beam radiotherapy. IOHDR applicators can be flexible, rigid, or molds. Catheters are placed 1 cm apart and covered volumes that ranged from 4 to 96 cm^3 (Nag et al. 2001). Nearby normal tissues were excluded from treatment by retraction or shielded with pliable lead sheets. These studies suggest that IOHDR is a useful adjunct to external beam radiotherapy and provides good local control and survival with fewer side effects than external beam treatment in selected patients.

Fractionated HDR brachytherapy (HDRBT) has also been used by Nag, Tippin, and Ruymann (2003) to take advantage of reoxygenation and redistribution of residual cancer cells by delivering 36 Gy in 12 fractions. Long-term morbidity of this treatment in 15 patients was reviewed. External beam radiotherapy was not part of the primary treatment although 4 patients needed it for salvage. After a median follow-up of 10 years, 12 patients (80%) were alive without evidence of disease. Four patients (20%) had grade 3 to 4 brachytherapy-related late effects including trismus, osteonecrosis, vaginal stenosis, and periurethral fibrosis. Viani et al. (2008) reported on 18 children treated with interstitial or intracavitary HDRBT for pediatric STS. Eight were treated with HDRBT alone (21 to 30 Gy) and 10 with HDRBT (18 to 24 Gy) plus external beam radiotherapy (41.4 to 50 Gy). Treatment sites included the extremities, head and neck, and pelvis. The dose was delivered in 3 to 5 Gy fractions twice per day, 8 hours apart, 0.5 cm beyond the plane(s) of the implant. HDRBT alone was used after complete tumor resection with negative margins, and combined therapy was used otherwise. Five-year overall survival was 84.5% and the local control rate was 94.5% with 80 months median follow-up (Viani et al. 2008).

Permanent I-125 seeds have also been used for pediatric STS, as reported by Healey et al. (1995) for 15 patients, including patients with brain tumors, sarcomas, neuroblastoma, hepatoblastoma, and adenocarcinoma of the pancreas. Low activity permanent I-125 seeds spaced at 1 cm intervals in vicryl suture were directly sewn into target tissue with the seeds extended at least 1 cm beyond the clinical target volume. Strand spacing was 0.75 to 1.0 cm, depending on seed activity. Forty to 50 Gy was delivered. All of six evaluable patients were locally controlled with no long-term complications (Healey et al. 1995).

8.4 Proton Therapy

Proton therapy has been reported for pediatric periorbital (Hug et al. 2000; Miralbell et al. 2000; Yock et al. 2005), parameningeal RMS (Kozak et al. 2009) and other STS sites (Lee et al. 2005; Timmermann et al. 2007; Fogliata et al. 2009). Nearly all of these papers compare proton therapy dose distributions with either 3DCRT or IMRT plans. For the periorbital patients, proton therapy was able to reduce the dose to the lacrimal glands, lenses, and pituitary gland below that which is thought to cause dysfunction. Hug et al. (2000) described a multiple field patch technique to produce the normal tissue sparing. Yock et al. (2005) reported on five additional patients treated at the same institution and compared proton plans to 3DCRT plans, calculating the dose spared to normal structures with the proton plan, noting

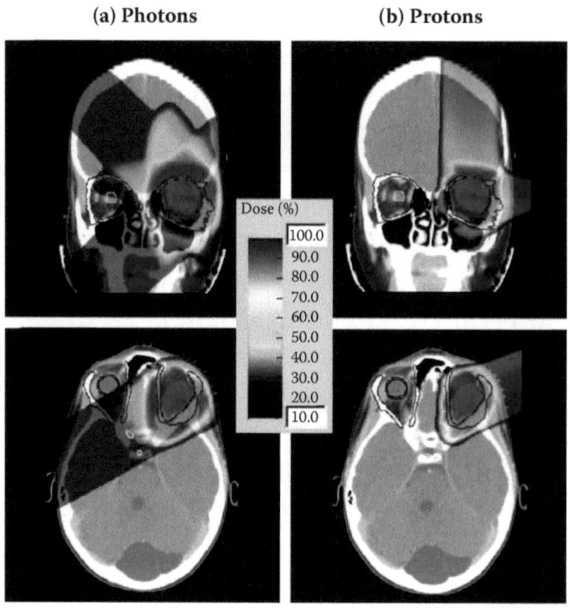

FIGURE 8.9 Protons versus 3DCRT photons for orbital RMS. (a) Coronal and axial photon plan, (b) coronal and axial proton plan. (From Yock, T., et al., *International Journal of Radiation Oncology • Biology • Physics* 63 (4):1161–8, 2005. With permission.)

substantial sparing (Figure 8.9). Six of 6 patients treated with intact orbits had good vision in the treated eye although they all had mild to moderate bony asymmetry. None had any neuroendocrine effects from proton therapy. The contralateral structures were already well spared with photons but the ipsilateral lens, orbital bone, and lacrimal gland received a lower dose with protons that could translate into clinical benefit. If IMRT had been used, it is likely that only the lens would have been spared to a clinically significant level. For the parameningeal proton paper, the pituitary/hypothalamus and ipsilateral parotid were the only sites that could receive a clinically relevant lower dose with intensity-modulated proton treatment (IMPT) than IMRT. The other sites were well protected with either technique (Figure 8.10). In most other cases where protons were compared to IMRT, the proton plans almost always produced plans with nontarget doses less than the IMRT plans but since both were able to keep normal organ doses well below tolerance, this dose reduction may not have any clinical benefit to the patient. In many cases, the IMRT plan that was compared could have been a better plan. A potential benefit of otherwise clinically insignificant reductions in normal tissue doses is the ability to escalate the tumor dose should that be deemed beneficial. In fact, the ability to conformally treat the target and spare contralateral tissues, by any technique, may cause asymmetric growth of bones and soft tissues compared to less conformal techniques (Kozak et al. 2009).

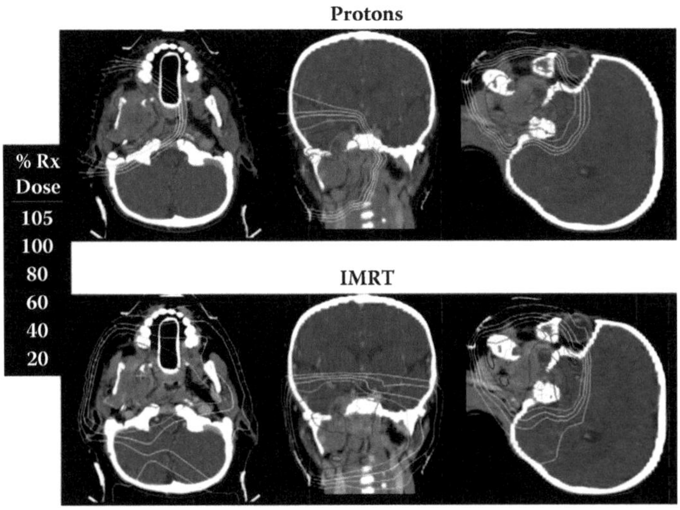

FIGURE 8.10 (See color insert.) Protons versus IMRT for parameningeal RMS. Dose distributions in the axial, coronal, and sagittal planes. Upper panel is the three-field proton plan, lower panel is the five-field IRMT plan. (After Kozak, K. R., et al., *International Journal of Radiation Oncology • Biology • Physics* 74 (1):179–86, 2009.)

In the paper by Folgliata et al. (2009), intensity modulated proton therapy was compared to helical TomoTherapy and rapid arc (Figure 8.11).

8.5 Organ-at-Risk Doses and Late Effects

Although most of the normal organ late effects have been discussed in preceding chapters, a brief review of the literature on these effects specifically in the context of STS will be given here. In the head and neck region, late effects from treatment of STS include xerostomia, dental abnormalities, facial growth retardation, neuroendocrine dysfunction, and visual and hearing abnormalities. For treatments in the pelvis, musculoskeletal growth delay, bowel obstruction, and gonadal dysfunction can occur while in the extremities, fractures, growth disturbance, fibrosis, atrophy, peripheral nerve damage, and limitations in range of motion can occur. Most of these effects are as a result of radiation therapy, chemotherapy, and surgery. Kroll et al. (1994) tabularized the late effects of radiotherapy for all treatment sites and diagnoses. For RMS, 80% of patients had bone and soft-tissue deformity, 36% endocrine deficiency, and 16% skin atrophy (Kroll et al. 1994). The Childhood Cancer Survivor Study (CCSS) report on the relative risk of various late effects of treatment for childhood RMS is another good reference (Punyko, Mertens, Gurney, et al. 2005). Treatment using advanced technologies such as IMRT can reduce this toxicity profile but not eliminate it. Krasin et al. (2010) found that 47% of their NRMS patients had grade II or

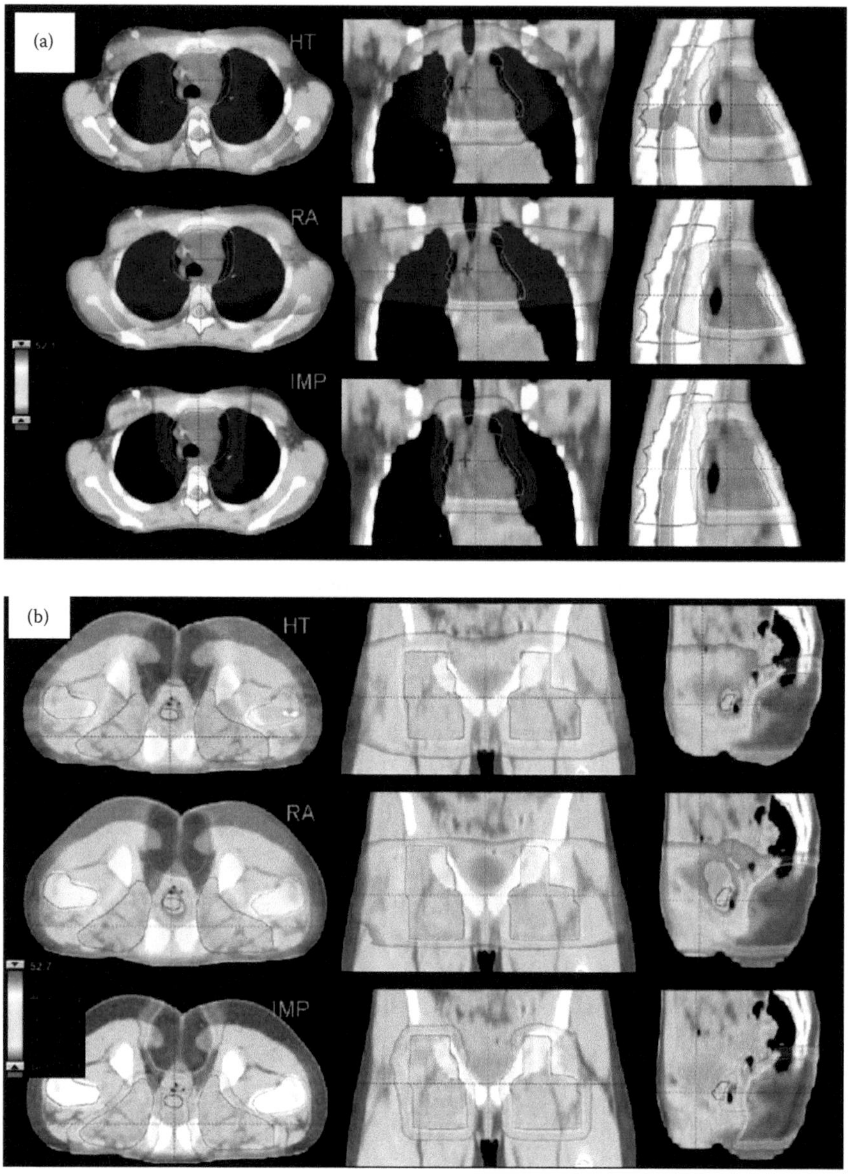

FIGURE 8.11 **(See color insert.) Helical TomoTherapy (HT) versus rapid arc (RA) versus intensity modulated protons (IMP) for (a) mediastinal RMS, (b) RMS of anus with metastases to lymph nodes. (After Fogliata, A., et al., *Radiation Oncology* 4:2, 2009.)**

greater late soft-tissue fibrosis and 31% had moderate stiffness or reduction of joint range of motion related to radiotherapy.

8.5.1 Effects Due to Irradiation of the Head and Neck

Several studies have examined the late effects from head and neck RMS. Irradiation of the face was found to cause both soft-tissue and bone growth deficits. A dose–response relationship was found for asymmetry for soft tissues at about 34 Gy and for bone at about 46 Gy, but doses as low as 4 Gy caused soft tissue changes. Irradiation of the orbit was found to cause growth deficits in the lower face as well as in the region actually irradiated (Guyuron et al. 1983; Larson et al. 1990). Paulino et al. (2000) reported on 30 children treated for head and neck RMS and found that late effects were seen in all patients. Most occurred by 10 years after treatment. Without the benefit of 3D planning for these patients, only approximate dose data could be presented for hypothalamic/pituitary, vision, hearing, and dental late effects. A timeline for type of late effect and time of occurrence shows that many of the effects do not occur until 5 years posttreatment and some take decades to become apparent. Median doses to the hypothalamic/pituitary region were 45 Gy (range 20–64.6) and 40.1 Gy (range 27.6–56.0) for those with and without decreased growth velocity, indicating the lack of a well-defined dose–response. A similar scatter of doses was seen for the other documented late effects. In the 15 children that had a portion of the face in the field, 11 developed facial asymmetry and had received between 44-60 Gy (Paulino et al. 2000). In the much larger study of 213 patients (median age at diagnosis was 5 years old) treated on IRS-II and III protocols between 1978 and 1987, 77% of the patients had one or more problems during the 7-year median length of follow-up. Forty-eight percent had subnormal statural growth, which is usually considered a result of pituitary irradiation. Many of these patients were given growth hormone injections, which did help restore growth. Patients less than 10 years old at diagnosis were more than twice as likely to have decreased growth as those aged 10 to 14. It was noted that there was no dose–response relationship and not all of the patients who received greater than 40 Gy to their pituitary became growth hormone deficient. Of the 76 patients with information about facial symmetry, 74 had asymmetry or hypoplasia of tissues in the irradiated region and several needed reconstructive surgery. About one-third had dental problems including malformed teeth. Impaired vision and cataracts were noted in about 20% of patients. Hearing loss was also seen although much of this can be attributed to cisplatin-containing chemotherapy. Dose–volume toxicity data was presented for pituitary, thyroid, and vision although the range of doses was wide and again could only be estimated due to the era in which these patients were treated (Raney et al. 1999). Retinal complications as a function of percentage of the retina receiving a dose for adult nasal, paranasal, and periorbital sarcomas have been reported. Retinas that developed late complications received doses of between 54 and 75 Gy and more than half of those that received more than 50 Gy to more than 60% of the retina developed severe retinal complications. No patients who received less than 50 Gy to the retina developed retinal complications (Takeda et al. 1999).

8.5.2 Effects Due to Irradiation of the Extremities

Late effects of pediatric extremity sarcoma were reported for 15 survivors with 20 years follow-up. Late effects were seen in all patients and included atrophy, fibrosis, bone growth abnormalities, impairment of extremity function, edema, and peripheral nerve injury. Eight of the children required surgical procedures to correct the late effects (Paulino 2004).

8.5.3 Irradiation of the Orbital Region

Late effects of treatment for orbital RMS have been reported by several groups including the IRSG (Heyn et al. 1986; Fiorillo et al. 1999; Raney et al. 2000; Skowronska-Gardas et al. 2000; Oberlin et al. 2001). Orbital RMS has been associated with excellent survival when treated with both chemotherapy and radiation therapy, but a significant number of patients suffer late effects. In the earliest report covering treatments given before 1980 when most patients received 50 to 60 Gy to the orbit, 89% of the patients had a decrease or absence of vision. Structural deficits in the bone or soft tissues were seen in more than half. Cataracts were seen in 90% of the patients (Heyn et al. 1986). Abramson and Notis (1994) analyzed the long-term visual status in 32 orbital RMS patients treated between 1960 and 1992. All patients received 55 to 60 Gy over 5.5 to 6 weeks, delivered by a combination of lateral photons and anterior electrons or anterior photons. Twenty-one of 32 eyes (66%) were functionless, and only 6 patients (19%) retained vision better than 20/200 in the treated eye (Abramson and Notis 1994).

Oberlin et al. (2001) summarized the experience of four international cooperative groups with a total of 306 patients with an overall survival of 87%. The most frequent late effects were dry eye (72%), impaired vision in the treated eye (54%), cataract (51%), ptosis (36%), and orbital hypoplasia (29%). These occurred almost exclusively in irradiated patients. In the IRSG study by Raney et al. (2000), overall survival was 96% and the eye was preserved in 86% of patients, but vision was impaired in 70% of them. The current IRS-V study recommends decreasing the dose of irradiation and using conformal techniques to minimize these effects. Fiorillo et al. (1999) reported on 19 patients treated in the Cobalt-60 era of the 1980s and 1990s. All patients were treated to 60 Gy with an anterior field and a lateral field, weighted 4:1 with a 1 cm lead cylindrical lens block on the anterior field. Doses of 40 to 60 Gy were recorded to the lacrimal gland, anterior retina, boney orbit, pituitary gland, facial bones, and portions of the optic nerves. Although all children showed at least one complication, only about half the patients experienced a late effect related to the dose to each organ at risk. Cataracts occurred in 11 of 19 patients after a dose of 5 to 12 Gy. Optic nerve damage was seen in just one patient despite most patients partially receiving 60 Gy. They also evaluated the patient's height over time and found it to be normal. All four patients treated under the age of 3 suffered facial asymmetry but none treated after puberty had this toxicity. Lacrimal duct stenosis was seen just a few months after irradiation with nearly all seen by 2 years after radiotherapy but could almost always be resolved by a surgical procedure (Fiorillo et al. 1999).

8.5.4 Irradiation of the Genitourinary System

About 20% of STSs occur in the genitourinary system and late effects of treatment in these sites for girls and boys are common and can be severe. In a review of 26 female patients treated for pelvic RMS treated before 1997, 22 received radiotherapy. Late effects occurred in 24 patients, 23 of which had grade 3 or 4 effects. There was a median of 9.5 late effects per patient who received radiotherapy. The most common grade 3-5 late effects were ovarian failure, pubertal delay, intestinal stricture, disfiguring fibrosis, removal of the uterus, vaginal stenosis, fistula, acute kidney infection, urinary obstruction, and removal of the bladder, but patients were affected in every listed organ system. Endocrine late effects were seen in 77%; GI, 69%; musculoskeletal, 69%; gynecologic, 58%; renal, 54%; and psychological, 38%. More than half of the patients required surgical intervention to correct the late effect. Three second malignancies (SMs) were found, all within the radiation field. It was noted that the burden of late effects in this population seemed to be greater than in adults treated similarly. Interestingly, no evidence was found of an association between age at diagnosis (less than 3 years of age versus older than 3 years of age) and number of late effects (Spunt 2005).

For patients with bladder or prostate STS, total or partial surgical removal of the bladder or prostate, once performed regularly, is now avoided in most patients. Incontinence and sexual dysfunction after therapy is of course dependent on the type of surgery performed and to some degree on the type and duration of chemotherapy, but the growth retardation and fibrosis of the bladder and prostate in irradiated children is worse than those not receiving radiotherapy (Yeung et al. 1994). In 109 children with RMS of the bladder or prostate, 39 operative procedures were required for those receiving chemotherapy plus radiation compared to just 6 for patients not receiving radiation. Thirty of 38 patients who did not have impaired bladder function were irradiated to a median dose of 41 Gy (range 8–60 Gy) but 13 of 14 patients with bladder dysfunction received about the same median dose. A similar lack of dose–response was seen for pubertal development. Thirteen out of 17 normally developed patients received about the same median dose of 45 Gy as 7 of 7 patients with abnormal development. Hematuria occurred at a median dose of 44 Gy (Raney et al. 1993). It should be noted that 44 Gy to the whole bladder in adults would not generally cause this level of dysfunction.

8.5.5 Irradiation of Very Young Children

Infants and very young children with RMS represent a particularly challenging group. Approximately 5% to 10% of all cases occur before the age of 1 year. Several studies have shown that infants have significantly lower survival, 55% compared to 80% for children over 1 year old (Ferrari et al. 2003). One reason for this decreased survival could be reduced use of radiotherapy for fear of late effects. In the Italian study reported by Ferrari et al. (2003), 10 out of 50 infants with RMS received 40 to 45 Gy external beam radiation therapy and 6 of them were long-term survivors but with significant bone and soft tissue growth retardation. The local failure rate was particularly high in patients whose stage of disease warranted radiotherapy but did not receive it. In a study of 20 patients less than 36 months old, a reduced dose of ≤36 Gy external beam radiotherapy (mostly IMRT) was used after gross total resection. Patients with tumors that could not be resected received 50.4 Gy. There were just two local failures in the entire cohort. With a median follow-up of about 3 years, only five patients had late effects, all mild. Use of treatment techniques that reduce the dose to normal structures, such as IMRT, IORT, IOHDR, and protons should improve functional outcomes (Puri et al. 2006). An international clinical trial group studying 59 children also found radiotherapy increased survival (Defachelles et al. 2009). They and others support the concept that despite the concerns about late effects, local control should be the goal for RMS patients regardless of age to maximize chances of long-term survival and the use of radiation therapy is necessary to maximize the chance for cure.

Reluctance to radiate very young children even extends to patients treated on protocol. In a study specifically reviewing noncompliance of radiation therapy protocol guidelines for IRS studies I through IV, 55% of patients with operative bed recurrences did not receive the intended radiation (Million et al. 2010). Three-fourths of the children die when local-regional disease is not controlled, emphasizing the importance of radiation therapy for many patients. The most common age group with recurrence of disease was less than 5 years old and the majority of those were less than 2 years old. In another IRS group report, salvage of local failures was also found to be less than 20% (Pappo et al. 1999).

8.5.6 Second Malignancy

Second malignancies after treatment of STS has been studied by Heyn et al. (1993) for IRS I and II and by Spunt et al. (2001) for IRS I to IV. The estimated cumulative incidence of SMs was 1.7% at 10 years and 3.5% at 20 years. Twenty-seven of the 67 SMs were leukemias/lymphomas, 27 were soft-tissue or bone sarcomas (osteogenic sarcoma was most common), and 13 were other cancers. Almost all of the SMs occurred in the radiation field. It is anticipated that the incidence of SM due to radiation therapy will decrease due to reductions in radiotherapy dose in patients with favorable initial response to chemotherapy. It also can be hypothesized that reduction of margins (field size) in new studies compared to those of 20 years ago will also contribute to a reduction of SMs.

References

Abramson, D. H., and C. M. Notis. 1994. Visual acuity after radiation for orbital rhabdomyosarcoma. *American Journal of Ophthalmology* 118 (6):808–9.

Blank, L. E., K. Koedooder, B. R. Pieters, et al. 2009. The AMORE protocol for advanced-stage and recurrent nonorbital rhabdomyosarcoma in the head-and-neck region of children: A radiation oncology view. *International Journal of Radiation Oncology • Biology • Physics* 74 (5):1555–62.

Blank, L. E. C. M., K. Koedooder, H. N. B. van der Grient, et al. 2010. Brachytherapy as part of the multidisciplinary treatment of childhood rhabdomyosarcomas of the orbit. *International Journal of Radiation Oncology • Biology • Physics* 77 (5):1463–9.

Breneman, J., J. Meza, S. S. Donaldson, et al. 2012. Local control with reduced-dose radiotherapy for low-risk rhabdomyosarcoma: A report from the Children's Oncology Group D9602 Study. *International Journal of Radiation Oncology • Biology • Physics* 83 (2):720–6.

Calvo, F. A., L. Sierrasesumaga, I. Martin, et al. 1989. Intraoperative radiotherapy in the multidisciplinary treatment of pediatric tumors. A preliminary report on initial results. *Acta Oncologica* 28 (2):257–60.

Chan, M. F., K. Schupak, C. Burman, C.-S. Chui, and C. C. Ling. 2003. Comparison of intensity-modulated radiotherapy with three-dimensional conformal radiation therapy planning for glioblastoma multiforme. *Medical Dosimetry* 28 (4):261–5.

Chen, C., H. K. Shu, J. W. Goldwein, R. B. Womer, and A. Maity. 2003. Volumetric considerations in radiotherapy for pediatric parameningeal rhabdomyosarcomas. *International Journal of Radiation Oncology • Biology • Physics* 55 (5):1294–9.

Combs, S. E., W. Behnisch, A. E. Kulozik, P. E. Huber, J. Debus, and D. Schulz-Ertner. 2007. Intensity modulated radiotherapy (IMRT) and fractionated stereotactic radiotherapy (FSRT) for children with head-and-neck-rhabdomyosarcoma. *BMC Cancer* 7:177.

Crist, W. M., J. R. Anderson, J. L. Meza, et al. 2001. Intergroup rhabdomyosarcoma study-IV: Results for patients with nonmetastatic disease. *Journal of Clinical Oncology* 19 (12):3091–102.

Defachelles, A. S., A. Rey, O. Oberlin, D. Spooner, and M. C. Stevens. 2009. Treatment of nonmetastatic cranial parameningeal rhabdomyosarcoma in children younger than 3 years old: Results from international society of pediatric oncology studies MMT 89 and 95. *Journal of Clinical Oncology* 27 (8):1310–5.

Donaldson, S. S., and J. Anderson. 1997. Factors that influence treatment decisions in childhood rhabdomyosarcoma. Intergroup Rhabdomyosarcoma Study Group of the Children's Cancer Group, the Pediatric Oncology Group, and the Intergroup Rhabdomyosarcoma Study Group Statistical Center. *Radiology* 203 (1):17–22.

Donaldson, S. S., J. Meza, J, C. Breneman, et al. 2001. Results from the IRS-IV randomized trial of hyperfractionated radiotherapy in children with rhabdomyosarcoma—A report from the IRSG. *International Journal of Radiation Oncology • Biology • Physics* 51 (3):718–28.

Ferrari, A., M. Casanova, G. Bisogno, et al. 2003. Rhabdomyosarcoma in infants younger than one year old. *Cancer* 97 (10):2597–604.

Fiorillo, A., R. Migliorati, P. Vassallo, et al. 1999. Radiation late effects in children treated for orbital rhabdomyosarcoma. *Radiotherapy & Oncology* 53 (2):143–8.

Fogliata, A., S. Yartsev, G. Nicolini, et al. 2009. On the performances of Intensity Modulated Protons, RapidArc and Helical TomoTherapy for selected paediatric cases. *Radiation Oncology* 4:2.

Gerbaulet, A., X. Panis, F. Flamant, and D. Chassagne. 1985. Iridium afterloading curietherapy in the treatment of pediatric malignancies: The Institut Gustave Roussy experience. *Cancer* 56 (6):1274–9.

Gillespie, M. B., D. T. Marshall, T. A. Day, A. O. Mitchell, D. R. White, and J. C. Barredo. 2006. Pediatric rhabdomyosarcoma of the head and neck. *Current Treatment Options in Oncology* 7 (1):13–22.

Goodman, K. A., S. L. Wolden, M. P. LaQuaglia, K. Alektiar, D. D'Souza, and M. J. Zelefsky. 2003. Intraoperative high-dose-rate brachytherapy for pediatric solid tumors: A 10-year experience. *Brachytherapy* 2 (3):139–46.

Guyuron, B., A. P. Dagys, I. R. Munro, and R. B. Ross. 1983. Effect of irradiation on facial growth: A 7- to 25-year follow-up. *Annals of Plastic Surgery* 11 (5):423–7.

Haase, G. M., D. P. Meagher, Jr., L. K. McNeely, et al. 1994. Electron beam intraoperative radiation therapy for pediatric neoplasms. *Cancer* 74 (2):740–7.

Healey, E. A., R. C. Shamberger, H. E. Grier, J. S. Loeffler, and N. J. Tarbell. 1995. A 10-year experience of pediatric brachytherapy. *International Journal of Radiation Oncology • Biology • Physics* 32 (2):451–5.

Heyn, R., A. Ragab, R. B. Raney, Jr., et al. 1986. Late effects of therapy in orbital rhabdomyosarcoma in children. A report from the Intergroup Rhabdomyosarcoma Study. *Cancer* 57 (9):1738–43.

Heyn, R., V. Haeberlen, W. A. Newton, et al. 1993. Second malignant neoplasms in children treated for rhabdomyosarcoma. Intergroup Rhabdomyosarcoma Study Committee. *Journal of Clinical Oncology* 11 (2):262–70.

Hua, C., J. M. Gray, T. E. Merchant, L. E. Kun, and M. J. Krasin. 2008. Treatment planning and delivery of external beam radiotherapy for pediatric sarcoma: The St. Jude Children's Research Hospital experience. *International Journal of Radiation Oncology • Biology • Physics* 70 (5):1598–1606.

Hug, E. B., J. Adams, M. Fitzek, A. De Vries, and J. E. Munzenrider. 2000. Fractionated, three-dimensional, planning-assisted proton-radiation therapy for orbital rhabdomyosarcoma: A novel technique. *International Journal of Radiation Oncology • Biology • Physics* 47 (4):979–84.

Kozak, K. R., J. Adams, S. J. Krejcarek, N. J. Tarbell, and T. I. Yock. 2009. A dosimetric comparison of proton and intensity-modulated photon radiotherapy for pediatric parameningeal rhabdomyosarcomas. *International Journal of Radiation Oncology • Biology • Physics* 74 (1):179–86.

Krasin, M. J., A. M. Davidoff, X. Xiong, et al. 2010. Preliminary results from a prospective study using limited margin radiotherapy in pediatric and young adult patients with high-grade nonrhabdomyosarcoma soft-tissue sarcoma. *International Journal of Radiation Oncology • Biology • Physics* 76 (3):874–8.

Kroll, S. S., S. Y. Woo, A. Santin, et al. 1994. Long-term effects of radiotherapy administered in childhood for the treatment of malignant diseases. *Annals of Surgical Oncology* 1 (6):473–9.

Larson, D. L., S. Kroll, N. Jaffe, A. Serure, and H. Goepfert. 1990. Long-term effects of radiotherapy in childhood and adolescence. *American Journal of Surgery* 160 (4):348–51.

Lee, C. T., S. D. Bilton, R. M. Famiglietti, et al. 2005. Treatment planning with protons for pediatric retinoblastoma, medulloblastoma, and pelvic sarcoma: How do protons compare with other conformal techniques? *International Journal of Radiation Oncology • Biology • Physics* 63 (2):362–72.

Martelli, H., C. Haie-Meder, S. Branchereau, et al. 2009. Conservative surgery plus brachytherapy treatment for boys with prostate and/or bladder neck rhabdomyosarcoma: A single team experience. *Journal of Pediatric Surgery* 44 (1):190–6.

Matuszak, M. M., D. Yan, I. Grills, and A. Martinez. 2010. Clinical applications of volumetric modulated arc therapy. *International Journal of Radiation Oncology • Biology • Physics* 77 (2):608–16.

McDonald, M. W., N. Esiashvili, B. A. George, et al. 2008. Intensity-modulated radiotherapy with use of cone-down boost for pediatric head-and-neck rhabdomyosarcoma. *International Journal of Radiation Oncology • Biology • Physics* 72 (3):884–91.

Meazza, C., A. Ferrari, M. Casanova, et al. 2006. Rhabdomyosarcoma of the head and neck region: Experience at the pediatric unit of the Istituto Nazionale Tumori, Milan. *Journal of Otolaryngology* 35 (1):53–9.

Merchant, T. E. 1997. Conformal therapy for pediatric sarcomas. *Seminars in Radiation Oncology* 7 (3):236–45.

Merchant, T. E., N. Parsh, P. L. del Valle, et al. 2000. Brachytherapy for pediatric soft-tissue sarcoma. *International Journal of Radiation Oncology • Biology • Physics* 46 (2):427–32.

Merchant, T. E., M. J. Zelefsky, J. M. Sheldon, M. B. LaQuaglia, and L. B. Harrison. 1998. High-dose rate intraoperative radiation therapy for pediatric solid tumors. *Medical and Pediatric Oncology* 30 (1):34–9.

Michalski, J. M., J. Meza, J. C. Breneman, et al. 2004. Influence of radiation therapy parameters on outcome in children treated with radiation therapy for localized parameningeal rhabdomyosarcoma in Intergroup Rhabdomyosarcoma Study Group trials II through IV. *International Journal of Radiation Oncology • Biology • Physics* 59 (4):1027–38.

Million, L., J. Anderson, J. Breneman, et al. 2010. Influence of noncompliance with radiation therapy protocol guidelines and operative bed recurrences for children with rhabdomyosarcoma and microscopic residual disease: A Report from the Children's Oncology Group. *International Journal of Radiation Oncology • Biology • Physics* 80 (2):333–8.

Miralbell, R., L. Cella, D. Weber, and A. Lomax. 2000. Optimizing radiotherapy of orbital and paraorbital tumors: Intensity-modulated X-ray beams vs. intensity-modulated proton beams. *International Journal of Radiation Oncology • Biology • Physics* 47 (4):1111-9.

Nag, S., D. Tippin, and F. B. Ruymann. 2001. Intraoperative high-dose-rate brachytherapy for the treatment of pediatric tumors: The Ohio State University experience. *International Journal of Radiation Oncology • Biology • Physics* 51 (3):729–35.

Nag, S., D. Tippin, and F. B. Ruymann. 2003. Long-term morbidity in children treated with fractionated high-dose-rate brachytherapy for soft tissue sarcomas. *Journal of Pediatric Hematology Oncology* 25 (6):448–52.

Nag, S., D. Tippin, S. Smith, C. Bauer, and F. B. Ruymann. 2003. Intraoperative electron beam treatment for pediatric malignancies: The Ohio State University experience. *Medical & Pediatric Oncology* 40 (6):360–6.

Oberlin, O., A. Rey, J. Anderson, et al. 2001. Treatment of orbital rhabdomyosarcoma: Survival and late effects of treatment—Results of an international workshop. *Journal of Clinical Oncology* 19 (1):197–204.

Oertel, S., A. G. Niethammer, R. Krempien, et al. 2006. Combination of external-beam radiotherapy with intraoperative electron-beam therapy is effective in incompletely resected pediatric malignancies. *International Journal of Radiation Oncology • Biology • Physics* 64 (1):235–41.

Pappo, A. S., J. R. Anderson, W. M. Crist, et al. 1999. Survival after relapse in children and adolescents with rhabdomyosarcoma: A report from the Intergroup Rhabdomyosarcoma Study Group. *Journal of Clinical Oncology* 17 (11):3487–93.

Paulino, A. C. 2004. Late effects of radiotherapy for pediatric extremity sarcomas. *International Journal of Radiation Oncology • Biology • Physics* 60 (1):265–74.

Paulino, A. C., and M. F. Okcu. 2007. Rhabdomyosarcoma. *Current Problems in Cancer* 32 (1):7–34.

Paulino, A. C., J. H. Simon, W. Zhen, and B. C. Wen. 2000. Long-term effects in children treated with radiotherapy for head and neck rhabdomyosarcoma. *International Journal of Radiation Oncology • Biology • Physics* 48 (5):1489–95.

Punyko, J. A., A. C. Mertens, K. S. Baker, K. K. Ness, L. L. Robison, and J. G. Gurney. 2005. Long-term survival probabilities for childhood rhabdomyosarcoma. A population-based evaluation. *Cancer* 103 (7):1475–83.

Punyko, J. A., A. C. Mertens, J. G. Gurney, et al. 2005. Long-term medical effects of childhood and adolescent rhabdomyosarcoma: A report from the Childhood Cancer Survivor Study. *Pediatric Blood & Cancer* 44 (7):643–53.

Puri, D. R., L. H. Wexler, P. A. Meyers, M. P. La Quaglia, J. H. Healey, and S. L. Wolden. 2006. The challenging role of radiation therapy for very young children with rhabdomyosarcoma. *International Journal of Radiation Oncology • Biology • Physics* 65 (4):1177–84.

Raney, R. B., J. R. Anderson, J. Kollath, et al. 2000. Late effects of therapy in 94 patients with localized rhabdomyosarcoma of the orbit: Report from the Intergroup Rhabdomyosarcoma Study (IRS)-III, 1984–1991. *Medical & Pediatric Oncology* 34 (6):413–20.

Raney, R. B., L. Asmar, R. Vassilopoulou-Sellin, et al. 1999. Late complications of therapy in 213 children with localized, nonorbital soft-tissue sarcoma of the head and neck: A descriptive report from the Intergroup Rhabdomyosarcoma Studies (IRS)-II and -III. IRS Group of the Children's Cancer Group and the Pediatric Oncology Group. *Medical & Pediatric Oncology* 33 (4):362–71.

Raney, B., Jr., R. Heyn, D. M. Hays, et al. 1993. Sequelae of treatment in 109 patients followed for 5 to 15 years after diagnosis of sarcoma of the bladder and prostate. A report from the Intergroup Rhabdomyosarcoma Study Committee. *Cancer* 71 (7):2387–94.

Raney, R. B., H. M. Maurer, J. R. Anderson, et al. 2001. The Intergroup Rhabdomyosarcoma Study Group (IRSG): Major lessons from the IRS-I through IRS-IV studies as background for the current IRS-V treatment protocols. *Sarcoma* 5 (1):9–15.

Ries, L. A. G., M. A. Smith, J. G. Gurney, et al. 1999. *Cancer Incidence and Survival among Children and Adolescents: United States SEER Program 1975–1995*. NIH Pub. No. 99-4649. Bethesda, MD: National Cancer Institute.

Schomberg, P. J., L. L. Gunderson, C. R. Moir, G. S. Gilchrist, and W. A. Smithson. 1997. Intraoperative electron irradiation in the management of pediatric malignancies. *Cancer* 79 (11):2251–6.

Skowronska-Gardas, M. A., M. Chojancka, K. Pedziwiatr, and B. Zawadzka. 2000. Late complication of radiotherapy in children treated for orbital rhabdomyosarcoma. *Radiotherapy and Oncology* 57 (Suppl. 1):S20.

Spunt, S. L., J. L. Meza, and J. R. Anderson. 2001. Second malignant neoplasms (SMN) in children treated for rhabdomyosarcoma: A report from the Intergroup Rhabdomyosarcoma Studies (IRS) I-IV. *Proceedings of the American Society of Clinical Oncology* 20:369a.

Spunt, S. L., S. X. Skapek, and C. M. Coffin. 2008. Pediatric nonrhabdomyosarcoma soft tissue sarcomas. *Oncologist* 13 (6):668–78.

Spunt, S. L., T. A. Sweeney, M. M. Hudson, C. A. Billups, M. J. Krasin, and A. L. Hester. 2005. Late effects of pelvic rhabdomyosarcoma and its treatment in female survivors. *Journal of Clinical Oncology* 23 (28):7143–51.

Sterzing, F., E. Stoiber, S. Nill, et al. 2009. Intensity modulated radiotherapy (IMRT) in the treatment of children and adolescents—A single institution's experience and a review of the literature. *Radiation Oncology* 4 (1):37.

Takeda, A., N. Shigematsu, S. Suzuki, et al. 1999. Late retinal complications of radiation therapy for nasal and paranasal malignancies: Relationship between irradiated-dose area and severity. *International Journal of Radiation Oncology • Biology • Physics* 44 (3):599–605.

Timmermann, B., A. Schuck, F. Niggli, et al. 2007. Spot-scanning proton therapy for malignant soft tissue tumors in childhood: First experiences at the Paul Scherrer Institute. *International Journal of Radiation Oncology • Biology • Physics* 67 (2):497–504.

Viani, G. A., P. E. Novaes, A. A. Jacinto, et al. 2008. High-dose-rate brachytherapy for soft tissue sarcoma in children: A single institution experience. *Radiation Oncology* 3:9.

Walterhouse, D., and A. Watson. 2007. Optimal management strategies for rhabdomyosarcoma in children. *Paediatric Drugs* 9 (6):391–400.

Wolden, S. L. 2005. Radiation therapy for non-rhabdomyosarcoma soft tissue sarcomas in adolescents and young adults. *Journal of Pediatric Hematology Oncology* 27 (4):212–4.

Yeung, C. K., H. C. Ward, P. G. Ransley, P. G. Duffy, and J. Pritchard. 1994. Bladder and kidney function after cure of pelvic rhabdomyosarcoma in childhood. *British Journal of Cancer* 70 (5):1000–3.

Yock, T., R. Schneider, A. Friedmann, J. Adams, B. Fullerton, and N. Tarbell. 2005. Proton radiotherapy for orbital rhabdomyosarcoma: Clinical outcome and a dosimetric comparison with photons. *International Journal of Radiation Oncology • Biology • Physics* 63 (4):1161–8.

Chapter 9

Bone Sarcomas (Osteosarcoma and Ewing's Sarcoma)

9.1 Clinical Overview

Primary bone tumors are the most common type of solid tumor in adolescents and young adults, trailing only leukemias and lymphomas in incidence. They are infrequent in very young children. Half of childhood bone tumors are malignant. Osteosarcoma comprises about 60% of the malignant childhood bone tumors, with a peak incidence between ages 10 and 25 years. Approximately 400 cases are diagnosed in children below the age of 20 in the United States annually. It usually arises in the metaphyseal regions of the long bones of the extremities: the distal femur, proximal tibia (Figure 9.1), proximal or mid-femur, and proximal humerus, in descending order. Less commonly, it can occur in other bones in the body, for example, the pelvis (Figure 9.2) (Ries et al. 1999). The incidence of osteosarcoma of the cranial bones is markedly elevated in children with the heritable form of retinoblastoma, particularly among those treated with radiation therapy (Wong et al. 1997). Osteosarcoma is one of the more common secondary cancers occurring after radiation therapy for other childhood sarcomas (Hawkins et al. 1996). Ewing's sarcoma is the second most common malignant childhood bone tumor, comprising about 20% of the total. It usually occurs in the pelvis (Figure 9.3), ribs, or diaphyseal portions of the long bones of the lower extremity, but like osteosarcoma, can occur in virtually any bone in the body. About half of tumors are located in the extremities.

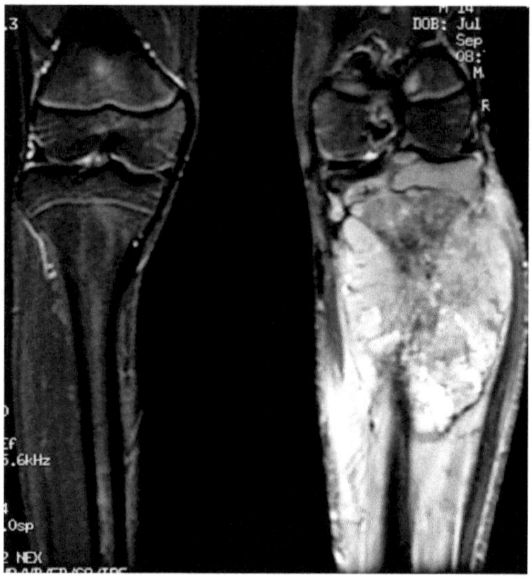

FIGURE 9.1 Coronal MRI image of osteosarcoma of proximal tibia.

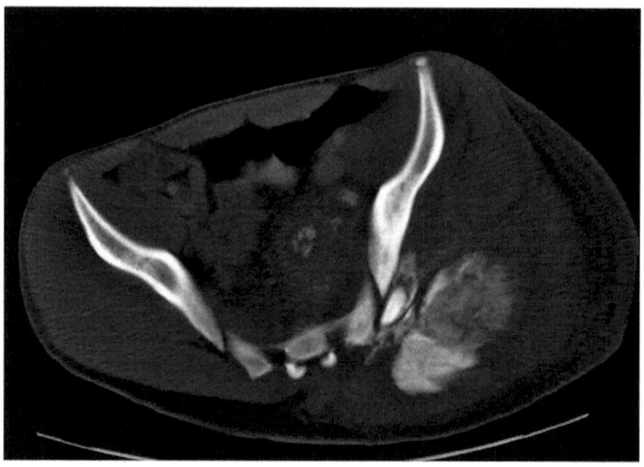

FIGURE 9.2 Axial CT image of pelvic osteosarcoma.

It is more common in males than females and in Caucasians than Africans Americans. The peak incidence of Ewing's sarcoma is between ages 10 and 20 years, with 30% of cases occurring in children under age 10 and only 5% of cases in adults over 25 years of age. About 200 new cases are diagnosed each year (Ries et al. 1999).

9.1.1 Osteosarcoma

Osteosarcoma patients usually present with pain in the involved bone, often accompanied by swelling due to extension of the tumor into soft tissues adjacent to the bone. A biopsy is required for diagnosis. The extent of the tumor is best assessed by MRI scan. There is no generally utilized staging

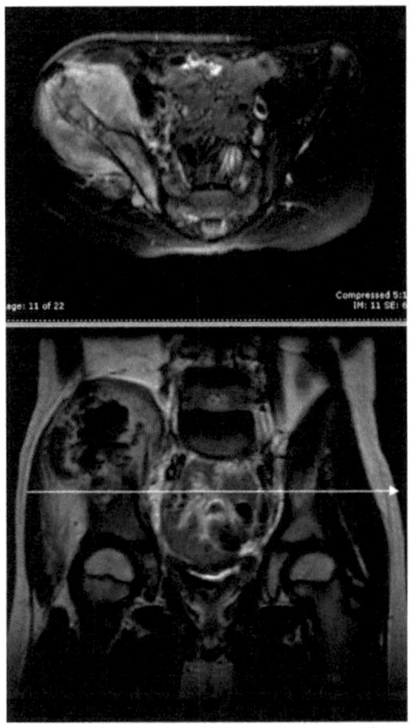

FIGURE 9.3 Axial and coronal MRI of Ewing's sarcoma of the right hip. Horizontal line on coronal shows location of axial image.

system for osteosarcoma. Osteosarcoma is usually simply classified as localized or metastatic. Metastases are detected at diagnosis in 15% to 20% of patients. The most common metastatic site is the lung, followed by bones. Radionuclide bone scan and chest CT scan are important to evaluate for metastatic disease. Unless effective systemic chemotherapy is given, lung metastases will develop in approximately 50% of patients within 6 months and in 80% within 5 years of their diagnosis. For patients presenting with localized disease, multiagent chemotherapy has been demonstrated to decrease the incidence of metastasis and improve the probability of cure. The addition of chemotherapy to surgical resection of the primary tumor has increased the 5-year survival rate for localized osteosarcoma from approximately 20% to 50%–75% (DeLaney et al. 2005).

Chemotherapy is generally given for 2 to 3 months prior to surgical resection, as the initial treatment of either localized or metastatic osteosarcoma. Several months of multiagent chemotherapy are given postoperatively to combat or prevent the development of macroscopic metastatic disease. Neoadjuvant chemotherapy was introduced to give the surgeon time to plan for intricate surgical procedures intended to completely resect the primary tumor while preserving the affected limb. Several other reasons have emerged for giving chemotherapy prior to surgical resection of the primary tumor. First, it addresses the metastatic disease that is present in the majority of patients, even if undetectable, when the metastatic tumor burden is lowest. Second, it often induces extensive necrosis and shrinkage of the tumor, making limb-sparing surgery more feasible. Third, histologic analysis of the extent of tumor necrosis following chemotherapy provides prognostic information about the probability of cure and may influence the selection of chemotherapy following surgery. Despite these theoretical considerations, clinical trials have found no improvement in survival with the use of neoadjuvant chemotherapy in addition to postoperative chemotherapy.

The traditional treatment of osteosarcoma of a limb was amputation of the limb above the level affected by the tumor or disarticulation at the joint if the tumor involved the proximal aspect of the extremity bone. Modern surgical and reconstructive techniques permit the complete resection of many extremity osteosarcomas with preservation of a functional limb. When a wide margin is obtained, limb-sparing procedures can result in a 90% local control rate. The individual tumor needs to be carefully imaged using MRI, CT, or angiography and evaluated by an experienced oncologic surgeon to determine the suitable surgical procedure. Amputation with an endoprosthesis may produce a better functional result than preservation of the limb with neurological, musculoskeletal, or vascular compromise. Surgery should be planned to produce a complete resection with a margin, even if that means sacrifice of the limb, as the probability of relapse is significantly higher after marginal or incomplete resections. Complete resection may not be feasible in some locations, such as the spine, sacrum, or base of the skull. In circumstances in which gross total resection cannot be performed or the patient refuses surgery, external beam radiation therapy is indicated.

Osteosarcoma has the reputation of being relatively insensitive to radiation therapy. In a series of patients treated with radiation therapy preoperatively, persistent viable cells were present in all tumors resected after 50 Gy, but absent after 80 to 100 Gy (Caceres and Zaharia 1972). The limited experience with radiation therapy as the definitive local treatment modality suggests that 60 to 70 Gy can provide a reasonable chance of obtaining long-term tumor control. Patients must be made aware that surgery is the preferred modality to obtain local control and that long bones treated with radiation doses above 50 Gy are susceptible to pathological fracture. A trend toward a higher local control rate was found for doses >55 Gy (71%) compared to <55 Gy (54%) (DeLaney et al. 2005). The radiation dose mandated on the current Children's Oncology Group (COG) protocol for unresectable disease is 70 Gy in once daily fractions of 2 Gy each. Definitive radiation therapy, unlike surgery, is delayed until all chemotherapy has been completed in order to avoid interrupting the chemotherapy schedule. Postoperative radiation therapy is mandated for patients in whom the resection did not yield wide surgical margins. The prescribed dose on the current COG protocol for gross residual disease is 66 Gy and for microscopically involved margins is 60 Gy in 2 Gy daily fractions.

9.1.1.1 Target Definition

The clinical target volume (CTV) margin is 2 cm for nonextremity tumors and up to 4 to 5 cm for tumors of the extremities. The presurgical tumor extension is considered when drawing the CTV. A planning target volume (PTV) margin of 0.5 cm should be sufficient when reasonably good immobilization is used and less may be justified if daily image-guided radiation therapy (IGRT) is performed. Normal tissue tolerance limits of adjacent structures, such as the spinal cord, may necessitate reduction of the dose to at least a portion of the target volume.

In patients with metastatic disease and limited life expectancy, palliative radiation therapy may be given to the primary tumor following chemotherapy to avoid a surgical resection that would disfigure and entail a long rehabilitation period. A relatively high radiation dose is usually prescribed for palliation of either primary disease or treatment of bone metastases.

Osteosarcoma bone metastases are usually osteoblastic and show increased uptake on 99m-technitium radionuclide bone scan. Treatment of these bone metastases with the bone-seeking radiopharmaceutical ^{153}Samarium-EDTMP has been reported in a small number of osteosarcoma patients by multiple institutions. Some partial responses were noted, but they have been transient (Anderson et al. 2005; Loeb et al. 2009).

9.1.2 Ewing's Sarcoma

The Ewing sarcoma family of tumors (ESFT) is a group of small, blue, round cell tumors with shared immunologic and genetic traits consisting of Ewing's sarcoma of bone, extraskeletal Ewing's sarcoma, primitive neuroectodermal tumor, and Askin tumor of the thoracic wall. The cell of origin is unknown. Other small, blue, round cell tumors include neuroblastoma, lymphoblastic lymphoma, rhabdomyosarcoma, and synovial cell sarcoma. A reciprocal chromosomal translocation between chromosomes 11 and 22 involving the *fli*1 and *ews* genes is present in 85% of ESFT patients and is pathognomonic for the disease.

ESFT usually arises in a bone but may also originate from soft tissues. Patients usually present with local pain, often accompanied by swelling. The primary tumor is generally best delineated by MRI scan. A biopsy is required for diagnosis. Approximately 25% of patients have distant metastases at diagnosis, usually to the lung or bones. Bone marrow or lymph node involvement is less common. There is no generally accepted staging system. As with osteosarcoma, disease is classified as localized or metastatic, and the majority of patients with localized disease will develop distant metastases if chemotherapy is not given. The addition of chemotherapy to definitive local therapy has raised the cure rate for localized disease from approximately 15% to 70%. Extremity tumors and smaller tumors have a higher survival rate. The prognosis for those who present with or develop metastatic disease remains poor.

The multimodality treatment strategy of localized ESFT is similar to that for osteosarcoma. Multiagent chemotherapy is given for approximately 12 weeks prior to the initiation of and for several months following definitive local therapy. A good response to induction chemotherapy usually makes surgical removal more feasible. If the tumor can be completely resected with adequate margins and a reasonable functional result, the preferred local treatment is surgical resection. Complete resection may require extensive reconstruction of the chest wall or pelvis and use of an allograft and endoprosthesis. Postoperative radiation therapy providing 45 to 55 Gy to the tumor bed is indicated if an adequate surgical margin has not been obtained. Postoperative radiation therapy has also been shown to improve the local control rate for completely resected tumors that demonstrate a poor histological response to induction chemotherapy, defined as >10% viable cells in the resection specimen. The local control rate after surgery, with or without postoperative radiation therapy, is over 90% (Schuck et al. 2003).

Tumors that remain bulky after induction chemotherapy or are located in surgically difficult sites, such as the spine, skull, facial bones, pelvis, or acetabulum, may be treated with definitive radiation therapy to a dose of 50 to 55 Gy. Chemotherapy is continued during and following the course of radiation therapy. The local control rate with radiation therapy alone is 75% to 80%, demonstrating that ESFT is more radiosensitive than osteosarcoma. The local control rate is higher for tumors less than 8 cm in diameter than for larger tumors. Incomplete tumor resection prior to radiation therapy does not increase the local control rate compared to radiation therapy alone. Therefore, resection should only be attempted if a complete tumor extirpation is anticipated. Radiation therapy has been used preoperatively for tumors that appear to be marginally resectable after induction chemotherapy. The radiation therapy dose given for preoperative therapy is the same as for definitive treatment. Donaldson (2004) summarized the results of 10 large ESFT patient cohorts. The 5-year disease-free, event-free, and relapse-free survival rates following multidisciplinary therapy were 60% to 69% and local control was 74% to 93%.

Whole lung irradiation following a complete response to conventional chemotherapy is a standard treatment for patients with metastatic disease limited to the lungs. The dose given to the whole lungs is 15 to 18 Gy in 1.5 Gy fractions. A randomized trial found that incorporating consolidative whole lung irradiation doubled the 5-year event-free survival to 47% in patients with isolated lung metastases (Paulussen et al. 1998). Prophylactic whole lung irradiation has been used for patients with localized disease in the past but has been replaced by more aggressive chemotherapy.

9.1.2.1 Target Definition

There are commonly two image-based gross tumor volumes (GTVs). GTV1 is the extent of disease prior to surgery and chemotherapy, and GTV2 is defined by the extent of disease after chemotherapy with or without surgery. For unresected or partially resected tumors, GTV2 includes the pretreatment abnormalities in bone and the gross residual tumor in soft tissue (and the tumor bed) after induction chemotherapy. CTV1 and 2 are the expansions of GTV1 and 2 by 1.5 cm tailored by considering anatomic barriers to disease extension. These CTVs include lymph nodes adjacent to the GTV when appropriate. PTV1 and 2 are three-dimensional geometric expansions of CTV1 and 2 by at least 0.5 cm but depends on the immobilization and image guidance being used. In the case of a primary tumor, which is resected with adequate margins but nodal disease is present, only the nodal disease is targeted as GTV1. Only unresected lymph nodes are defined as GTV2 and include the nodes remaining after chemotherapy. GTV2 is not used for resected lymph nodes or when there has been a complete response to chemotherapy. Target margins for Ewing's sarcoma are shown in Table 8.1 in Chapter 8.

9.2 Treatment Planning

The treatment planning considerations for osteosarcoma or ESFT are similar so will be discussed together. Treatments can be relatively simple to extremely complex, depending on the site of disease. For example, treatment of a limb is typically done using opposing fields, which irradiate the entire section of the affected limb while sparing peripheral lymphatics where possible. If opposed anterior and posterior fields are used and the patient is treated on a carbon fiber sandwich couch top (with or without fixation devices), consideration must be given to the posterior skin dose for these relatively large tumor doses, because most of the skin sparing can be lost (Butson et al. 2007). Also, skin dose reactions can be enhanced due to the chemotherapy given. Treatment of targets in the head, neck, and trunk are usually better treated by multibeam intensity-modulated radiation therapy (IMRT) than opposing fields to spare nearby normal structures. One difference between osteosarcoma and ESFT is that the dose for osteosarcoma is significantly higher than for ESFT; up to 70 Gy for osteosarcoma compared to about 55 Gy for ESFT. In fact, this dose is the highest for any pediatric tumor type. Depending on the location of the target, this dose can make organs at risk (OARs) sparing nearly impossible without localized underdosage of the PTV. Due to the high doses given, the volume, location, and magnitude of the maximum dose is an especially important consideration. Ten percent hot spots that may be acceptable for lower prescribed doses amount to 77 Gy in this context, a dose that is not tolerated by most normal tissues.

In the case of Figure 9.4, prior to radiotherapy, a surgical excision of the proximal femur was performed, leaving a metal femoral prosthesis in place. One challenge for treatment planning of bone tumor patients is the fairly common occurrence of metal structural supports and prostheses required after the surgical resection. CT artifacts, both low and high Hounsfield unit (HU) number regions, must be dealt with, usually by contouring these regions and manually overriding the HU value to 0 (water) or the density to 1. The actual metal regions (as best visualized as possible using a wide open level and window range) should be contoured and a bulk density consistent with the actual material used should be assigned. Care should be taken to ensure that the CT number to relative electron density table in the planning system has been extended to account for materials with a relative electron density of at least 7. Unlike the case with hip prostheses in prostate cancer patients, where beam angles can be chosen that avoid the metal, in the case of bone tumors, the metal is inside the target volume and cannot be avoided. The presence of large amounts of metal will increase the uncertainty of the dose distribution. The plan in Figure 9.4 uses an eight coplanar 6 MV x-ray beam IMRT field arrangement with no beams entering from the left lateral 90 degrees region since the PTV was right-sided. However, all beams pass through the femoral prosthesis before exiting the PTV, so the impact of the high-density metal is present for each beam. A 0.5 cm planning-at-risk volume (PRV) margin was used around the bladder visible on CT. The small bowel was a secondary OAR. Both the bladder and small bowel could be adequately spared while at least 95% of the PTV received the 66 Gy prescribed dose. Although the water equivalent depth was up to 12 cm longer for rays passing through the prosthesis, total monitor units (MUs) per beam were only moderately changed compared to when no metal is present. This is because only about 10% to 20% of each beam's area contains the metal; portions of the modulated field that do not pass through the metal are unchanged in fluence, whereas those portions that do may require 30% to 50% more MUs. For an open beam passing through the prosthesis, a significant dose perturbation will be seen for calculation to a point behind the metal (Figure 9.5).

In the unusual case when an osteosarcoma occurs in structures in the head, such as the sinus location in Figures 9.6 and 9.7, it can present a situation where the PTV cannot be fully covered due to nearby normal tissue dose constraints. In this case, the PTV is adjacent to optic nerves, optic chiasm, and the right eye. The optic structures have a dose limit of 54 Gy

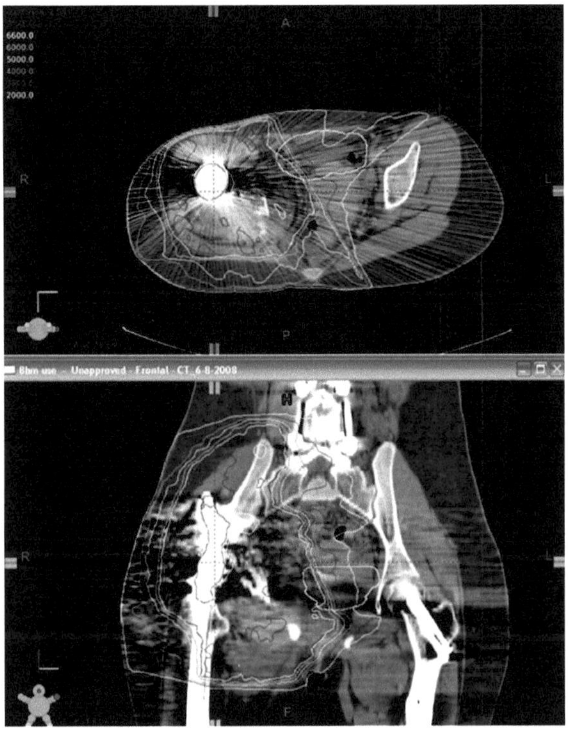

FIGURE 9.4 (Top) Ewing's sarcoma of right femur with femoral prosthesis. IMRT plan with metallic prosthesis inside the PTV. (Bottom) Coronal image of same patient.

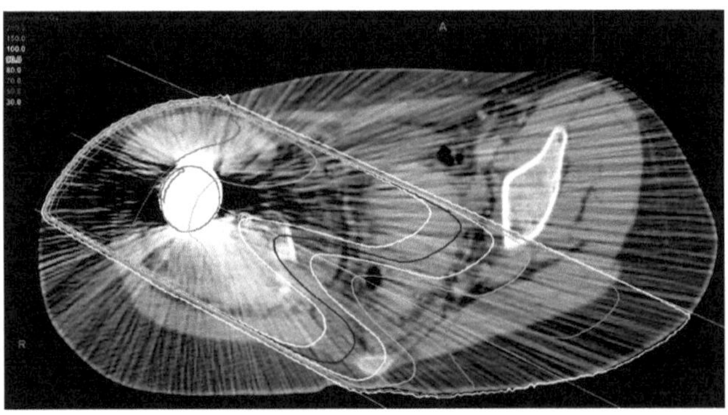

FIGURE 9.5 Single-beam dosimetry through metallic prosthesis.

maximum while the retina has a dose limit of 45 Gy. Planning for this case used nine noncoplanar, mostly anteriorly oriented intensity-modulated beams (Figure 9.7). PTV dose had to be compromised along the corridors adjacent to the critical structures but could be fully realized elsewhere. Tolerance doses were applied to PRVs created by three-dimensional expansion by 2 mm for all these critical structures to account for daily setup error and uncertainties in isocenter verification on orthogonal megavoltage images. Only about 90% of the PTV received 66 Gy, while all normal structure PRVs were maintained

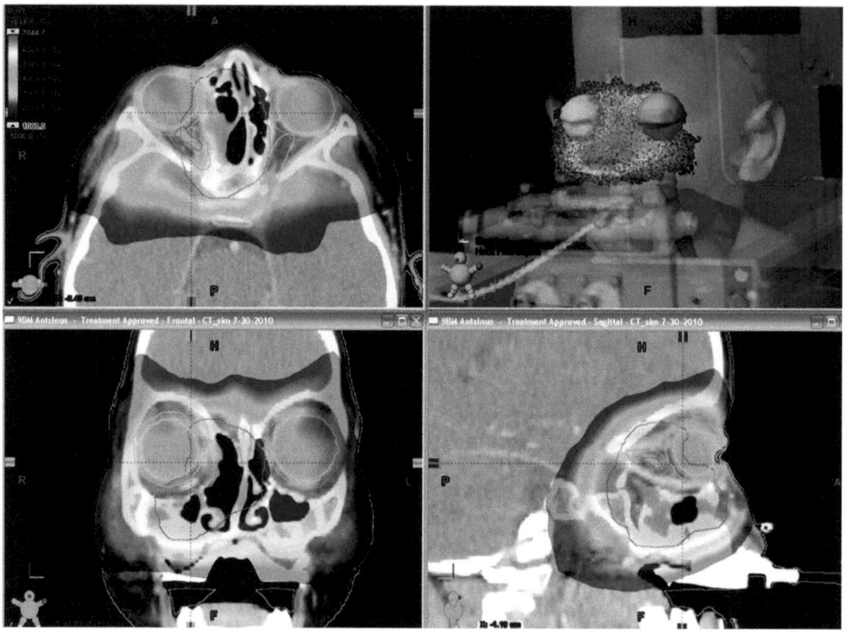

FIGURE 9.6 (See color insert.) IMRT for osteosarcoma (66 Gy) of the sinus with color wash doses.

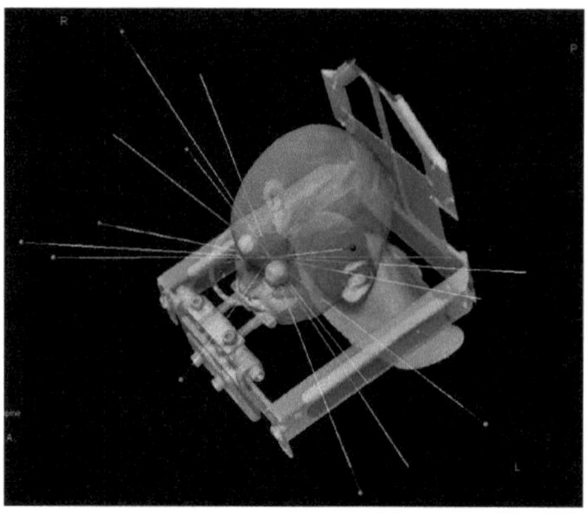

Gantry [deg]	Couch [deg]
270.00	40.00
262.00	348.00
292.00	348.00
323.00	36.00
49.00	326.00
73.00	13.00
101.00	13.00
90.00	320.00
321.00	90.00

FIGURE 9.7 Noncoplanar beams for osteosarcoma of the sinus.

at or below their tolerance dose. Figure 9.6 shows a color wash of the dose distribution with sparing of the optic structures and little dose to the normal brain tissue.

ESFT is more often found near the spine than is osteosarcoma, presenting planning challenges for protection of the spinal cord, small bowel, and kidneys. Figure 9.8 shows a paraspinal Ewing's sarcoma that envelops the spinal cord and is adjacent to the left kidney. The superior half of the target volume includes the spinal cord, but the inferior extent includes cauda equina, a less sensitive structure. The prescribed dose of 55.8 Gy exceeds cord tolerance, so where there is spinal cord, a PRV was made using a 0.5 cm expansion around

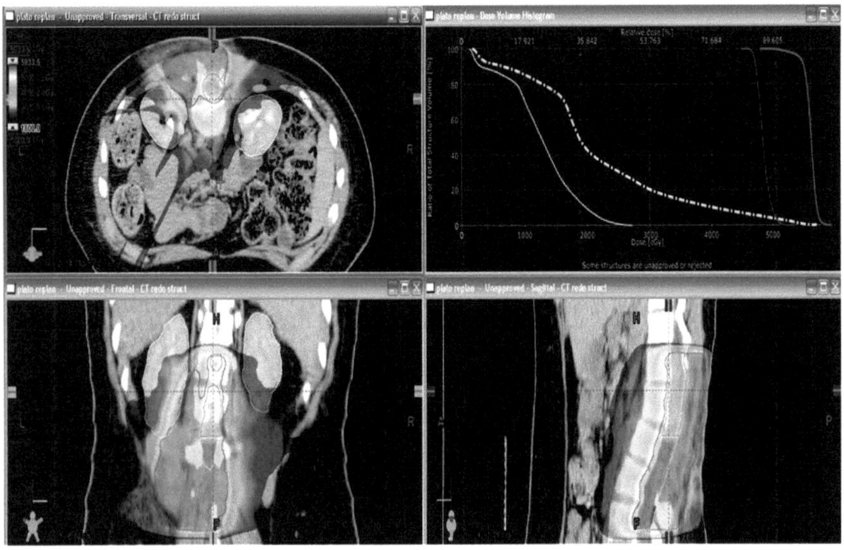

FIGURE 9.8 (See color insert.) IMRT for paraspinal Ewing's sarcoma. DVHs are shown: red, cauda equina PTV; purple, spinal cord target/OAR; dotted yellow, left kidney; solid yellow, spinal cord target.

the spinal cord with a goal of restricting the maximum dose to a few percent of the included section of the cord to less than 50 Gy. The cauda equina section was allowed to receive the full prescribed dose with the restriction that only a few percent of it could receive more than 55 Gy. A nine coplanar, nonopposing beam IMRT plan was used for treatment. International Electrotechnical Commission (IEC) gantry angles ranged from 95 to 340 degrees, omitting the right lateral anterior beam entrances for this posterior and somewhat lateral target. The left kidney was understandably less well spared than one achieves in a neuroblastoma case, for example, where the prescribed dose is only 21.6 Gy. Here, about 40% of the left kidney received 20 Gy, close to the limit set in recent COG protocols. The right kidney, being over 1 cm away from the PTV, is much better spared, receiving 20 Gy to only 10%. Due to the proximity of the PTV to the posterior skin surface in such cases, skin dose can be a concern for the supine patient, especially since carbon fiber couch tops and body immobilization devices can all but eliminate skin sparing. The patient could be treated prone to avoid this problem, but a less stable and reproducible patient position could result and the posterior structures will rise and fall with breathing, causing the need for gating or breath-hold techniques to be employed. For the supine position, this target motion problem is much less pronounced.

Either type of sarcoma can occur in the pelvis or the sacrum, and challenge the planner to create a plan that avoids bladder, rectal, and intestinal toxicity. Figure 9.9 shows a Ewing's sarcoma of the sacrum adjacent to the bladder. The treatment was delivered with the patient having a full bladder

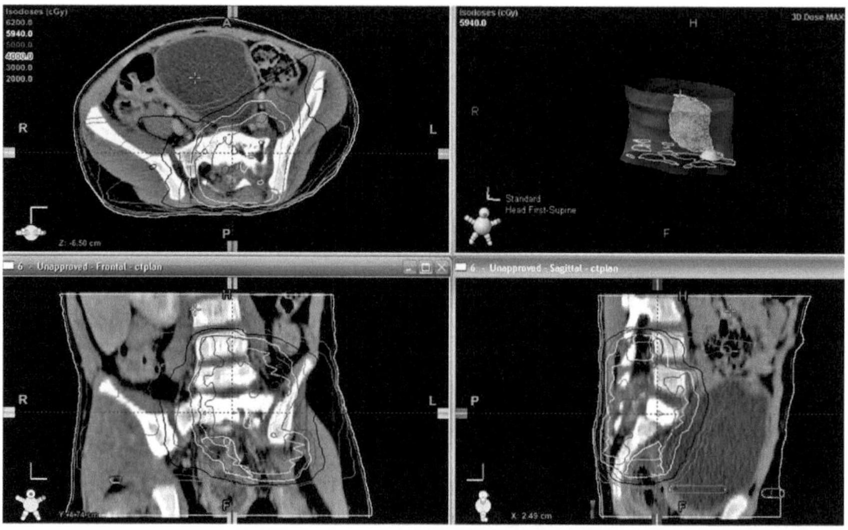

FIGURE 9.9 IMRT for sacral Ewing's sarcoma, 59.4 Gy.

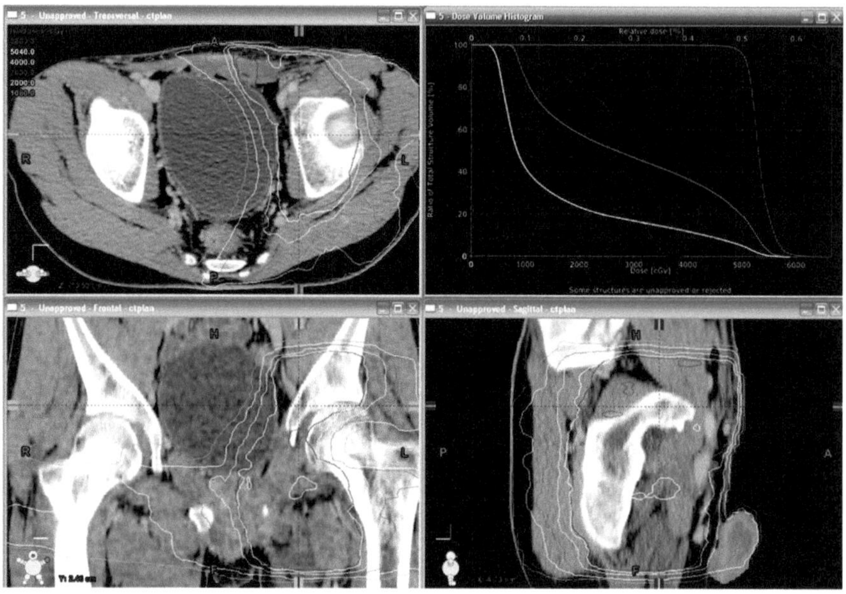

FIGURE 9.10 IMRT for pelvic Ewing's sarcoma with bladder sparing.

to push as much bladder away from the PTV as possible. Figure 9.10 shows a left-sided pelvic Ewing's sarcoma. Although the prescribed dose of less than 60 Gy is under the tolerance dose for severe late bladder effects, doses above 50 Gy to large volumes of bladder can cause severe acute toxicity that can and should be avoided. In cases where the PTV overlaps the bladder, either at a lateral or posterior surface, IMRT can generally produce the necessary dose sculpting inside the bladder to achieve the necessary sparing of the bladder wall, as seen in the isodoses in Figure 9.9 and Figure 9.10.

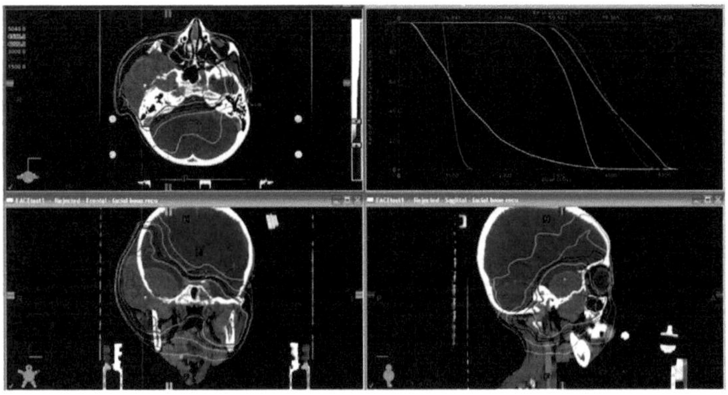

Gantry [deg]	Couch [deg]
266	19
345	94
303	0
310	80
36	337
80	0
39	304
271	39
329	27

FIGURE 9.11 (See color insert.) Parameningeal Ewing's sarcoma involving facial bones and soft tissues and brain. Nine-beam noncoplanar IMRT plan. DVHs: red, PTV; yellow, brain; green, right eye; dark blue, optic chiasm; turquoise blue, right lens; pink, left lens; light blue, left cochlea.

Head and neck or parameningeal Ewing's sarcoma present near or around critical structures can result in treatment plans that can cause permanent injury with the doses employed, more so if hot spots occur in those structures. Because of the location, complete surgical resection (or any surgical resection) is often not possible, and chemotherapy may leave a large mass remaining for radiation therapy. Figure 9.11 shows a large parameningeal Ewing's sarcoma of the facial bones spreading inferiorly to invade the ipsilateral mandible and crossing the midline to invade the contralateral mandible and the intervening soft tissues. In addition, the tumor closely approached the skin surface of the right side of the face so bolus was used in that region to bring the full prescribed dose to the skin (skin reaction must be monitored and skin conditioner should be applied). The optic nerves and chiasm were inside or adjacent to the PTV but could be kept below the 54 Gy tolerance dose by using a nine-beam noncoplanar IMRT technique to ensure that no hot spot occurs in those structures. The pituitary was inside the CTV and could not be spared, and the right lens received well over the dose to create a cataract. Normal brain dose was a concern because the PTV enclosed a large volume of the right temporal lobe, which could not be spared, but also surrounded a large volume of normal brain tissue. IMRT dose constraints were selected to keep the high dose volume as conformal to the PTV as possible, carving away dose from normal brain tissue so only about 15% received 20 Gy.

Because of the variety of locations in the body where bone tumors can occur and the high doses that they require, any method that will both preserve normal tissues and adequately treat the tumor is justified. In a dosimetric analysis of 72 sarcoma patients, including 17 bone tumors, treated at St. Jude Children's Research Hospital in Memphis, Tennessee, they concluded that it was not uncommon for the dose–volume constraints of critical normal structures, such as spinal cord, visual apparatus, and kidneys to compromise the delivery of the intended dose to the entire target. Dose–volume statistics for the PTV and all critical structures for their cases were tabulated (Hua et al. 2008).

In recent COG trials for all types of sarcoma, intraoperative radiotherapy (IORT) and brachytherapy have been allowed, and protons have been allowed on the ESFT and soft-tissue sarcoma trials; presumably protons will begin to be allowed for osteosarcoma trials in the near future. Preliminary results from proton treatment of ESFT and osteosarcoma patients have been reported (Ciernik et al. 2011; Rombi et al. 2012). Proton therapy treatments reviewed in Chapter 8 can generally apply to treatment of bone tumors.

9.3 Late Effects

At least 50% of patients receiving radiotherapy for treatment of Ewing's sarcoma and osteosarcoma suffer late effects, including inhibition of bone growth, muscular hypoplasia, lymphedema, fibrosis, and second malignancies. This compares to at least 25% for surgery alone (Paulino 2007). Treatment should avoid irradiation of the entire circumference of an extremity and limit coverage of epiphyseal growth plates and joint spaces if feasible. A wide range of chronic health conditions are seen in these patients and are similar to those seen for soft-tissue sarcoma (see Chapter 8). Disease recurrence or progression of the primary ESFT accounted for 60% of deaths compared to 8% for causes other than the disease. Other causes of death included subsequent malignant neoplasms (standardized mortality ratio [SMR] = 20.0), cardiac disease (SMR = 12.0), and other medical causes (SMR = 4.0). Cumulative mortality due to causes other than the disease was similar for survivors treated with and without radiation therapy (Ginsberg et al. 2010).

9.3.1 Secondary Malignancy

ESFT patients treated with high-dose radiation therapy have been reported to have a high incidence of secondary malignancies (SMs), particularly osteosarcoma. The risk of secondary malignancies is dose dependent, with the highest risk associated with doses above 60 Gy (Tucker et al. 1987; Kuttesch et al. 1996). It may be that the risk of SM is decreasing over time as dose and volume is reduced. In a study from 1979, the 10-year cumulative risk of SM was reported to be 35% (Strong et al. 1979). A 1987 study reported the 20-year risk to be 22% (Tucker et al. 1987). More recent studies reported 15 to 25 year cumulative incidence rates of 5% to 9% (Hawkins et al. 1996; Kuttesch et al. 1996; Dunst et al. 1998; Ginsberg et al. 2010; Sultan et al. 2010). Notably, breast and thyroid cancer represented 32% and 12% of the second cancers, respectively (Ginsberg et al. 2010). In a study comparing radiation therapy to surgery alone, the second malignancy risk within 15 years was 6.1% after 36 to 46 Gy postoperative radiation therapy compared to 0.9% after surgery alone (Ahrens et al. 1997).

References

Ahrens, S., J. Dunst, C. Rübe, M. Paulussen, C. Hoffmann, and H. Jürgens. 1997. Second malignancies after treatment for Ewing's sarcoma. *International Journal of Radiation Oncology • Biology • Physics* 39 (2, Suppl. 1):142.

Anderson, P. M., G. A. Wiseman, L. Erlandson, et al. 2005. Gemcitabine radiosensitization after high-dose samarium for osteoblastic osteosarcoma. *Clinical Cancer Research* 11 (19 Pt 1):6895–900.

Butson, M. J., T. Cheung, P. K. Yu, M. J. Butson, T. Cheung, and P. K. N. Yu. 2007. Megavoltage x-ray skin dose variation with an angle using grid carbon fibre couch tops. *Physics in Medicine & Biology* 52 (20):N485–92.

Caceres, E., and M. Zaharia. 1972. Massive preoperative radiation therapy in the treatment of osteogenic sarcoma. *Cancer* 30 (3):634–8.

Ciernik, I. F., A. Niemierko, D. C. Harmon, et al. 2011. Proton-based radiotherapy for unresectable or incompletely resected osteosarcoma. *Cancer* 117 (19):4522–30.

DeLaney, T. F., L. Park, S. I. Goldberg, et al. 2005. Radiotherapy for local control of osteosarcoma. *International Journal of Radiation Oncology • Biology • Physics* 61 (2):492–8.

Donaldson, S. S. 2004. Ewing sarcoma: Radiation dose and target volume. *Pediatric Blood & Cancer* 42 (5):471–6.

Dunst, J., S. Ahrens, M. Paulussen, et al. 1998. Second malignancies after treatment for Ewing's sarcoma: A report of the CESS-studies. *International Journal of Radiation Oncology • Biology • Physics* 42 (2):379–84.

Ginsberg, J. P., P. Goodman, W. Leisenring, et al. 2010. Long-term survivors of childhood Ewing sarcoma: Report from the childhood cancer survivor study. *Journal of the National Cancer Institute* 102 (16):1272–83.

Hawkins, M. M., L. M. Wilson, H. S. Burton, et al. 1996. Radiotherapy, alkylating agents, and risk of bone cancer after childhood cancer. *Journal of the National Cancer Institute* 88 (5):270–8.

Hua, C., J. M. Gray, T. E. Merchant, L. E. Kun, and M. J. Krasin. 2008. Treatment planning and delivery of external beam radiotherapy for pediatric sarcoma: The St. Jude Children's Research Hospital experience. *International Journal of Radiation Oncology • Biology • Physics* 70 (5):1598–1606.

Kuttesch, J. F., Jr., L. H. Wexler, R. B. Marcus, et al. 1996. Second malignancies after Ewing's sarcoma: Radiation dose-dependency of secondary sarcomas. *Journal of Clinical Oncology* 14 (10):2818–25.

Loeb, D. M., E. Garrett-Mayer, R. F. Hobbs, et al. 2009. Dose-finding study of 153Sm-EDTMP in patients with poor-prognosis osteosarcoma. *Cancer* 115 (11):2514–22.

Paulino, A. C., T. X. Nguyen, and W. Y. Mai. 2007. An analysis of primary site control and late effects according to local control modality in non-metastatic Ewing sarcoma. *Pediatric Blood & Cancer* 48 (4):423–9.

Paulussen, M., S. Ahrens, S. Burdach, et al. 1998. Primary metastatic (stage IV) Ewing tumor: Survival analysis of 171 patients from the EICESS studies. European Intergroup Cooperative Ewing Sarcoma Studies. *Annals of Oncology* 9 (3):275–81.

Ries, L. A. G., M. A. Smith, J. G. Gurney, et al. 1999. *Cancer Incidence and Survival among Children and Adolescents: United States SEER Program 1975–1995.* NIH Pub. No. 99-4649. Bethesda, MD: National Cancer Institute.

Rombi, B., T. F. DeLaney, S. M. MacDonald, et al. 2012. Proton radiotherapy for pediatric Ewing's sarcoma: Initial clinical outcomes. *International Journal of Radiation Oncology • Biology • Physics* 82 (3):1142–8.

Schuck, A., S. Ahrens, M. Paulussen, et al. 2003. Local therapy in localized Ewing tumors: Results of 1058 patients treated in the CESS 81, CESS 86, and EICESS 92 trials. *International Journal of Radiation Oncology • Biology • Physics* 55 (1):168–77.

Strong, L. C., J. Herson, B. M. Osborne, and W. W. Sutow. 1979. Risk of radiation-related subsequent malignant tumors in survivors of Ewing's sarcoma. *Journal of the National Cancer Institute* 62 (6):1401–6.

Sultan, I., R. Rihani, R. Hazin, and C. Rodriguez-Galindo. 2010. Second malignancies in patients with Ewing Sarcoma Family of Tumors: A population-based study. *Acta Oncologica* 49 (2):237–44.

Tucker, M. A., G. J. D'Angio, J. D. Boice, Jr., et al. 1987. Bone sarcomas linked to radiotherapy and chemotherapy in children. *New England Journal of Medicine* 317 (10):588–93.

Wong, F. L., J. D. Boice, Jr., D. H. Abramson, et al. 1997. Cancer incidence after retinoblastoma. Radiation dose and sarcoma risk. *JAMA* 278 (15):1262–7.

Chapter 10
Retinoblastoma

10.1 Clinical Overview

Retinoblastoma (RB) is a congenital malignancy that arises in the retina of the eye and in rare cases also in the pineal gland (trilateral RB). Retinoblastoma was originally called "glioma of the retina" by the famous German pathologist Rudolph Verchow in 1864. It was not until 1926 when the term *retinoblastoma* was widely adopted. It is the most common intraocular tumor in children and is diagnosed in infancy (even prenatally) or early childhood, with a median age at diagnosis of about 2 years old. It occurs with an incidence of 1 case in about 15,000 live births amounting to about 300 new cases in the United States per year. This represents only about 3% of cancers in children younger than 15 years of age but 11% of cancers developing in the first year of life (Ries et al. 1999). RB is the first diagnosis found to be the result of a genetic defect. The RB1 gene 13q14, found in the early 1980s, was the first tumor suppressor gene to be discovered and when mutated, RB can occur. RB patients with a positive family history of retinoblastoma and those with bilateral retinoblastoma are carriers of a germline mutation of the RB1 gene and are classified as hereditary (30%–40%). Unilateral patients without a family history (60%–70%) are classified as having nonhereditary disease, where retinoblastoma is presumably caused by somatic mutations of the RB1 gene. Approximately one-fourth of all retinoblastomas are bilateral. All bilateral cases are hereditary but just 15% of patients with unilateral RB also harbor the germinal mutation, the rest are as a result of a new mutation, usually in the father's sperm. The incidence rate for bilateral RB falls to nearly zero after age 2, but the rate of unilateral RB remains elevated until after age 7. Survival for children with RB is excellent with about 93% alive at 5 years after diagnosis (Ries et al. 1999). Trilateral RB (originally called pineoblastoma), which is bilateral RB plus intracranial involvement frequently of the pineal gland, occurs in about 6% of bilateral cases and 10% in those with a family history of RB. Trilateral RB is nearly always fatal (Blach et al. 1994).

Diagnosis of RB was based on the Reese-Ellsworth (R-E) staging system, first published in the 1960s. It divided eyes into five groups, I to V, each with subgroups, a and b. These divisions were according to the extent and location of disease as determined by ophthalmoscopy. The relative risk of losing an eye treated with primary external beam radiotherapy (EBRT) served as the index that clinically separated the groups. Group I included eyes with the lowest risk of enucleation and group V eyes the highest. In 2005, Murphree proposed a new classification system for intraocular retinoblastoma that ranks tumors for the risk of treatment failure and enucleation or EBRT by specific morphologic features, and the extent of disease in the eye at initial diagnosis. Its groups follow the natural history of intraocular retinoblastoma from early disease (group A) to late disease (group E). Group A eyes have the lowest risk of treatment failure and group E eyes have the highest risk with groups B, C, and D having intermediate risk. The presence of vitreous or subretinal seeding is the major ophthalmologic feature that separates eyes with a low risk for enucleation (groups A and B) from those containing advanced tumors (groups C and D) with a higher risk (Figure 10.1).

Systemic chemotherapy and focal therapies used by ophthalmologists, including cryotherapy, laser therapy, and transpupillary thermocoagulation, have changed the role of external beam radiotherapy from the primary modality to one used for eye preservation following intraocular progression after chemotherapy and focal ophthalmologic therapy or for orbital recurrence after enucleation. External beam radiation therapy is no longer considered first-line treatment because of its important long-term side effects such as radiation-induced chronic dry eye, retinopathy, optic neuropathy, cataract, induction of second malignancy, and midfacial malformation secondary to poor orbital development. Cryo- and laser therapy are generally ineffective in treating tumors greater than 4 mm in diameter or those with vitreous seeds.

10.2 Treatment Planning

The primary objective of treatment of RB is to preserve life. The secondary objective is to preserve vision. As with any treatment, one must balance the

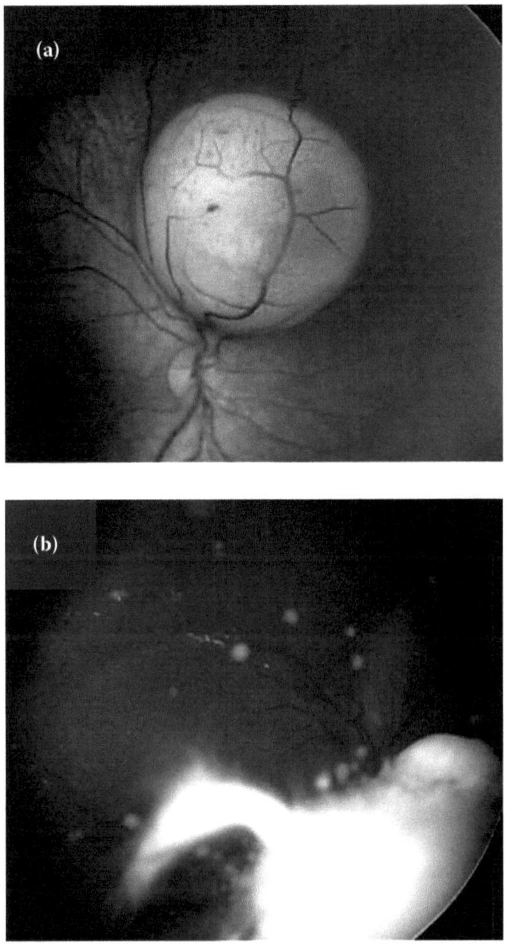

FIGURE 10.1 (a) Retcam image of Group B (low risk) solitary tumor larger than 3 mm. (b) Retcam image of Group D (high risk) large tumor with vitreous seeding. (From Murphree, L. A., *Ophthalmology Clinics of North America* 18 (1):41–53, 2005. With permission.)

risk of normal tissue damage with the risk of persistent or new tumors that cannot be cured with local therapies.

There has been a remarkable volume of literature on treatment planning techniques for RB considering the apparently straightforward targeting of the globe. Hilgartner (1903) reported the successful treatment of retinoblastoma with x-rays in 1903. In 1958, Reese and colleagues reported a 90% cure rate using up to 100 Gy with orthovoltage x-rays through temporal and nasal ports. Radon seeds were used by Moore and Stallard in the 1920s and 1930s (Moore et al. 1931). By the 1940s and 1950s, both the x-ray and radon seed clinical results showed that much lower doses, as low as 35 Gy, could be used effectively against retinoblastoma, although the most typical doses since the 1960s ranged from about 45 Gy to 60 Gy. Prior to the

1960s and the advent of the megavoltage era, orthovoltage treatments were given. In the megavoltage era starting in the 1960s, reports on treatment techniques for retinoblastoma began to appear with a large number in the 1980s and 1990s. Most of these dealt with variations on plans with lateral fields, with or without anterior-posterior (AP) fields. Technique papers continue to be published as interest increases in intensity-modulated radiation therapy (IMRT), stereotactic radiosurgery (SRS), and proton treatments. Radioactive plaque therapy has also played an important role in the treatment of retinoblastoma, with Co-60 sources replacing radon seeds starting in the 1960s followed by Iridium-192 (Ir-192), Iodine-125 (I-125), and Ruthenium-106 (Ru-106) later. As will be seen, the structures to include in the target volume drive the choice of treatment technique.

10.2.1 Target and Organ-at-Risk Volumes

The planning of retinoblastoma critically depends on whether the lens can be spared. When vitreous seeding is present, then the whole eye must be treated. When foci of RB are active in the retina and not in the vitreous, then a judgment must be made as to whether the entire retina needs treatment. There has been much controversy on this subject. If a single focus of tumor is in the posterior retina (posterior to the equator of the globe), then some have deemed it safe to treat just that small volume allowing sparing of the lens and other nearby tissues. When there are multiple foci of tumor in the posterior hemisphere of the globe, some have elected to treat the whole retina with sparing of the lens, whereas others have treated the entire globe. By sparing the lens, it is fairly certain that the ora serrata, the anterior-most aspect of the retina, ending between 1 to 1.5 mm behind the lens, will be underdosed. This can be understood by the illustration of the structures of the eye seen in Figure 10.2. If lens sparing is attempted, one must consider if this is worth the risk of tumor recurrence. As we will see, techniques that purport to both spare the lens and fully treat the ora serrata have been proposed in several publications. Offsetting the risk of a small anterior recurrence with lens-sparing radiotherapy is the relatively easier approach for subsequent laser or plaque therapy of the anterior globe.

The clinical target volume (CTV) for RB localized to the globe is the whole or partial retina plus 0.5 to 1 cm of the optic nerve as it enters the eye. The planning target volume (PTV) margin should be no larger than 0.3 cm and could be 0.2 cm with daily kilovoltage (KV) imaging; a smaller margin than that may conflict with the uncertainties of the verification imaging acquisition and registration process. To the extent one can reduce the PTV margin below 3 mm, one can spare an increasing volume of orbital bone with IMRT, and the V20 decreases from 20% to 10% for a 36 Gy tumor dose with a 1 mm instead of a 3 mm PTV.

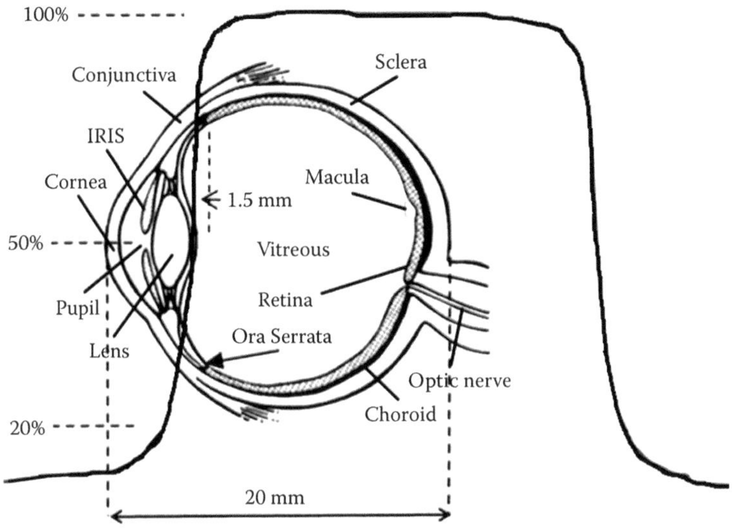

FIGURE 10.2 Relationship of single lateral 6 MV photon field with the anterior beam edge at the posterior aspect of the lens (approximately at bony canthus) and structures of the eye. Dose profile of a lateral D-shaped 26 mm × 32 mm 6 MV field along the anterior–posterior axis of the eye. Note that the field edge is inside the 1.5 mm space between the ora serrata and the lens. (Adapted from Schipper, J., *Radiotherapy and Oncology* 1 (1):31–41, 1983. With permission.)

Structures to protect include the lens if applicable, pituitary, bony orbit, brain, and the lacrimal gland if there is localized RB at least 1 cm from the gland. Late effects from excessive doses to these structures will be discussed later, but generally, cataractogenesis occurs above about 8 Gy, pituitary dysfunction above about 20 Gy, bone growth arrest above 20 Gy, cognitive deficits above about 18 Gy to a large volume of brain, retinal damage above 45 Gy, and lacrimal gland dysfunction above 40 Gy. For small structures such as the lens, the mean dose is probably a better measure than dose–volume metrics. The accuracy with which the lens is drawn as well as the image and dose calculation resolution and penumbral accuracy of the TPS beam model will greatly affect the accuracy of any dosimetric measure for the lens.

RB patients are all very young, generally less than 2 years old, so the dose–volume relationship for these effects are more uncertain and depending on the late effect, may cause a greater effect than for older children. The lacrimal gland is in contact with the superior-lateral aspect of the retina, and sparing it necessarily spares the adjacent retina. Minimizing the dose–volume histogram (DVH) for the bony orbit is important for avoiding the devastating bony and soft-tissue growth deformity seen in RB patients later in life known as the "hourglass" deformity, where the temporal and orbital bones cease to grow while less irradiated bones outside the target volume grow normally. To define the orbital bones in the planning system, one can create a 3D region of interest around the treated orbit. Automatic bone segmentation works well to generate the bone contours. Approximate boundaries for this volume are 2 cm superior and inferior to the globe, the medial extent is midline, and the lateral extent is the lateral skin surface. DVH metrics for orbital bone will vary greatly depending on how the bone is contoured, and, therefore, comparisons of orbital bone DVH between centers and across reports in the literature should be considered with caution.

10.3 Treatment Techniques

10.3.1 External Beam Treatment

One of the earliest reports on treatment technique was by Reese et al. in 1949 in which an ipsilateral temporal and contralateral oblique nasal pair of orthovoltage beams was described. In 1969, Cassady et al. published results of 230 cases, some who were treated with the orthovoltage technique but the more recent patients were treated with a new megavoltage technique using a 22.5 MeV 4 cm × 3 cm lateral x-ray field with 1 cm of bolus on the lateral entrance. The anterior border of the lateral field was placed at the bony lateral canthus with the intention to spare the lens, when appropriate, more easily done with the sharper penumbra of the megavoltage beam than the orthovoltage beam. The ora serrata is anterior to the bony canthus and about 1 mm posterior to the lens. To fully treat the ora serrata, the lens dose would have to be 70% to 80% of the tumor dose depending on the beam penumbra. The beam was angled posteriorly to avoid irradiation of the contralateral lens. An anterior open 22.5 MeV or Co-60 beam was used for whole globe irradiation (Cassady et al. 1969). By 1975, most centers treating RB used the single lateral field lens-sparing technique of Cassady but reports of recurrence in the ora serrata and other peripheral regions of the retina led many to believe that the entire retina was at risk and should be treated (Weiss et al. 1975; Reese 1976; Salmonsen et al. 1979; Messmer et al. 1990). In 1975, Weiss and colleagues described a modification of the lateral field technique whereby the anterior border of the lateral field was placed at the equator of the eye and an anterior field was added that had a divergent lens block hung along the central axis, projecting a 1 cm shadow on the lens. The 4 MV x-ray beams were weighted 1 to 4.5 anterior to lateral. Anesthesia and body cast immobilization was used, and it was noted that the eyes were seen to remain in an "eyes-front" position throughout the period of anesthesia. Forty-two to 45 Gy was delivered to the retinal surface in 3 to 4 fractions per week. They estimated the lens dose to be about 3%, or 144 cGy based on thermoluminescent dosimeter (TLD) measurements in an eye phantom (Weiss et al. 1975). From the dose distribution shown in this paper, the 80% isodose covered the ora serrata, but the paper did not state the isodose line to which the dose was prescribed.

In 1981, Abramson et al. published dosimetry and clinical results for a number of different lens-sparing treatment techniques, some fairly sophisticated for the time, which addressed unilateral or bilateral disease. For unilateral treatment, isodoses were shown for the commonly used lateral 6 MV photon beam angled 5 degrees posteriorly and for a mixed beam lateral technique giving 15 Gy from 6 MV photons added to 20 Gy with 11 MeV electrons. A more complex technique was also described and isodoses presented for 5 degree posterior oblique left and right lateral 6 MV photon beams weighted 1:5 for 32.5 Gy along with 12.5 Gy from an anterior 13 MeV electron beam with the lens block mounted on a low vacuum contact lens for the right eye (Figure 10.3a). In some cases, the contralateral photon

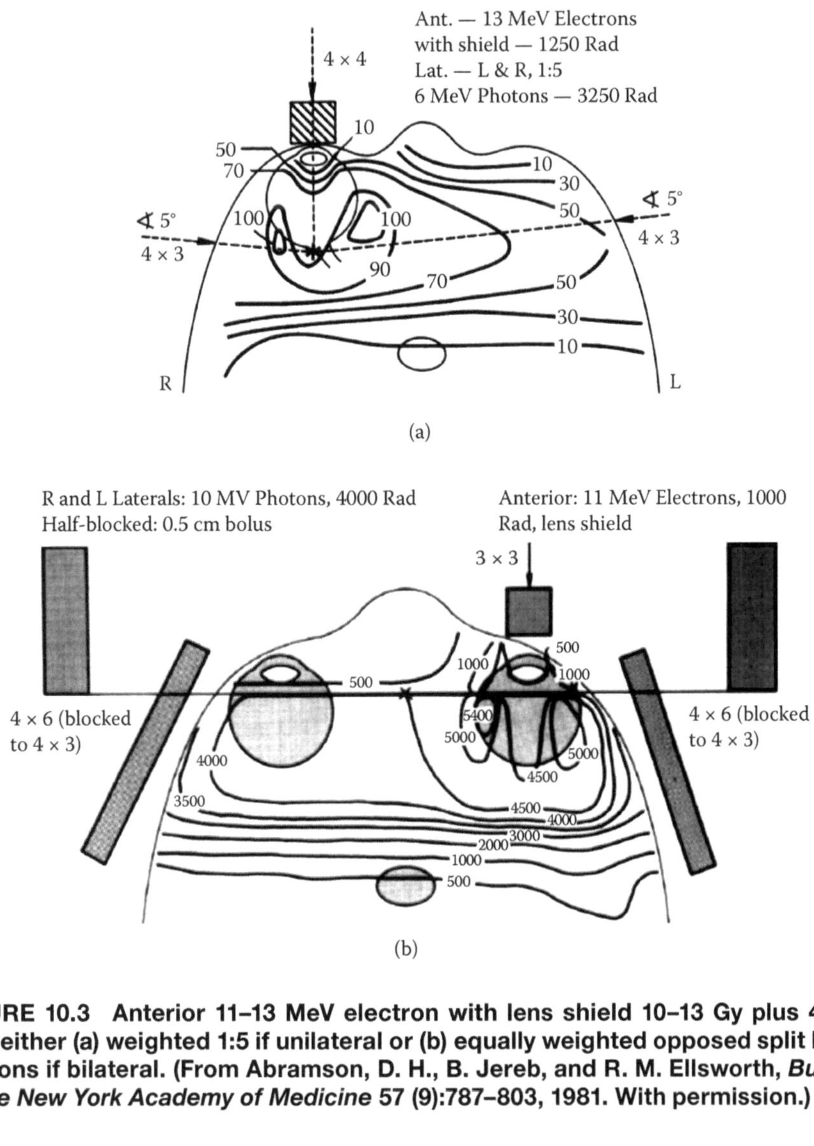

FIGURE 10.3 Anterior 11–13 MeV electron with lens shield 10–13 Gy plus 40 Gy with either (a) weighted 1:5 if unilateral or (b) equally weighted opposed split beam photons if bilateral. (From Abramson, D. H., B. Jereb, and R. M. Ellsworth, *Bulletin of the New York Academy of Medicine* 57 (9):787–803, 1981. With permission.)

beam was omitted altogether. For bilateral disease, isodoses were shown for right and left lateral 5 degree posterior oblique 6 MV photon beams giving 45 Gy with an open anterior 6 MV field giving 9 Gy to the more advanced eye. A variation on that bilateral treatment utilized opposing half-beam 10 MV photons to 40 Gy with 10 Gy from an anterior 11 MeV electron beam with a 1 cm lens block (Figure 10.3b). They point out that the lens block can cause underdosage of the posterior retina if not compensated properly by the other fields. These techniques were widely used for many years and generated much interest and study into perfecting these mixed beam techniques (Abramson et al. 1981).

In 1983, Schipper, at the University Hospital Utrecht, reported on a novel lens-sparing technique billed as both simple and precise that could replace the Abramson methods. The technique used a 26 mm × 32 mm D-shaped lateral x-ray field, produced with a special SRS-type collimator (17 cm from end of collimator to isocenter) to sharpen the beam penumbra. Ultrasound was used to accurately determine the intraocular geometry for each patient. The anterior edge of the field was positioned tangent to the posterior lens surface. A plastic beam spoiler was also used to raise the superficial dose to adequately treat the lateral aspect of the eye. Also, to ensure precise daily positioning of this sharp beam with respect to the eye, a contact lens was fixed to the cornea by means of a vacuum. A steel pin protruding from the outside of the contact lens was held by magnets in a fixture that was connected to the collimator housing. The fixture was adjustable for customization to any patient. The stated precision of the system was 0.3 mm. If both eyes required treatment, then parallel opposed lateral fields were used; if just one eye needed treatment, then a superior oblique beam (40 degree couch rotation) was used to avoid irradiation of the healthy contralateral eye. Others have used the lateral beam angled a few degrees posteriorly instead of the superior oblique beam. The superior oblique approach considers the possibility that at a later time, the contralateral eye will also need treatment and will have received some dose with the posterior oblique lateral but not with the superior oblique lateral. The 50% isodose line was aligned to the back of the lens while the 95% dose covered the ora serrata (Schipper 1983; Schipper et al. 1985) (Figure 10.2). Dosimetry was performed with this system, which showed that the penumbral distance from the 15% to the 85% dose was 2 mm at a depth of 2 cm (compared to 6 mm for standard collimation). Schipper's method resulted in saving 100% of the RE I and II eyes, and about 80% of the RE III and IV eyes. This system was later replicated by Toma et al. (1995) at St. Bartholomew's in London and Phillips et al. (2003) at Peter MacCallum in Melbourne (Figure 10.4). Phillips placed the anterior edge of the beam at the ora serrata (unless there was disease there), posterior to the placement by Schipper, and reported that the 50% isodose falls

FIGURE 10.4 Contact lens technique using lateral field. (After Phillips, C., et al., *Australasian Radiology* 47 (3):226–30, 2003.)

at the level of the ora serrata, the 90% isodose is approximately 2 to 3 mm behind it while the lens receives less than 10% of the prescribed dose. A few years later, Schipper reported on the clinical results of their method that delivered 45 Gy in 3 Gy fractions. In some of their early patients, the anterior border of the field had been placed at the bony canthus, which corresponded to a position about 2 mm posterior to the back of the lens. Tumor recurrences were found in the underdosed region of the ora serrata. Based on this finding, they routinely placed the anterior field edge at the posterior surface of the lens for tumors in the posterior half of the eye. If there were any tumors or seeding anterior to the equator of the eye, they moved the anterior border 1 to 3 mm into the lens. Their cataract incidence data indicated that the posterior 1 mm of the lens could be inside their sharply collimated field edge without inducing an opacity that impaired vision or permitted development of new tumors (Schipper et al. 1985). By the 1990s, several more reports had been published demonstrating that (1) underdosage of the anterior retina due to aggressive lens sparing was associated with a higher risk of anterior recurrences (Schipper et al. 1985; Chin et al. 1988; Toma et al. 1995), (2) whole eye treatments with an anterior beam rarely resulted in anterior recurrences but resulted in nearly all patients getting cataracts (Egbert et al. 1978; Hungerford et al. 1995), and (3) local therapy resulted in an equal percentage of recurrences in the anterior and posterior eye (Messmer et al. 1990). Although there may be a higher incidence of new anterior tumors when the ora was underdosed, the overall ocular survival rate was not reduced compared to whole eye irradiation due to early detection and treatment with focal salvage techniques (Toma et al. 1995). Schipper's study of 73 eyes resulted in saving 100% of the eyes with lower stage tumors and 80% with higher stage (Schipper et al. 1985). The Utrecht group published an update of their results in 1997 on 128 eyes in which 73% of eyes were saved, 50 with radiation therapy alone, and 93% of the saved eyes retained useful vision (Schipper et al. 1997).

For eyes with higher stage disease where the entire globe must be treated, anterior field treatments were performed with Co-60 and other megavoltage beams when the fellow eye was present or a lateral beam including the whole eye when not (Bedford et al. 1971; Egbert et al. 1978; Hungerford et al. 1995) or with a wedged anterior and lateral pair of fields with the lateral angled behind the fellow lens if that eye were present (Amendola et al. 1990). Few tumors were found after these treatments with 80% eye preservation with additional local therapy in RE stage I to III eyes, but as one would guess, all patients had cataracts within 2 years (Egbert et al. 1978; Hungerford et al. 1995).

In the late 1980s through the 1990s, a host of creative techniques for sparing the lens without underdosage of the ora serrata were developed and published. These included x-ray beam techniques as well as electron beam and mixed beam techniques. By 2000, IMRT techniques and by 2005, proton beam techniques, had been published. The dilemma these techniques tried to overcome was how to "thread the needle": treat the whole retina (and vitreous in many cases) but spare the lens, which can only tolerate 20% of the target dose but was just 1 to 2 mm anterior to the target volume. These treatment techniques would have to account for the penumbral width of the beam edge defined either by the jaws or blocks and geometric precision of the treatment.

A few years after the Schipper publications, arc therapy techniques were developed to spare the lens while fully treating the retina. One technique consisted of three pairs of noncoplanar 4 MV arcs, each with its isocenter 1 cm anterior to the lens and each with a block covering the lens with a margin at every angle. Couch angles of 0 degrees and plus and minus 60 degrees were used. The gantry rotated 60 degrees for each partial arc with a 20 degree overlap so that 80 degrees was traversed for each of the three arcs. The maximum dose was noted to be 115% while the lens received about 30% of the target dose (13 Gy at 0.5 Gy per fraction) after consideration of the effects of setup variation. The lens dose would have been 10% with a perfect setup. Figure 10.5 shows the treatment geometry and the isodose distribution achieved. The report on this technique highlighted the critical nature of the positional setup in any technique where the objective was to place the high dose gradient perfectly between the lens and the ora serrata. A refinement of this technique involved designing a special U-shaped collimator insert for a stereotactic radiosurgery-style collimator and also added a fluence modulator to compensate for dose inhomogeneity in the target volume, which

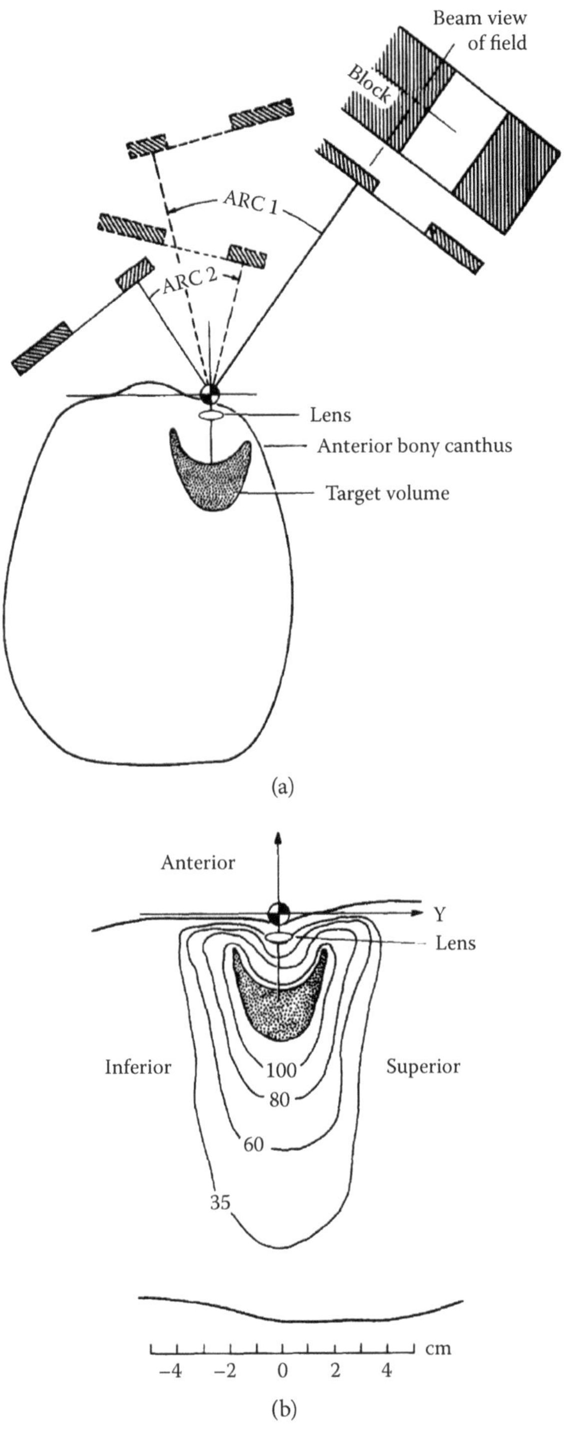

FIGURE 10.5 Arc therapy for RB. (a) Arc arrangement, (b) isodoses in sagittal plane. (From Chin, L. M., et al., *International Journal of Radiation Oncology • Biology • Physics* 15 (2):455–60, 1988. With permission.)

(a)

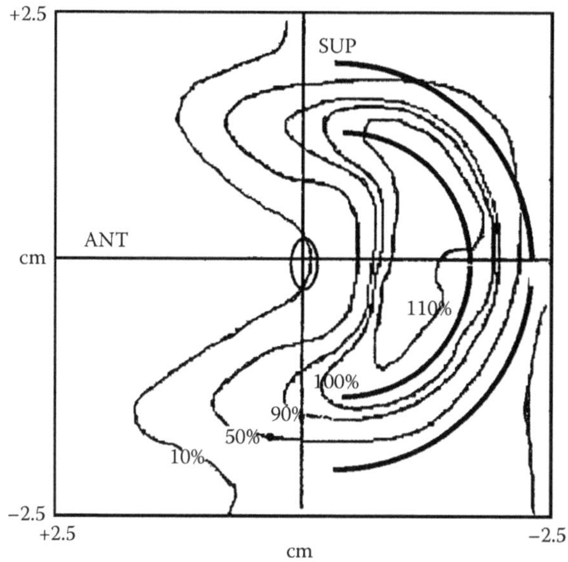

(b)

FIGURE 10.6 (a) Custom lead collimator with compensator. (b) Sagittal dose distribution of arc treatment with collimator in panel a. (From Cormack, R. A., et al., *Medical Physics* 25 (8):1438–42, 1998. With permission.)

was the entire retina excluding the lens (Figure 10.6a). Three noncoplanar arcs, one sagittal (couch 90 degrees, gantry 60–120 degrees) and two lateral obliques (couch 30 degrees, gantry 60–120 degrees; and couch 330 degrees, gantry 240–300 degrees), were used with the lens positioned at isocenter. The collimator always blocks the lens and the brass fluence modulator integrated into the collimator compensated for the otherwise reduced dose to the posterior retina. The lens received 10% of the dose while the 100% isodose fell 5 mm behind the lens and wrapped in a concave fashion around the entire retina (Figure 10.6b) Orbital bones were generally outside the 50% isodose surface. Film measurements confirmed their planning system calculations of this very sophisticated treatment technique, which was designed to be carried out with SRS patient positioning accuracy (Cormack et al. 1998).

An obvious technique for treatment of the whole eye, a very superficial structure, is an anterior electron beam, which indeed has been used. Electrons are attractive because of their ability to spare the contralateral side (for either an AP or lateral beam) and brain tissue beyond the treated orbit (for an anterior beam) due to their limited range. But sparing of the lens was also attempted with electron beams as early as the 1960s by the French, who employed a lateral D-shaped 28 MeV electron beam with the same positioning relative to the bony canthus as had been used for lateral photon beams. An anterior and lateral electron beam weighted 4:1 lateral to anterior was used for more anterior tumors (Haye et al. 1985). Griem et al. (1968) experimented with an anterior 10 MeV electron beam with a copper lens shield mounted to a scleral contact lens to avoid neutron production. This produced a dose to the lens of less than 10% but a dose depression of 40% at the posterior retina. By arcing the electron beam by 20 degrees, the posterior retinal dose could be raised. Extensive dosimetry, computed (pencil-beam algorithm) and measured, determined that a 10 MeV electron beam was optimal with a 1 cm diameter lens block in contact with the surface of the eye. The lens dose was less than 10% of the dose outside the block. The need for 3D heterogeneity corrections and scatter calculations was noted based on the presence of the surrounding bones. A 4.8 cm wide field was found to give better coverage of the posterior retina than a 3.8 cm wide field due to increased lateral scatter, but a 30% to 40% underdosage of the posterior retina was still observed (Figure 10.7). The trade-off for increasing the dose to the posterior retina was the increased volume of normal tissue irradiated with the larger field (Kirsner et al. 1987). This study showed that the retinal dose coverage of the anterior electron beam with lens block was at least as good as the lateral photon beam lens-sparing technique; each has a characteristic low dose zone, anterior for the lateral beam and posterior for the electron beam. Following up on this work, Al-Beteri and Raeside (1992) investigated the use of electron beams using a 3D Monte Carlo-based treatment planning system with heterogeneity corrections. They also used a 10 MeV anterior electron field, centrally blocked by a 1 cm diameter lens block, but theirs was a 6 cm tall divergent Lucite column whose distal end was positioned within 1 cm from the skin surface to sharpen the penumbra. In addition to this field, a 16 MeV D-shaped lateral electron beam with its anterior border at the equator of the eye was used, equally weighted with the anterior field. This combination was found to produce a dose distribution where the prescribed dose was given to the 90% isodose covering the entire retina including the ora serrata with a small volume of retina receiving 110% (122% of prescription) and less than 10% of the dose given to the lens and brain tissue. Al-Beteri and Raeside noted a complex dose perturbation

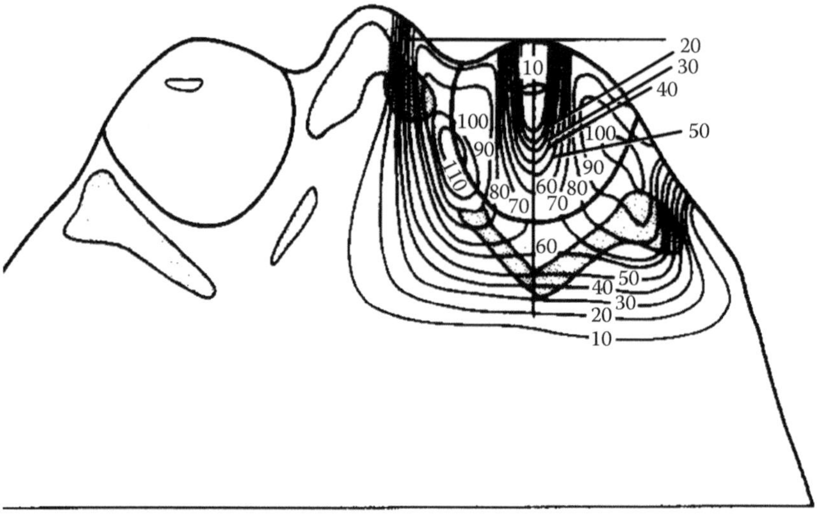

FIGURE 10.7 Calculated doses for a 10 MeV electron beam, 4.8 cm diameter, 96 cm SSD. (From Kirsner, S. M., et al., *Medical Physics* 14 (5):772–9, 1987. With permission.)

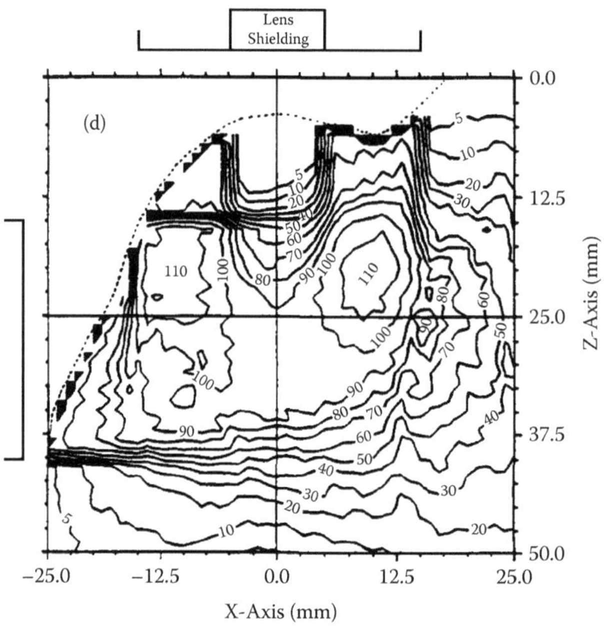

FIGURE 10.8 10 MeV AP electron beam with lead lens block and open 16 MeV lateral electron beam. (From Al-Beteri, A. A., and D. E. Raeside, *Medical Physics* 19 (1):125–35, 1992. With permission.)

pattern caused by the interplay of the body surface obliquity and the presence of heterogeneities, bone and air, around the orbit, which could not be appreciated with the pencil-beam algorithm (Figure 10.8). Inspection of their composite dose distribution shows coverage of the nasal aspect of the ora serrata by only the 70% isodose. Several years later, Steenbakkers et al.

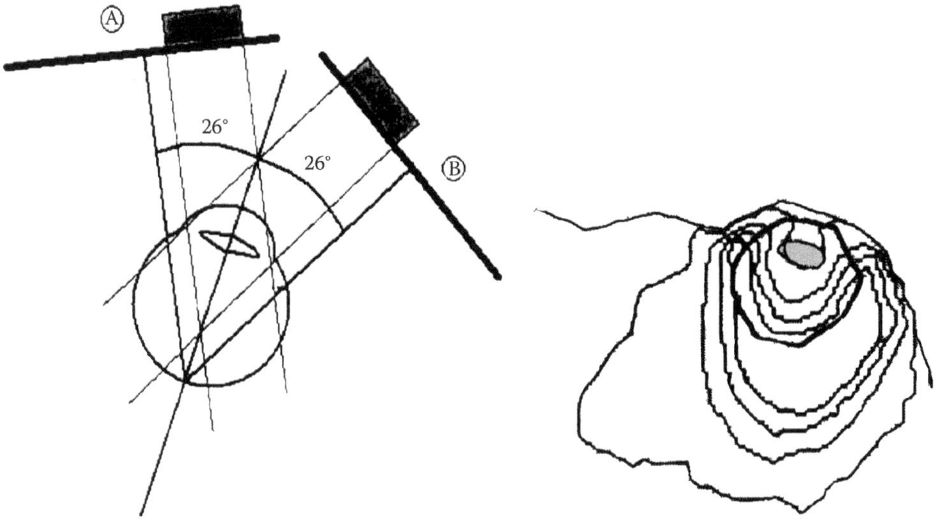

(a) (b)

FIGURE 10.9 (a) Two 9 MeV electron beams ±26 degrees from the axis of the optic nerve. (b) Isodose curves are 10%, 50%, 80%, and 100%. (From Steenbakkers, R. J. H. M., et al., *International Journal of Radiation Oncology • Biology • Physics* 39 (3):589–94, 1997. With permission.)

(1997) described their electron beam technique, either for whole eye or lens sparing. For whole eye, they recommended a 3.5 cm × 3.5 cm 9 MeV electron beam oriented along the axis of the optic nerve. For lens sparing, their best plan consisted of two 9 MeV beams angled 26 degrees medially and laterally to the optic nerve axis, with the isocenter of both beams at the border between the globe and the origin of the optic nerve. Lens blocks were used on both beams with a 2.5 mm margin (Figure 10.9a). The 100% isodose was used for treatment with the lens dose being about 10% (Figure 10.9b). Steenbakkers et al. found this technique was applicable to both infants and young children. Their dose distribution calculations were verified to be correct within 3% or 2mm with TLD in a phantom.

Following the publication of the various photon and electron techniques and nearly a decade of their use at Memorial Sloan Kettering, McCormick et al. (1988) compared their clinical results using an anterior electron beam with lens block plus a lateral photon field, referred to as the anterior lens sparing (ALS) technique (first described by Abramson) to the use of a lateral electron beam with a superior and inferior lateral oblique split beam wedged photon pair (weighted 2:1 superior:inferior) referred to as the modified lateral beam (MLB) technique (Figure 10.10a). In this latter three-field technique, they moved the anterior border of the fields forward relative to the original lateral beam technique, raising the lens dose but better covering the ora serrata (Figure 10b). They stated that this approach placed the lens at the 30% isodose, although others

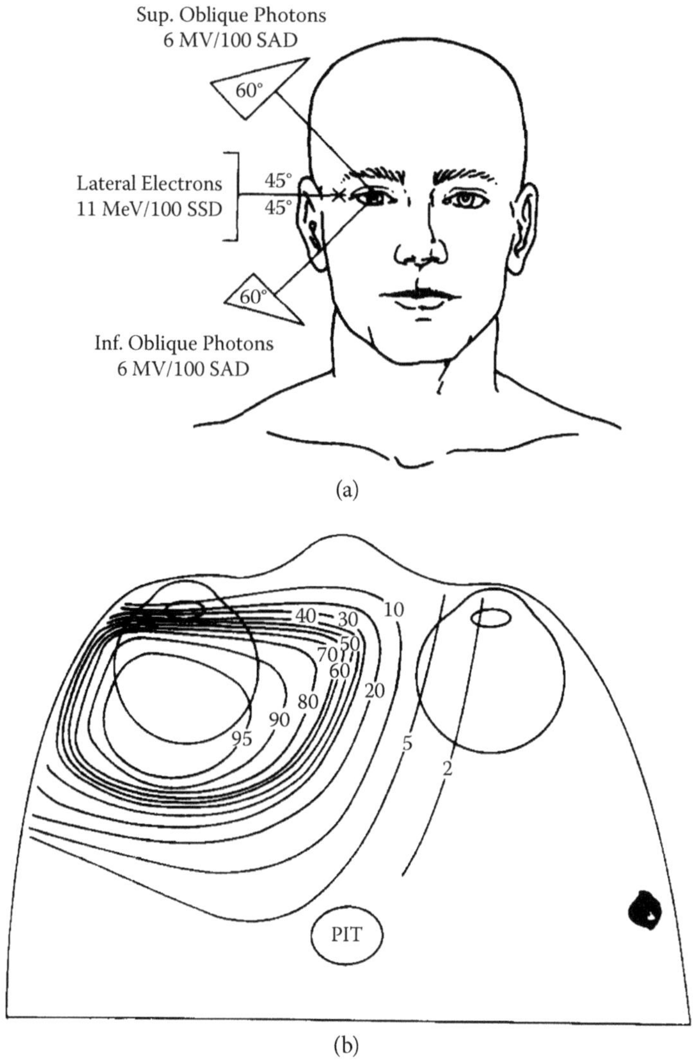

FIGURE 10.10 (a) Contact lens mounted eye shield with three-field technique, superior and inferior oblique 6 MV 60 degree wedged photon, and lateral 11 MeV electron beam. (b) Resulting dose distribution. (From McCormick, B., et al., *International Journal of Radiation Oncology • Biology • Physics* 15 (3):567–74, 1988. With permission.)

have reported that this value is at least 50%. Forty-four Gy was given in 2 Gy fractions to the 90% isodose, but the anterior nasal aspect of the retina was somewhat underdosed (McCormick et al. 1988). The lens-sparing technique resulted in a higher than expected local failure rate due to dosimetric underdosage of the retina. The new MLB technique reduced local relapse but did not significantly improve eye survival. This study was updated in 1996 by Blach et al. (1996) who reported on the clinical results of treatment of 180 eyes in 123 children, the largest cohort of children treated as a single institution at the time. They found significantly better local control in RE group I to III eyes with the MLB technique, 84% versus 37%, but no difference for higher stage eyes due to the successful use of various eye-sparing salvage techniques such as cryo- and laser therapy. Overall survival was 87% at 8 years. Blach et al. presented an analysis of nine studies between 1955 and 1991 with radiation doses between 35 and 60 Gy showing eye survival of 70% and cataract rate of 5% to 66%. However, smaller studies have reported the ALS technique to be superior (Foote et al. 1989).

Scott et al. (1999) performed a comparison of two methods, the MLB versus a technique that more thoroughly treated the globe while still providing some lens sparing, and found the same eye conservation rate but a better tumor control rate without salvage in the latter method due to a higher frequency of anterior recurrences in the former.

In a recent comprehensive review of 13 treatment techniques, Reisner et al. (2007) recomputed the dose distributions for each using three-dimensional treatment planning software according to the details given by each author and included their own IMRT technique. The plans were applied to the same CT data set, based on a 2-year-old patient with unilateral RB. They calculated dose statistics for the ora serrata, lens, orbit, vitreous, optic nerve, lacrimal gland, and cornea. All plans were calculated to give 45 Gy to a point in the posterior retina. The techniques that reached a minimum dose of 45 Gy to the ora serrata were the IMRT technique and those described by Haye et al. (1985), Cassady et al. (1969), Cormack et al. (1998), and Al-Beteri and Raeside (1992); however, none of these limited the lens dose to below 12 Gy although the IMRT and Cormack plans were best in limiting the volume of orbit getting 20 Gy (about 40%) and the IMRT plan was also best in limiting the lacrimal gland volume getting 34 Gy to just 14% (Reisner et al. 2007). It should be noted that the doses tabulated in the Reisner paper often differed substantially from those stated in the actual references, perhaps due to the recalculation.

Many of the techniques mentioned used a central lens block of approximately 1 cm diameter in either a photon or electron beam. The dosimetric result of this seemingly simple approach actually depends on several factors, including beam energy and distance from block to surface. Das et al. (1990) studied these effects for photon beams and concluded that the lowest lens dose was achieved by having the lens block about 0.5 cm from the surface. For electron beams, the lens block was generally in contact with the eye using a supportive device.

More recently, fractionated stereotactic radiosurgery techniques (SRT) have been reported for treatment of solitary locus of RB. Sahgal et al. (2006) used SRT, 40 Gy in 2 Gy fractions, to treat tumors of the posterior retina, referred to as perimacular or peripapillary tumors. The Tarbell–Loeffler–Cosman headframe was used to secure and position the patient. The planning CT was acquired with 1 mm thick slices. A 1 mm PTV margin was treated with four or five noncoplanar arcs using a 10 to 20 mm diameter circular collimator. The median lens dose was about 4 Gy and the median orbital bone dose was less than 7 Gy. Out of five patients treated, one patient recurred in the high dose volume and one outside (Sahgal et al. 2006). Stereotactic conformal therapy with fixed beams has been used as an alternative to plaque radiotherapy for small solitary tumors. CT planning with 2 mm slices and 3 mm PTV was used to plan the treatment and the Brainlab (Brainlab, Westchester, Illinois) mask system was used for immobilization. Six fixed noncoplanar beams shaped by a micromultileaf collimator (micro-MLC) were used to deliver 45 Gy to the 95% isodose surface in 1.8 Gy fractions. The maximum dose was 113%, the maximum lens dose was 6 Gy, the mean dose to the ipsilateral lacrimal gland was 14 Gy, and the V20 for the ipsilateral orbital bones was 22%. These doses compared favorably with the radioactive eye plaque plan (Eldebawy et al. 2010). Pica et al. (2011) also reported treatment of papillary or macular RB with stereotactic radiotherapy with a micromultileaf collimator. They gave 50.4 Gy in 1.8 Gy fractions after relapse or progression after chemotherapy by fixing the treated eye with a vacuum contact lens system integrated into the mask for maintaining the precise position of the eye (Figure 10.11, inset). CT with 1.25 mm slices was used for planning and a 2 to 2.5 mm PTV margin was used except for papillary tumors where a 4 mm margin into the optic nerve was applied. Similarly, good target coverage and low doses to the important structures were found as in the other reports where stereotactic techniques and small localized volumes were treated (Figure 10.11). Local control for the 15 mostly group B and C eyes treated was 87% (Pica et al. 2011).

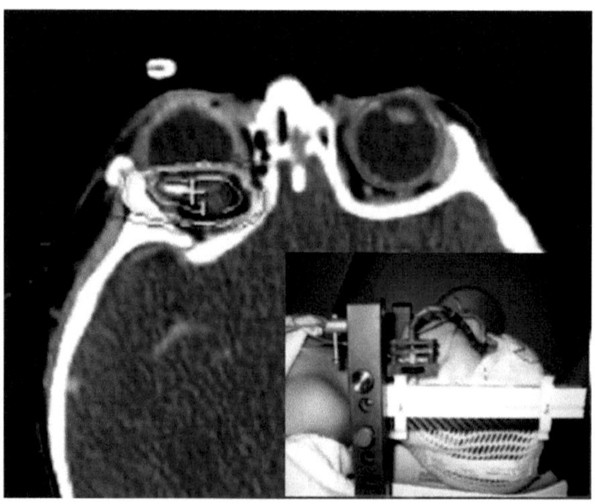

FIGURE 10.11 Vacuum lens and stereotactic mask immobilization (inset) for SRS treatment of a macular RB. (From Pica, A., et al., *International Journal of Radiation Oncology • Biology • Physics* 81 (5):13806, 2011. With permission.)

The normal tissue sparing properties of IMRT has been used to try to treat the entire eye while sparing all the relevant structures. Filion et al. (2006) and Krasin et al. (2004) reported on their experience with IMRT for RB. Either a 2 or 3 mm PTV margin was used with a margin of up to 1 cm around the optic nerve as it inserts into the back of the globe. Both groups compared their IMRT plans to various three-dimensional conformal radiation therapy (3DCRT) plans. Filion et al. used seven coplanar IMRT fields and compared these plans to an anterior oblique wedged pair or an SRS plan with five noncoplanar micro-MLC fields. Krasin et al. compared their four noncoplanar IMRT field plan to a wedged photon beam plan, anterior and lateral photon beam plan, or an en face electron plan. In both studies, IMRT was found to provide the greatest sparing of normal structures. Although Filion et al. reported that IMRT was superior to other techniques for brain sparing, each technique resulted in a V10 of less than 15%. Orbital bone sparing was marginally better with IMRT with a V25 of 75% versus 85% for 3DCRT. Krasin et al. focused on sparing of the bony orbit. They gave 45 Gy to the PTV (whole eye) and found a 23% to 33% reduction in the ipsilateral orbital bone V20 using IMRT with four noncoplanar beams (V20 was 60%) compared to the other techniques. They also found that the monitor units (MUs) given by IMRT was less than that of the plan with a pair of 60 degree wedges, and the integral dose was less for the IMRT plan than any of the other plans (Krasin et al. 2004). No lens sparing was attempted in either of these studies. In the Reisner et al. (2007) study, their IMRT plan did spare the lens and they also achieved substantial sparing of the orbital bone (V20 was 44%) compared to all but one other non-IMRT techniques.

At Children's Hospital Los Angeles, a novel eight-beam IMRT plan has been used since 2001 to treat whole globe, globe with lens sparing, or macular target volumes. The beam angles for a typical plan to treat the left eye are shown in Figure 10.12. These beams form an anterior cone whose axis is approximately aligned with the lens-to-optic nerve axis. A multileaf collimator with 0.5 cm leaf width is used. Smaller leaf widths would produce even better plans and a 1 cm leaf width is not recommended for treatment of this small volume. For a localized target without vitreous seeding, lens, pituitary, orbital bone, and possibly lacrimal gland sparing can be achieved. The planning CT scan is performed with 1.25 mm slices in the region of the orbit and 2 mm slices superior and inferior to that. The HeadFix (Elekta, Norcross, Georgia) vacuum-assisted mouthpiece immobilization system is used for daily reproduction of head position routinely with less than or equal to 1.2 mm deviations in any one direction (Olch and Lavey 2002). The PTV margin is 2 to 3 mm. Figure 10.13 shows the axial, coronal, and sagittal dose distributions along with the DVHs of each structure for a 36 Gy PTV dose. The mean lens dose was less than 9 Gy, the mean pituitary dose was less than 3 Gy, and the mean lacrimal gland dose (seen in the coronal view)

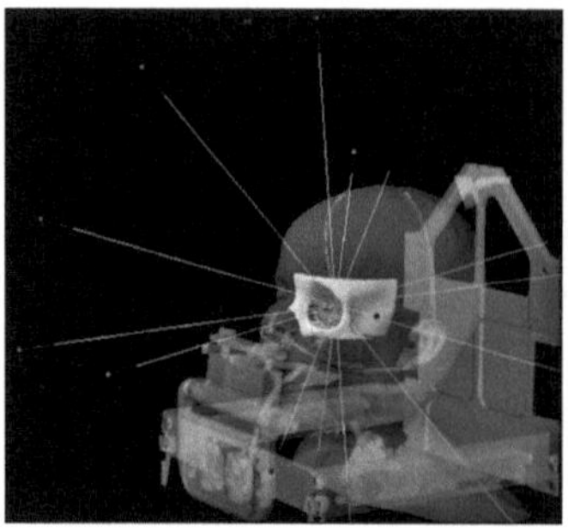

Beam Name	Gantry Angle (IEC)	Couch Angle (IEC)
LAIO1	47	24
LAIO2	13	25
LASO1	65	319
LASO2	77	281
LASO3	62	351
RAIO	330	344
RASO1	315	19
RASO2	300	60

FIGURE 10.12 Left eye eight-beam IMRT beam angles.

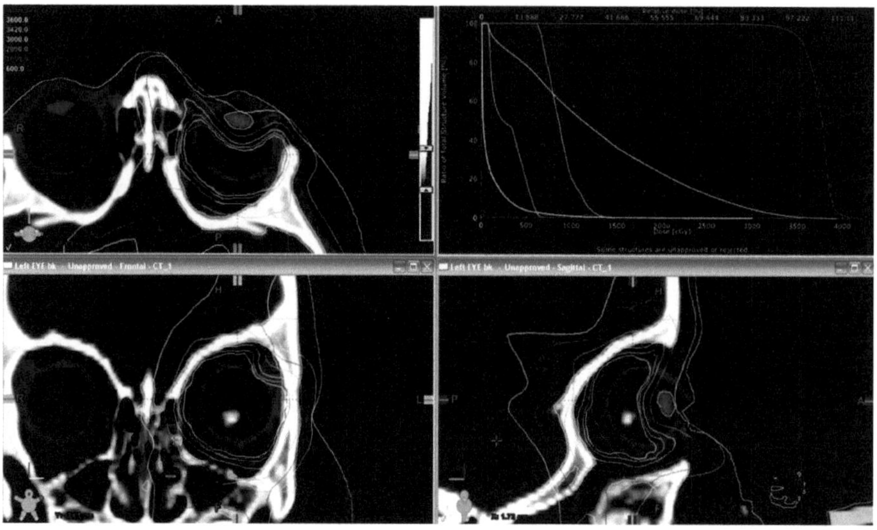

FIGURE 10.13 (See color insert.) Whole retina IMRT with lens and lacrimal gland sparing. 36 Gy PTV, mean lacrimal gland dose (blue) = 24 Gy, mean lens dose (pink) = 9 Gy, V20 orbital bone (tan) = 20%, V15 orbital bone (tan) = 35%, mean pituitary dose (light green) = 3 Gy.

was 24 Gy. The V15 and V20 for orbital bone was 35% and 20%, respectively. Only a very small volume of brain gets 10 Gy. The orbital bone that was contoured for a unilateral case is seen in Figures 10.12 and 10.15. Orbital bone DVH is greatly affected by exactly how this volume is drawn. When the whole globe is the target, custom bolus is created by applying a layer of dental impression material onto a thin plastic sheet that is placed over the region around the affected eye so the impression material does not get stuck in eyelashes, eyebrows, and on the skin. The dental mold material is applied to the region immediately surrounding the globe as seen in Figures 10.14 and 10.15. After a few minutes of hardening, a very conformal custom bolus is created. The patient is CT scanned with this bolus in place. The dose distribution can be seen in Figure 10.14. When bilateral eyes are treated, the same beam arrangement is used but with the isocenter on the midline between

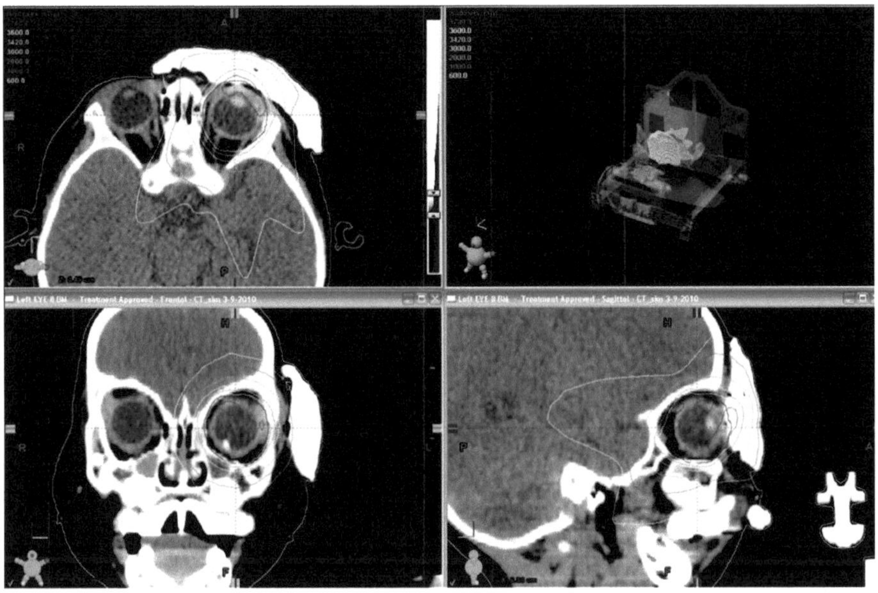

FIGURE 10.14 (See color insert.) Left whole eye IMRT to 36 Gy. Dental impression material was used as bolus.

the eyes. Figure 10.15 shows the comparison of the dose distribution for the eight-beam IMRT treatment and parallel opposed lateral beams conformally shaped to the same PTVs. The DVH for the PTVs are nearly identical but with the IMRT plan, the orbital bone dose is significantly reduced at the dose levels thought to cause bone growth arrest. Doses to the brain, optic chiasm, and pituitary are somewhat higher for the IMRT plan, but none are clinically significant. The decision to try to spare orbital bone that surrounds the PTV must be based on the trade-off between PTV coverage and the clinical significance of the orbital bone sparing achieved. An example of this trade-off is shown in Figure 10.16. A modest reduction in PTV coverage is required to produce a decrease in V20 from about 37% down to about 27%. Another way to look at this is to see that the D30 is reduced from 24 Gy down to 18 Gy. If lens sparing is to be attempted with IMRT or other highly conformal techniques, quality assurance measurements should be made with appropriate-sized detectors for both absolute dose and to determine if the dose distribution being delivered is as planned. Figure 10.17 is a coronal film in phantom through the location of the lens of the eye showing the lens sparing, which compared well to planning system dose calculations in that same plane.

Proton therapy has also been investigated as a normal tissue sparing technique for RB. Massachusetts General Hospital, Boston, evaluated proton plans using either a single lateral, anterolateral oblique, or anteromedial oblique beam to treat posterior, nasal, or temporal tumor locations to 46 CGE with lens sparing. A vacuum-assisted silicon corneal suction cup

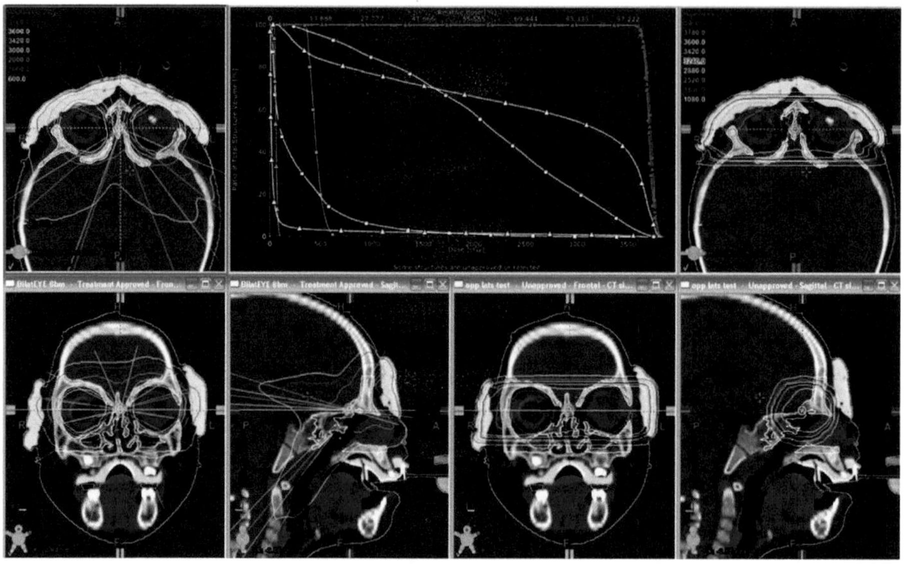

FIGURE 10.15 (See color insert.) Bilateral whole eye RB treatment with 36 Gy PTV dose. Opposed laterals (right panels) versus IMRT (left panels). DVH triangles are opposed laterals; red, PTV; white, orbital bone; yellow, brain tissue; pink, pituitary; blue, optic chiasm.

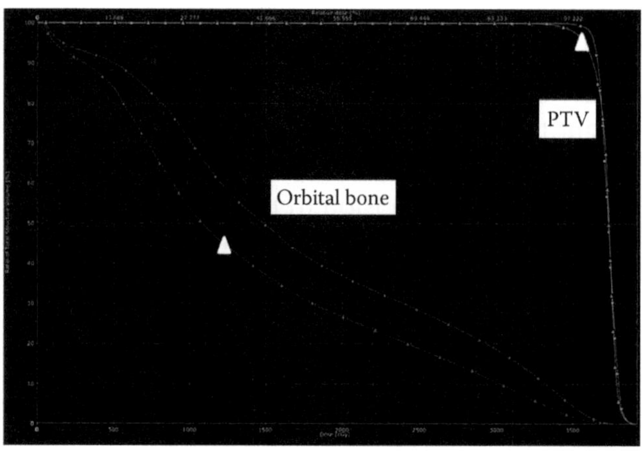

FIGURE 10.16 DVHs with triangles are for orbital bone sparing.

was used to immobilize the eye but also to rotate it so that the tumor could be positioned optimally relative to the beam and critical structures (Figure 10.18). The outlined organs at risk included the lens, orbital bone and soft tissues, optic nerve, lacrimal gland, contralateral eye, temporal and frontal lobes of the brain, and pituitary gland. Three bone growth centers within the orbit were identified. The mean lens dose was less than 3 Gy for all beam and tumor arrangements. The lacrimal gland could not be spared except for the nasal tumor location. Only a small volume of orbital bone received 20 Gy, but the doses to each of the three bone growth centers was calculated for

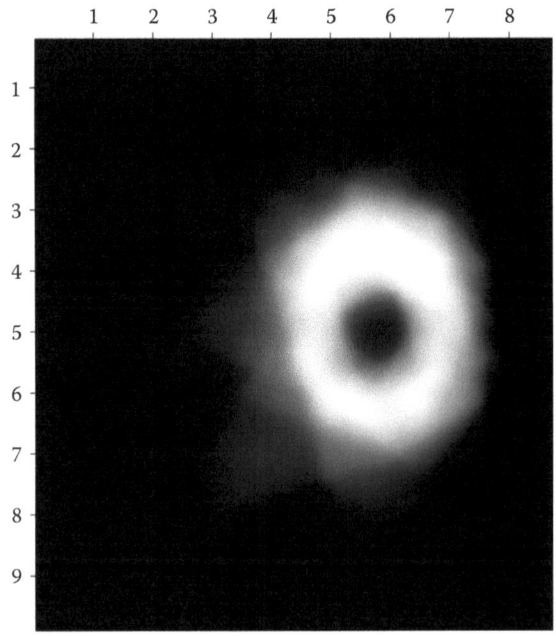

FIGURE 10.17 IMRT QA (true composite film irradiation) for eight-beam lens-sparing IMRT plan through the level of the lens.

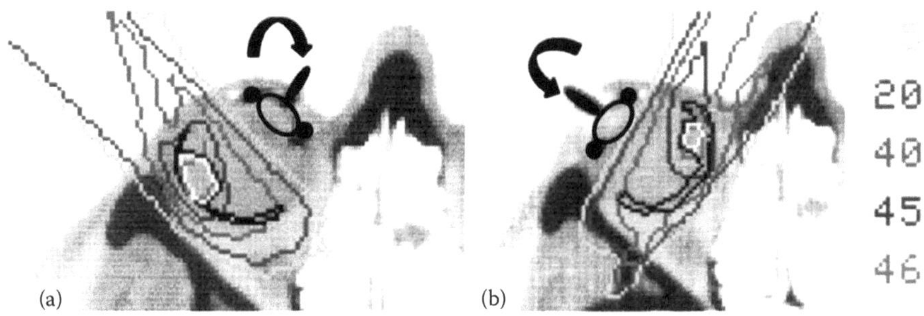

FIGURE 10.18 Eye rotated by movement of corneal suction cup to facilitate lens sparing for proton therapy. (a) Temporal tumor location treated with anterolateral oblique beam. (b) Nasal tumor location treated with anterolateral beam. (From Krengli, M., et al., *International Journal of Radiation Oncology • Biology • Physics* 61 (2):583–93, 2005. With permission.)

each plan and the anterolateral technique best spared bone. None of the plans delivered dose to any intracranial structure (Krengli et al. 2005). MD Anderson Cancer Center published their comparison of several photon beam plans with their proton beam plan for RB treated to 36 Gy without lens sparing. Their proton technique was a single anterior beam with up to a 10 degree angle with the vertical (Figure 10.19). Protons resulted in the best target coverage and the most orbital bone sparing, 3% of orbital bone volume received more than 20 Gy with protons versus 7% to 22% for photon techniques (Lee et al. 2005). Although V20 is commonly used as an index

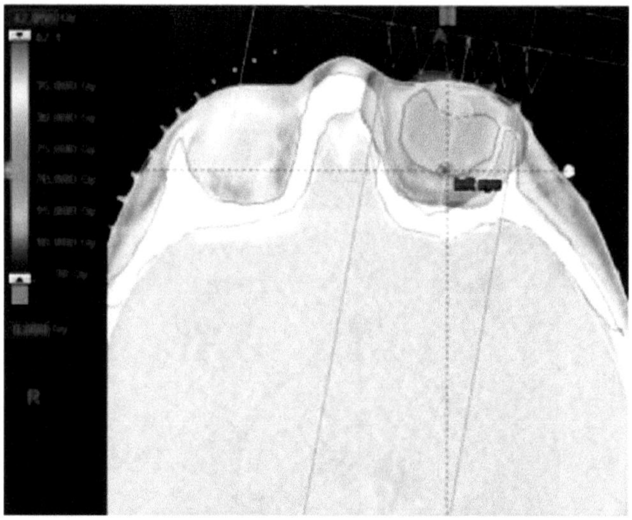

FIGURE 10.19 Dose distribution for an anterior oblique proton beam treatment of the whole eye. (From Lee, C. T., et al., *International Journal of Radiation Oncology • Biology • Physics* 63 (2):362–72, 2005. With permission.)

for bone growth in children, V5 was also calculated and was 10% for protons compared to 25%, 50%, and 69% for direct electrons, 3DCRT techniques, and IMRT, respectively. It is not clear that V5 is a meaningful metric for orbital bone growth cessation.

10.3.2 Episcleral Plaque Radiotherapy (EPRT)

As early as the 1920s, brachytherapy with radon seeds was used for small, localized RB. Up to five 1 to 5 mCi radon seeds were freely inserted into the retinal mass for 10 days. It was noted that frequently the seed would migrate beyond the tumor and float in the vitreous and have to be retrieved. Reduction in the size of the tumor was seen in most cases (Moore et al. 1931). By 1948, Stallard had designed special applicators or plaques, containing Co-60, which could be surgically fixed to the sclera at a location adjacent to the tumor. These plaques were later modified by Rosengren and Tengroth (1963). Total doses of 33 to 64 Gy at a depth of 2 mm were given and tumor control was obtained (Rosengren and Tengroth 1977). The use of 60Co persisted through the early 1990s by which time it became evident that the highly penetrating gamma rays were responsible for late effects of the critical ocular structures (Fass et al. 1991) as well as an increased radiation exposure to the patient and staff. However, much lower energy sources were becoming available that could address these problems. In Germany in the 1960s, a beta-emitter, Ru-106 deposited inside a silver disk, had become available (Lommatzsch and Vollmar 1966). It became more widely used after about 1980. In 1980, Sealy et al. (1980) published their techniques using

I-125 seeds in gold-backed plaque applicators for the treatment of ocular tumors including RB. It was apparent from dosimetric studies that as energy increased, scleral dose decreased but the ability to spare normal ocular structures also decreased. This observation motivated the use of lower energy radionuclides. In addition to I-125 seeds, Ir-192 and Pd-103 seeds have also been used for EPRT (Hernandez et al. 1993; Shields et al. 2001, 2002, 2006; Stannard et al. 2001, 2002; Merchant et al. 2004). Much of the work in EPRT has been applied to adult ocular melanoma as well as RB.

Indications for EPRT are (1) solitary tumor with diameter 6 to 15 mm, (2) tumor thickness 10 mm or less, and (3) location of lesion more than 3 mm from optic disc or fovea. EPRT has been used alone and as a boost after EBRT for advanced tumors where photocoagulation or cryotherapy cannot be used or has failed. In most cases, especially when treatment is for the child's only remaining eye, the goal is avoidance of enucleation and EBRT while maintaining vision.

10.3.2.1 Iodine-125 (I-125) Seeds

The majority of technical and clinical reports of EPRT for RB discuss the use of I-125 seeds inserted into plaques, but Ru-106 has also been of great interest and so applications with both of these sources will be discussed in more detail. The wide availability of low energy (28 keV photon) sealed I-125 sources prompted the development of gold-backed plaque applicators that shielded the normal periocular structures. In the paper by Sealy et al. (1980), I-125 seeds were glued to a gold-foil-backed applicator individually made for each patient. The applicator was positioned by the method of visualization by indirect ophthalmoscopy of the indentation made by pushing a forceps on the proposed site of the applicator. Forty Gy was given to the apex of the RB. More complex applicators were described that included wedge filters and acrylic filling to shape the isodoses away from critical structures, but no details were given about dose computations. An early paper on the dosimetry of I-125 seeds in eye plaques discussed the ability of a treatment planning computer to calculate the dose accurately in the short distances required for plaque therapy. TLD measurements were used to compare with calculations, both for a single seed and for an arrangement of seeds in the eye plaque, and reasonable agreement was found. An incidental finding was that the gold plaque reduced the dose at all measured depths by about 8% compared to seeds in paraffin due to the reduction in scatter from the plaque side of the seed array (Weaver 1986). This finding was confirmed a few years later by Luxton, Astrahan, and Petrovich (1988) at the University of Southern California (USC). The limitations of brachytherapy algorithms of the time were noted and led others to improve them. Luxton also reported on the relative dosimetry of I-125 as compared to Ir-192 and Co-60. TLD measurements showed that the I-125 plaque at the posterior pole was able to reduce the lens dose by 25% compared to the higher energy sources. This paper also described the USC plaques, which were made of gold alloy cast by the lost wax technique and had a pattern of grooves embedded in the concave surface into which seeds could be glued. A detailed dosimetric analysis based on measurements and calculations was presented (Luxton, Astrahan, Liggett, et al. 1988). Luxton also noted that the exposure rate to others near the patient was reduced by 97% for I-125 with the gold plaque present compared to no plaque. They also noted that a 0.5 mm thick lead eye patch placed over the affected eye was sufficient to reduce the exposure rate to less than 1 mr/hr at 1 m. A comprehensive exposure survey for I-125 eye plaque patients agreed with Luxton's findings and concluded that only about 3% of the monitored persons had registered a measureable whole body dose that averaged about .004 mSv (Al-Haj et al. 2004).

In 1990, Astrahan and colleagues first described his 3D treatment planning computer software for I-125 eye plaque dosimetry running on a Macintosh computer. A 3D rendering of the plaque positioned on a spherical globe with tumor volume, seed locations, and isodoses was displayed. Source activity, decay, anisotropy, and collimation of the primary photon flux by the gold shell were accounted for and later versions also accounted for geometric penumbra, fluorescent x-rays from the gold plaque, and secondary scatter from the plaque and interfaces, attenuation in the silicone seed carrier of COMS style plaques and slit collimation of individual seeds for the USC style plaques. The plaque and seed loading could interactively be adjusted with the dose calculation updated in near real time (Astrahan, Luxton, Jozsef, Kampp, et al. 1990). In another publication, Astrahan and colleagues described an improvement in the anatomical accuracy of his system by using patient-specific ocular anatomy instead of a spherical globe. The patient-specific information was derived from CT or MRI images, fundus photography, and ultrasound. Optimization of the source activity and position within the planning system for four test cases demonstrated that an improvement in tumor coverage and normal tissue sparing was possible. The potential for such optimization for plaques with I-125 seeds was an important advantage over Ru-106 plaques (Astrahan, Luxton, Jozsef, Liggett, et al. 1990). The details of the utilization of fundus photography in the planning of EPRT was described, showing that sub millimeter accuracy of plaque placement could be achieved. In addition, a polar diagram of the location of the suture holes of the plaque relative to the limbus, ora, and equator of the eye was developed to aid the surgeon in precise placement of the plaque at the time of surgery (Evans et al. 1993), an innovation that represented a major improvement over the method of visualization of the indentation of a forceps described by Sealy. Astrahan et al. (1997) further refined the slotted plaque concept by deepening the slots to produce a collimated beam

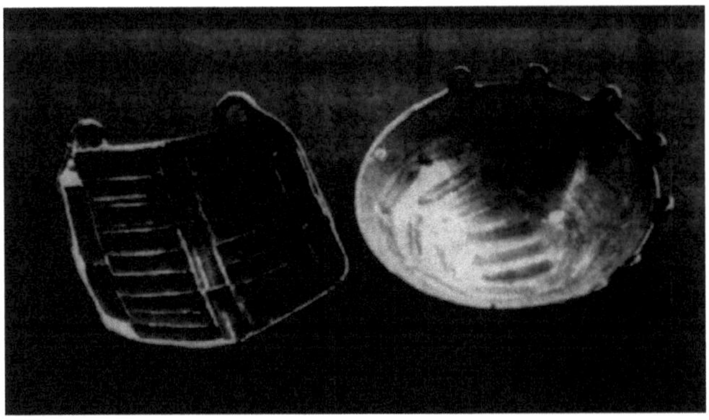

FIGURE 10.20 A deeply slotted plaque (left) compared to the more standard COMS plaque (right). (From Astrahan, M. A., G. Luxton, Q. Pu, and Z. Petrovich, *International Journal of Radiation Oncology • Biology • Physics* 39 (2):505–19, 1997. With permission.)

from each seed (Figure 10.20). This served to reduce the ratio of the scleral to apex dose from about 4:1 down to just 2:1 and for a posteriorly located tumor, the macular dose was reduced from 80% of the apical dose to just 20% (Figure 10.21). TLD measurements were made to confirm the accuracy of the dose calculation algorithm in estimating the effect of the collimation. The slotted plaque was also faster and easier to load than the conventional plaques (Astrahan et al. 1997).

In 1995, American Association of Physicists in Medicine (AAPM) Task Group 43 published new recommendations for dosimetry of I-125 seeds, which effectively lowered the dose by 9% to 17% for implants based on prior dosimetry methods (Nath et al. 1995). Ray et al. (1998) reviewed these changes specifically as they related to eye plaque dosimetry. Based on point and line source calculations for different plaque loading patterns, they also found that the new formalism would result in about 10% less dose being delivered than was prescribed. Later, an update to the TG43 report suggested further changes to the dose calculation formalism (Rivard et al. 2004). Using the modern dosimetry formalism, dose distribution tables were calculated for different seed loading patterns for Collaborative Ocular Melanoma Study (COMS) eye plaques using Pd-103, I-125, and Cesium-131 (Cs-131) sources. This report concluded that for a particular dose at the prescription point, lower photon energies resulted in higher scleral doses for a range of plaque sizes and tumor heights. However, for tumors less than 4 mm in height, Pd-103 produced better dose distributions than I-125 (Rivard et al. 2008). Looking at the effect of the presence of the plaque itself on the dose distribution, for COMS plaques, three TG43-based TPSs and two Monte Carlo (MC) systems were compared for ophthalmic plaque brachytherapy dose calculations. The TG43 TPS calculations agreed with MC on the central axis with or without heterogeneity corrections that accounted for the influence of the

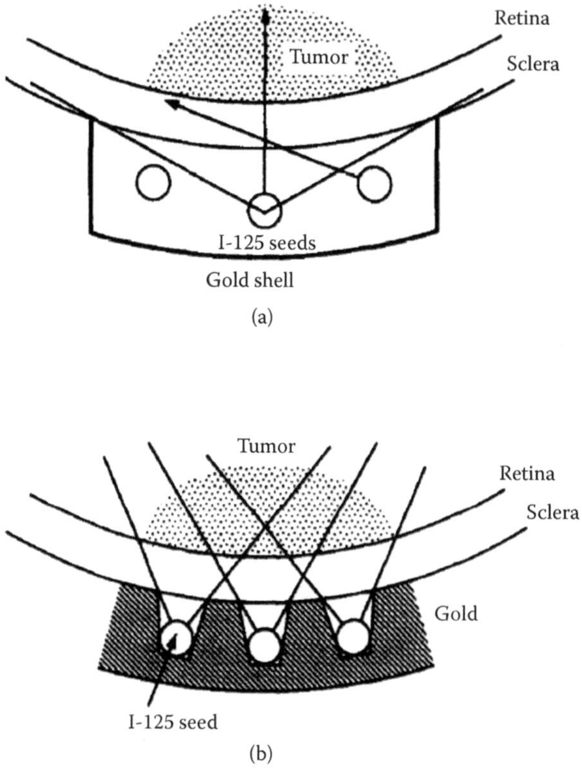

FIGURE 10.21 (a) Conventional plaque design, (b) slotted design that collimates radiation to avoid unwanted overlapping laterally directed radiation. (From Astrahan, M. A., G. Luxton, Q. Pu, and Z. Petrovich, *International Journal of Radiation Oncology • Biology • Physics* 39 (2):505–19, 1997. With permission.)

plaque itself on the dose. The difference between applying the heterogeneity corrections or not were substantial, up to 20% to 37% lower doses for I-125 and Pd-103, respectively. Much larger differences than that were found for off-axis points, and were particularly large in the plaque penumbra region. These data call into question the historical dose–response relationships for ocular critical structures (Rivard et al. 2011). Non-COMS plaques need to be studied separately. Astrahan et al. (2005) showed earlier that his slotted gold plaque had dosimetric as well as practical advantages over the silicone insert in the COMS plaque. Because the effective Z of silicone is about 11 compared to 7.4 for water, three times more photoelectric attenuation occurs in the silicone than water, resulting in about a 10% dose reduction compared to water at the tumor apex where the dose is prescribed (Astrahan et al. 2005).

Clinical results indicate that with doses of about 44 Gy to the apex, eye preservation and tumor control rates of 60% and nearly 100%, respectively, can be achieved although frequently additional whole eye treatment with EBRT is needed within about 1 year (Hernandez et al. 1993; Friedman et al. 2000; Stannard et al. 2001; Merchant et al. 2004; Shields et al. 2006).

10.3.2.2 Ruthenium-106

Ruthenium-106 (Ru-106), a beta emitter, has been seen as an attractive alternative to low-energy photon sources because of its ease of use; there are no seeds to arrange and adhere to the plaque. The ruthenium plaque is constructed with a 0.1 micron thick film of Ru-106 electrically deposited on a 0.2 mm thick silver foil, which is encapsulated by a 0.1 mm silver foil window on the concave side and 0.7 mm thick silver backing on the opposite side. The silver backing layer absorbs about 95% of the 3.5 MeV maximum (1.4 MeV mean) beta radiation. Ru-106 has a half-life of about 1 year so it potentially can be reused multiple times. The manufacturer, Bebig (Berlin, Germany), offers about 20 different sizes of plaques, which each cost about $5000. Because no one plaque can address every patient's tumor size and shape, a range of plaque sizes needs to be available. Radiation safety is even less of a concern for ruthenium than I-125 due to the short range of the beta particles. In fact, ruthenium is used for tumors of no more than about 6 mm thickness compared to a usual limit of about 10 mm for I-125 (Amendola et al. 1989; Fluhs et al. 1997; Schueler et al. 2006; Abouzeid et al. 2008). Lommatzsch and Vollmar were perhaps the first to publish their clinical experience with Ru-106 eye plaques for preservation of vision in 1966. Since then, several dosimetry and clinical papers have been published. Commercial treatment planning systems do not generally come equipped to calculate the dose from a ruthenium eye plaque, but most can do so for I-125 seeds. The manufacturer supplies a central axis depth dose table for each plaque and about 30 relative dose values in a plane 1 mm from the plaque's inner surface along with the calibration certificate but a 3D dose distribution is not supplied. The absolute dose rate uncertainty quoted by Bebig for their plaques was 20%. One approach to obtaining 3D dosimetric information used a linear array of 1 mm^3 plastic scintillators, which could be rotated and translated to measure a 3D dose distribution. This system was used to characterize and optimize plaque therapy (Fluhs et al. 1997) and was compared to Monte Carlo calculations. A higher resolution 3D measurement was performed more recently based on BANG gel dosimetry (Chan et al. 2001). Neither of these methods is practical for most people. An international intercomparison of the dosimetry of beta particle emitting eye plaques was performed and reported in 2001. Several types of dosimeters were used to measure the absolute dose rate to water at 1 mm from the concave applicator. This comparison revealed a range of 3% to 35% in measured dose and a standard deviation across measurement methods of about 10% (Soares et al. 2001). The detectors deemed most suitable for the measurement agreed within 3%. These data demonstrate that there can be a relatively high degree of uncertainty in knowing the dose at a point for Ru-106 applicator treatments, but due to the high dose gradient, this uncertainty is represented by just tenths of a millimeter in depth. Schueler et al. (2006) mentions that the change in calibration methods due to the International Commission on Radiation Units and Measurements (ICRU) report 72 in 2004 resulted in recalculation of their doses by up to a factor of 2 with the greatest changes occurring for the smaller plaques and for tumors with larger height. Based on these corrected doses, Schueler et al. gave a table of dose rates and total doses for the tumor height and diameters from 8 Bebig Ru-106 plaques used in his study.

Astrahan (2003) adapted his Plaque Simulator software to handle ruthenium dose calculations along with the other interactive features described earlier. This 2003 paper gives an excellent description of the source and the various plaque configurations available along with the dosimetric methodology developed for use by the Plaque Simulator. The method is based on subdividing the active surface into hundreds of hexagonal "patch sources" that are approximated as small discs and to which a beta-source analogue of the TG 43 formalism can be applied for dose calculation. The source strength of each patch can be adjusted by way of a simulated annealing process to closely reproduce the measured dose distribution of each Ru-106 plaque. Comparison to Monte Carlo calculations and radiochromic film measurements showed good agreement and was within the 10% error of measurement reported by Soares et al. (2001).

Two recent large patient cohort studies of ruthenium EPRT for RB have shown excellent local control and eye retention (Schueler et al. 2006; Abouzeid et al. 2008). The larger study from Essen, Germany, involved 134 patients and 140 eyes (Schueler et al. 2006). Their 5-year tumor control and eye preservation rates were 94% and 87%, respectively. They gave 138 Gy to the tumor apex but later reduced that to 88 Gy based on their good control rate and relatively high rate of late effects, including 54% incidence of cataract and greater than 20% rate of retinopathy and optic neuropathy. Another study reported on 41 eyes given a mean apex dose of 55 Gy (Abouzeid et al. 2008). It reported a 1-year local control and eye retention of 73% and 76%, respectively, and much fewer late effects than the Essen study. Perhaps the best dose is somewhere between 55 and 138 Gy.

10.4 Organ-at-Risk Doses and Late Effects

The most serious late effect for hereditary RB patients treated with radiotherapy is secondary malignancy, which is often fatal. Other serious quality of life late effects for all RB patients are orbital and facial bone deformity, and loss of vision due to optic neuropathy, retinal detachment, cataract, and other damage to ocular structures. Although the overall success rate for RB treatment is high, almost half of all patients develop some ocular or orbital complication (Anteby et al. 1998). Gordon et al. (1995) provided an excellent overview of visual late effects of radiation.

10.4.1 Second Malignancy (SM)

Hereditary RB patients have a genetic predisposition to develop secondary nonocular malignancies due to the mutation of their RB1 gene, whether or not they receive radiation. Radiation and alkylating agents and anthracycline chemotherapy, alone or combined with radiation, have been shown to increase the risk of second malignancy (SM) (Newton et al. 1991; Hawkins et al. 1996; Le Vu et al. 1998). Thirty-five different histological types of SMs have been reported in patients treated for RB (Moll et al. 1997). Osteosarcomas are the most common, but sarcomas of any type account for over 60% of SMs (Kleinerman et al. 2005), in some studies 90% (Smith et al. 1989). Malignant melanomas are also common. Some of the other sites of SMs include brain, female breast, lung, and uterus; leukemia rarely occurs (Kleinerman et al. 2005).

SM data from 1600 RB patients treated between 1914 and 1984 at New York Hospital was analyzed by Wong et al. (1997) and Abramson and Frank (1998), who reported that the cumulative incidence of SM 50 years after diagnosis for bilateral RB patients was 53% for irradiated and 22% for unirradiated patients. The SM rate was just 5% for nonhereditary RB patients (Wong et al. 1997; Abramson and Frank 1998). That same data revealed that the excess cancer risk was mostly due to high doses given in the 1930s to 1960s with orthovoltage beams where the orbital bone dose was over 100 Gy (Wong et al. 1997). Other studies of SM in hereditary RB patients reported a cumulative incidence of 19% to 38% at 30 to 40 years after diagnosis (Smith et al. 1989; Moll et al. 1997; Mohney et al. 1998). The SM data also suggests that the SM rate was two- to fourfold higher for hereditary RB patients irradiated before the age of 12 months than after 12 months (Abramson and Frank 1998; Moll et al. 2001).

More recently, Kleinerman et al. (2005) reanalyzed the updated New York data and reported that the cumulative incidence 50 years after diagnosis was a lower but still large rate; for hereditary RB, 38% for irradiated patients and 21% for nonirradiated patients. The nonhereditary rate remained at about 5%. Excess absolute risk per 10,000 person-years was 106 for irradiated and 45 for nonirradiated hereditary RB patients. The excess absolute risk for subsequent cancer after hereditary RB was 60 times greater than nonhereditary. For orthovoltage era hereditary RB patients, the cumulative incidence at 40 years was 33% compared to 26% for patients treated after 1960 with megavoltage beams, consistent with a British study showing a very high incidence for patients treated before 1950 (Fletcher et al. 2004). The overall risk of SM after radiation and chemotherapy for hereditary RB was about 30% greater than for radiation alone (Kleinerman et al. 2005). In another analysis of cumulative mortality from SM 50 years after diagnosis, the rate was 25.5% for hereditary and 1% for nonhereditary RB (Yu et al. 2009). Other recent analyses demonstrate similar findings (Marees et al. 2008).

Because the risk of SM is dose dependent, there has been great interest in finding the lowest dose that can effectively treat RB. Doses of 30 to 36 Gy have been found to be successful, well below the historic 45 Gy dose (Merchant et al. 2002).

10.4.2 Orbital Bone

Because their orbital bone and facial growth is still in progress, children treated for retinoblastoma are at risk of functionally and cosmetically significant bony orbital abnormalities. These effects were particularly pronounced for patients treated with orthovoltage radiation due to the increased absorption in bone. This growth retardation results in the "hour glass" appearance, with depression of the temporal bones, deep orbits, and other facial asymmetries. These conditions become evident by early adolescence, when orbital growth is mostly complete.

Several comprehensive studies of these effects can be found in the literature (Guyuron et al. 1983; Egawa et al. 1987; Guyuron 1990; Imhof et al. 1996; Yue and Benson 1996; Kaste et al. 1997). Orbital bone doses less than 14 Gy minimized growth defects in one study (Egawa et al. 1987) while another stated that 35 Gy was significantly worse than 22 Gy (Kaste et al. 1997). Bony changes were seen after higher doses (>23 Gy) than soft tissue changes (>6 Gy) (Guyuron et al. 1983). Data are inconsistent as to whether irradiation younger or older than 1 year of age is more harmful (Imhof et al. 1996; Kaste et al. 1997).

10.4.3 Cataract

Cataractogenic radiation damage was discussed in prior chapters in the context of cranial irradiation or periorbital sarcomas. One difference for RB patients is the very young age with the potential for disuse ambylopia (lazy eye). The other difference is the fact that the eye itself, with or without the lens, is the target of the radiotherapy. The risk of cataract formation is related to the dose to the germinative zone in the equatorial lens epithelium, which is the only place in the lens where mitosis occurs. The germinative zone is situated at an axial distance of one-third to one-half the lens thickness from the anterior pole. Although cataracts can be surgically removed and is usually successful (Portellos and Buckley 1998; Honavar et al. 2001), occasionally it can cause recurrence or extraorbital extension of the disease (Brooks et al. 1990). Also, an opacified lens makes it difficult to look into the eye and detect recurrent RB (Osman et al. 2010). Thus, avoiding cataract formation is a worthwhile goal if feasible. In healthy eyes, cataracts develop 1 to 3 years after irradiation (Hungerford et al. 1995; Anteby et al. 1998).

For all of the technical challenges in lens-sparing treatments, do they actually result in avoidance of cataract without an increased risk of recurrence? In whole eye studies, where the rate of new tumors of the anterior retina was low, the cataract rate was high (Hungerford et al. 1995; Toma et al. 1995). Looking across external beam series, those with extremely low cataract rates frequently but not always reported higher local recurrence rates (Egbert et al. 1978). The large variation in reported cataract rate may be explained by the fact that the dosimetry is uncertain for lens sparing where a high dose gradient (dose changes of 10% over 1 mm) is placed just behind the lens even when the eye was immobilized and the lack of precision of the dosimetry in the region of the dose gradient. Lens dose was roughly 30% of the prescribed dose in the lens-sparing studies reviewed by Reisner et al. (2007). About one-quarter to one-third of the saved eyes developed cataracts (Foote et al. 1989; Amendola et al. 1990; Blach et al. 1996; Schipper et al. 1997; Scott et al. 1999). With the contact lens technique, the lens dose was reported to be 5 Gy with less than a 10% cataract rate (Phillips et al. 2003). Schipper et al. (1997) noted that if no more than 1 mm of the posterior lens is included in the high gradient of a lateral beam, the child will rarely develop a cataract.

Eye plaque therapy also causes cataract. One confounding factor is that most eyes had prior EBRT. For Pd-103 plaques, the cataract rate was 90% for lens doses >20Gy, but 30% of patients receiving 10 to 19 Gy and half of the patients receiving between 4 and 9 Gy were projected to stay cataract-free. Shields et al. (2006) found that a 5 Gy lens dose using I-125 seed plaques after chemoreduction resulted in a 43% cataract rate, but a 10 Gy lens dose with plaque alone resulted in a 33% cataract rate (Shields et al. 2001).

10.4.4 Lacrimal Gland (Dry Eye)

Dry eye is rarely observed at external beam doses below 30 Gy but occurs at a rate of 5% to 25% with doses of 30 to 40 Gy (Parsons et al. 1994a). Although dry eye is common after EBRT for RB, it is seldom the cause of visual impairment at doses below 45 Gy even when combined with chemotherapy and when it occurs, is manageable.

10.4.5 Retinopathy

Radiation-induced retinopathy was discussed in Chapter 8 on soft tissue periorbital sarcomas, but for RB, the retina is the target tissue. Several adult studies established that below 45 Gy, retinopathy was rarely seen, but the risk increases rapidly as the total dose increases and for daily doses greater than 1.9 Gy (Coucke et al. 1993; Parsons et al. 1994a; Anteby et al. 1998; Takeda et al. 1999). The time from irradiation to retinopathy was approximately 2 years (Coucke et al. 1993; Parsons et al. 1994a). Radiation retinopathy occurred in at least 50% of patients who received 60 Gy to a large fraction of the retina (Coucke et al. 1993; Parsons et al. 1994a; Takeda et al. 1999). There appears to be a decrease in risk of retinopathy for cases where less than half the retina was in the high dose region (Parsons et al. 1994a; Takeda et al. 1999).

Retinopathy for plaque therapy is a concern due to the fact that the radioactive sources are placed directly on the sclera, just 1 mm from the retina. EPRT frequently follows EBRT or other local treatments, each of which can damage the retina. It has been observed that 40 Gy to the retina immediately after chemotherapy is too toxic, so a 6 to 8 week interval or a dose reduction to 35 Gy should be considered (Stannard et al. 2001). Retinopathy generally occurs at a significantly higher rate after chemotherapy or EBRT than with plaque therapy alone (Shields et al. 2006). Retinopathy was reported in less than 20% of cases treated with I-125 seed plaques to a 40 Gy apex dose (Shields et al. 2001). For Ru-106 with an apex dose of 55 Gy, the retinopathy rate was 2.4% (Abouzeid et al. 2008). The rate rose to 22% for patients given 138 Gy apex dose (Schueler et al. 2006). Even in combination with other therapies, most patients with localized RB receiving EPRT with I-125 or Ru-106 can avoid enucleation with excellent vision preservation (Friedman et al. 2000; Shields et al. 2001; Merchant et al. 2004; Shields et al. 2006).

References

Abouzeid, H., R. Moeckli, M.-C. Gaillard, et al. 2008. ^{106}Ruthenium brachytherapy for retinoblastoma. *International Journal of Radiation Oncology • Biology • Physics* 71 (3):821–8.

Abramson, D. H., and C. M. Frank. 1998. Second nonocular tumors in survivors of bilateral retinoblastoma: A possible age effect on radiation-related risk. *Ophthalmology* 105 (4):573–9; discussion 579–80.

Abramson, D. H., B. Jereb, and R. M. Ellsworth. 1981. External beam radiation for retinoblastoma. *Bulletin of the New York Academy of Medicine* 57 (9):787–803.

Al-Beteri, A. A., and D. E. Raeside. 1992. Optimal electron-beam treatment planning for retinoblastoma using a new three-dimensional Monte Carlo-based treatment planning system. *Medical Physics* 19 (1):125–35.

Al-Haj, A. N., A. M. Lobriguito, and C. S. Lagarde. 2004. Radiation dose profile in 125I brachytherapy: An 8-year review. *Radiation Protection Dosimetry* 111 (1):115–9.

Amendola, B. E., F. R. Lamm, A. M. Markoe, et al. 1990. Radiotherapy of retinoblastoma. A review of 63 children treated with different irradiation techniques. *Cancer* 66 (1):21–6.

Amendola, B. E., A. M. Markoe, J. J. Augsburger, et al. 1989. Analysis of treatment results in 36 children with retinoblastoma treated by scleral plaque irradiation. *International Journal of Radiation Oncology • Biology • Physics* 17 (1):63–70.

Anteby, I., N. Ramu, L. Gradstein, H. Miskin, J. Pe'er, and D. Benezra. 1998. Ocular and orbital complications following the treatment of retinoblastoma. *European Journal of Ophthalmology* 8 (2):106–11.

Astrahan, M. A. 2003. A patch source model for treatment planning of ruthenium ophthalmic applicators. *Medical Physics* 30 (6):1219–28.

Astrahan, M. A., G. Luxton, G. Jozsef, T. D. Kampp, et al. 1990. An interactive treatment planning system for ophthalmic plaque radiotherapy. *International Journal of Radiation Oncology • Biology • Physics* 18 (3):679–87.

Astrahan, M. A., G. Luxton, G. Jozsef, P. E. Liggett, and Z. Petrovich. 1990. Optimization of 125I ophthalmic plaque brachytherapy. *Medical Physics* 17 (6):1053–7.

Astrahan, M. A., G. Luxton, Q. Pu, and Z. Petrovich. 1997. Conformal episcleral plaque therapy. *International Journal of Radiation Oncology • Biology • Physics* 39 (2):505–19.

Astrahan, M. A., A. Szechter, and P. T. Finger. 2005. Design and dosimetric considerations of a modified COMS plaque: The reusable "seed-guide" insert. *Medical Physics* 32 (8):2706–16.

Bedford, M. A., C. Bedotto, and P. A. Macfaul. 1971. Retinoblastoma. A study of 139 cases. *British Journal of Ophthalmology* 55 (1):19–27.

Blach, L. E., B. McCormick, and D. H. Abramson. 1996. External beam radiation therapy and retinoblastoma: Long-term results in the comparison of two techniques. *International Journal of Radiation Oncology • Biology • Physics* 35 (1):45–51.

Blach, L. E., B. McCormick, D. H. Abramson, and R. M. Ellsworth. 1994. Trilateral retinoblastoma-incidence and outcome: A decade of experience. *International Journal of Radiation Oncology • Biology • Physics* 29 (4):729–33.

Brooks, H. L., Jr., D. Meyer, J. A. Shields, A. G. Balas, L. B. Nelson, and J. Fontanesi. 1990. Removal of radiation-induced cataracts in patients treated for retinoblastoma. *Archives of Ophthalmology* 108 (12):1701–8.

Cassady, J. R., R. H. Sagerman, P. Tretter, and R. M. Ellsworth. 1969. Radiation therapy in retinoblastoma. An analysis of 230 cases. *Radiology* 93 (2):405–9.

Chan, M. F., A. Y. Fung, Y. C. Hu, et al. 2001. The measurement of three dimensional dose distribution of a ruthenium-106 ophthalmological applicator using magnetic resonance imaging of BANG polymer gels. *Journal of Applied Clinical Medical Physics* 2 (2):85–9.

Chin, L. M., W. Harter, G. K. Svensson, and J. R. Cassady. 1988. An external beam treatment technique for retinoblastoma. *International Journal of Radiation Oncology • Biology • Physics* 15 (2):455–60.

Cormack, R. A., H. M. Kooy, M. R. Bellerive, J. S. Loeffler, R. A. Petersen, and N. J. Tarbell. 1998. A stereotactic radiation therapy device for retinoblastoma using a noncircular collimator and intensity filter. *Medical Physics* 25 (8):1438–42.

Coucke, P. A., C. Schmid, A. Balmer, R. O. Mirimanoff, and H. D. Thames. 1993. Hypofractionation in retinoblastoma: An increased risk of retinopathy. *Radiotherapy and Oncology* 28 (2):157–61.

Das, I. J., K. R. Kase, T. J. Fitzgerald, and D. A. Ligon. 1990. Study of dose perturbation parameters for eye shielding in megavoltage photon beam therapy. *International Journal of Radiation Oncology • Biology • Physics* 19 (2):461–7.

Egawa, S., I. Tsukiyama, Y. Akine, Y. Kajiura, and T. Ogino. 1987. Suppression of bony growth of the orbit after radiotherapy for retinoblastoma. *Radiation Medicine* 5 (6):207–11.

Egbert, P. R., S. S. Donaldson, K. Moazed, and A. R. Rosenthal. 1978. Visual results and ocular complications following radiotherapy for retinoblastoma. *Arch Ophthalmol* 96 (10):1826–30.

Eldebawy, E., H. Patrocinio, M. Evans, et al. 2010. Stereotactic radiotherapy as an alternative to plaque brachytherapy in retinoblastoma. *Pediatric Blood & Cancer* 55 (6):1210–2.

Evans, M. D., M. A. Astrahan, and R. Bate. 1993. Tumor localization using fundus view photography for episcleral plaque therapy. *Medical Physics* 20 (3):769–75.

Fass, D., B. McCormick, D. Abramson, and R. Ellsworth. 1991. Cobalt60 plaques in recurrent retinoblastoma. *International Journal of Radiation Oncology • Biology • Physics* 21 (3):625–7.

Filion, E., W. Parker, and C. Freeman. 2006. 171 Intensity modulated radiation therapy (IMRT) provides superior normal tissue sparing in the treatment of patients with paediatric retinoblastoma. *Radiotherapy and Oncology* 80 (Suppl. 1):S50.

Fletcher, O., D. Easton, K. Anderson, C. Gilham, M. Jay, and J. Peto. 2004. Lifetime risks of common cancers among retinoblastoma survivors. *Journal of the National Cancer Institute* 96 (5):357–63.

Fluhs, D., M. Bambynek, M. Heintz, et al. 1997. Dosimetry and design of radioactive eye plaques. *Frontiers of Radiation Therapy & Oncology* 30:26–38.

Foote, R. L., B. R. Garretson, P. J. Schomberg, S. J. Buskirk, D. M. Robertson, and J. D. Earle. 1989. External beam irradiation for retinoblastoma: Patterns of failure and dose-response analysis. *International Journal of Radiation Oncology • Biology • Physics* 16 (3):823–30.

Friedman, D. L., B. Himelstein, C. L. Shields, et al. 2000. Chemoreduction and local ophthalmic therapy for intraocular retinoblastoma. *Journal of Clinical Oncology* 18 (1):12–7.

Gordon, K. B., D. H. Char, and R. H. Sagerman. 1995. Late effects of radiation on the eye and ocular adnexa. *International Journal of Radiation Oncology • Biology • Physics* 31 (5):1123–39.

Griem, M. L., J. T. Ernest, M. L. Rozenfeld, and F. W. Newell. 1968. Eye lens protection in the treatment of retinoblastoma with high-energy electrons. *Radiology* 90 (2):351–2.

Guyuron, B. 1990. The hourglass facial deformity. *Journal of Cranio-Maxillofacial Surgery* 18 (5):187–91.

Guyuron, B., A. P. Dagys, I. R. Munro, and R. B. Ross. 1983. Effect of irradiation on facial growth: A 7- to 25-year follow-up. *Annals of Plastic Surgery* 11 (5):423–7.

Hawkins, M. M., L. M. Wilson, H. S. Burton, et al. 1996. Radiotherapy, alkylating agents, and risk of bone cancer after childhood cancer. *Journal of the National Cancer Institute* 88 (5):2708.

Haye, C., P. Schlienger, R. Calle, J. R. Vilcoq, and B. Blaszka. 1985. Conservative treatment of tumors of the retina at the Curie Institute. Long-term results of 129 cases treated with Stallard's disk and electrons. *Bulletin du Cancer* 72 (2):8798.

Hernandez, J. C., L. W. Brady, C. L. Shields, J. A. Shields, and P. DePotter. 1993. Conservative treatment of retinoblastoma. The use of plaque brachytherapy. *American Journal of Clinical Oncology* 16 (5):397–401.

Hilgartner, H. L. 1903. Report of a case of double glioma treated by x-rays. *Texas Medical Journal* 18:322–3.

Honavar, S. G., C. L. Shields, J. A. Shields, H. Demirci, and T. J. Naduvilath. 2001. Intraocular surgery after treatment of retinoblastoma. *Archives of Ophthalmology* 119 (11):1613–21.

Hungerford, J. L., N. M. Toma, P. N. Plowman, and J. E. Kingston. 1995. External beam radiotherapy for retinoblastoma: I. Whole eye technique. *British Journal of Ophthalmology* 79 (2):109-11.

Imhof, S. M., M. P. Mourits, P. Hofman, et al. 1996. Quantification of orbital and mid-facial growth retardation after megavoltage external beam irradiation in children with retinoblastoma. *Ophthalmology* 103 (2):263–8.

Kaste, S. C., G. Chen, J. Fontanesi, D. B. Crom, and C. B. Pratt. 1997. Orbital development in long-term survivors of retinoblastoma. *Journal of Clinical Oncology* 15 (3):1183–9.

Kirsner, S. M., K. R. Hogstrom, R. G. Kurup, and M. F. Moyers. 1987. Dosimetric evaluation in heterogeneous tissue of anterior electron beam irradiation for treatment of retinoblastoma. *Medical Physics* 14 (5):772–9.

Kleinerman, R. A., M. A. Tucker, R. E. Tarone, et al. 2005. Risk of new cancers after radiotherapy in long-term survivors of retinoblastoma: An extended follow-up. *Journal of Clinical Oncology* 23 (10):2272–9.

Krasin, M. J., Bruce T. Crawford, Y. Zhu, et al. 2004. Intensity-modulated radiation therapy for children with intraocular retinoblastoma: Potential sparing of the bony orbit. *Clinical Oncology* 16 (3):215–22.

Krengli, M., E. Hug, J. Adams, A. Smith, N. Tarbell, and J. Munzenrider. 2005. Proton radiation therapy for retinoblastoma: Comparison of various intraocular tumor locations and beam arrangements. *International Journal of Radiation Oncology • Biology • Physics* 61 (2):583–93.

Le Vu, B., F. de Vathaire, A. Shamsaldin, et al. 1998. Radiation dose, chemotherapy and risk of osteosarcoma after solid tumours during childhood. *International Journal of Cancer* 77 (3):370–7.

Lee, C. T., S. D. Bilton, R. M. Famiglietti, et al. 2005. Treatment planning with protons for pediatric retinoblastoma, medulloblastoma, and pelvic sarcoma: How do protons compare with other conformal techniques? *International Journal of Radiation Oncology • Biology • Physics* 63 (2):362–72.

Lommatzsch, P., and R. Vollmar. 1966. A new way in the conservative therapy of intraocular tumors by means of beta-irradiation (Ruthenium 106) with preservation of vision. *Klinische Monatsblatter fur Augenheilkunde* 148 (5):682–99.

Luxton, G., M. A. Astrahan, P. E. Liggett, D. L. Neblett, D. M. Cohen, and Z. Petrovich. 1988. Dosimetric calculations and measurements of gold plaque ophthalmic irradiators using iridium-192 and iodine-125 seeds. *International Journal of Radiation Oncology • Biology • Physics* 15 (1):167–76.

Luxton, G., M. A. Astrahan, and Z. Petrovich. 1988. Backscatter measurements from a single seed of 125I for ophthalmic plaque dosimetry. *Medical Physics* 15 (3):397–400.

Marees, T., A. C. Moll, S. M. Imhof, M. R. de Boer, P. J. Ringens, and F. E. van Leeuwen. 2008. Risk of second malignancies in survivors of retinoblastoma: More than 40 years of follow-up. *Journal of the National Cancer Institute* 100 (24):1771–9.

McCormick, B., R. Ellsworth, D. Abramson, et al. 1988. Radiation therapy for retinoblastoma: Comparison of results with lens-sparing versus lateral beam techniques. *International Journal of Radiation Oncology • Biology • Physics* 15 (3):567–74.

Merchant, T. E., C. J. Gould, N. E. Hilton, et al. 2002. Ocular preservation after 36 Gy external beam radiation therapy for retinoblastoma. *Journal of Pediatric Hematology/Oncology* 24 (4):246–9.

Merchant, T. E., C. J. Gould, M. W. Wilson, N. E. Hilton, Carlos Rodriguez-Galindo, and Barrett G. Haik. 2004. Episcleral plaque brachytherapy for retinoblastoma. *Pediatric Blood & Cancer* 43 (2):134–9.

Messmer, E. P., W. Sauerwein, T. Heinrich, et al. 1990. New and recurrent tumor foci following local treatment as well as external beam radiation in eyes of patients with hereditary retinoblastoma. *Graefes Archive for Clinical and Experimental Ophthalmology* 228 (5):426–31.

Mohney, B. G., D. M. Robertson, P. J. Schomberg, and D. O. Hodge. 1998. Second nonocular tumors in survivors of heritable retinoblastoma and prior radiation therapy. *American Journal of Ophthalmology* 126 (2):269–77.

Moll, A. C., S. M. Imhof, L. M. Bouter, and K. E. Tan. 1997. Second primary tumors in patients with retinoblastoma. A review of the literature. *Ophthalmic Genetics* 18 (1):27–34.

Moll, A. C., S. M. Imhof, A. Y. Schouten-Van Meeteren, D. J. Kuik, P. Hofman, and M. Boers. 2001. Second primary tumors in hereditary retinoblastoma: A register-based study, 1945–1997: Is there an age effect on radiation-related risk? *Ophthalmology* 108 (6):1109–14.

Moore, R. F., H. B. Stallard, and J. G. Milner. 1931. Retinal gliomata treated by radon seeds. *British Journal of Ophthalmology* 15 (12):673–96.

Murphree, L. A. 2005. Intraocular retinoblastoma: The case for a new group classification. *Ophthalmology Clinics of North America* 18 (1):41–53.

Nath, R., L. L. Anderson, G. Luxton, K. A. Weaver, J. F. Williamson, and A. S. Meigooni. 1995. Dosimetry of interstitial brachytherapy sources: Recommendations of the AAPM Radiation Therapy Committee Task Group No. 43. American Association of Physicists in Medicine. *Medical Physics* 22 (2):209–34.

Newton, W. A., A. T. Meadows, H. Shimada, G. R. Bunin, and G. F. Vawter. 1991. Bone sarcomas as 2nd malignant neoplasms following childhood-cancer. *Cancer* 67 (1):193–201.

Olch, A. J., and R. S. Lavey. 2002. Reproducibility and treatment planning advantages of a carbon fiber relocatable head fixation system. *Radiotherapy and Oncology* 65 (3):165–8.

Osman, I. M, H. Abouzeid, A. Balmer, et al. 2010. Modern cataract surgery for radiation-induced cataracts in retinoblastoma. *British Journal of Ophthalmology* 95 (2):227–30.

Parsons, J. T., F. J. Bova, C. R. Fitzgerald, W. M. Mendenhall, and R. R. Million. 1994a. Radiation retinopathy after external-beam irradiation: Analysis of time-dose factors. *International Journal of Radiation Oncology • Biology • Physics* 30 (4):765–73.

———. 1994b. Severe dry-eye syndrome following external beam irradiation. *International Journal of Radiation Oncology • Biology • Physics* 30 (4):775–80.

Phillips, C., M. Sexton, G. Wheeler, and J. McKenzie. 2003. Retinoblastoma: Review of 30 years' experience with external beam radiotherapy. *Australasian Radiology* 47 (3):226–30.

Pica, A., R. Moeckli, A. Balmer, et al. 2011. Preliminary experience in treatment of papillary and macular retinoblastoma: Evaluation of local control and local complications after treatment with linear accelerator-based stereotactic radiotherapy with micromultileaf collimator as second-line or salvage treatment after chemotherapy. *International Journal of Radiation Oncology • Biology • Physics* 81 (5):13806.

Portellos, M., and E. G. Buckley. 1998. Cataract surgery and intraocular lens implantation in patients with retinoblastoma. *Archives of Ophthalmology* 116 (4):449–52.

Ray, S. K., R. Bhatnagar, W. F. Hartsell, and G. R. Desai. 1998. Review of eye plaque dosimetry based on AAPM Task Group 43 recommendations. American Association of Physicists in Medicine. *International Journal of Radiation Oncology • Biology • Physics* 41 (3):701–6.

Reese, A. B. 1976. *Tumors of the Eye*. 3rd ed. Hagerstown, MD: Harper & Row.

Reese, A. B., G. A. Hyman, N. D. Tapley, and A. W. Forrest. 1958. The treatment of retinoblastoma by x-ray and triethylene melamine. *AMA Archives of Ophthalmology* 60 (5):897–906.

Reese, A. B., G. R. Merriam, and H. E. Martin. 1949. Treatment of bilateral retinoblastoma by irradiation and surgery: Report on 15-year results. *American Journal of Ophthalmology* 32 (2):175–90.

Reisner, M., C. Viegas, R. Grazziotin, et al. 2007. Retinoblastoma—comparative analysis of external radiotherapy techniques, including an IMRT technique. *International Journal of Radiation Oncology • Biology • Physics* 67 (3):933-941.

Ries, L. A. G., M. A. Smith, J. G. Gurney, et al. 1999. *Cancer Incidence and Survival among Children and Adolescents: United States SEER Program 1975–1995*. NIH Pub. No. 99-4649. Bethesda, MD: National Cancer Institute.

Rivard, M. J., S.-T. Chiu-Tsao, P. T. Finger, et al. 2011. Comparison of dose calculation methods for brachytherapy of intraocular tumors. *Medical Physics* 38 (1):306.

Rivard, M. J., B. M. Coursey, L. A. DeWerd, et al. 2004. Update of AAPM Task Group No. 43 Report: A revised AAPM protocol for brachytherapy dose calculations. *Medical Physics* 31 (3):633–74.

Rivard, M. J., C. S. Melhus, S. Sioshansi, and J. Morr. 2008. The impact of prescription depth, dose rate, plaque size, and source loading on the central axis using ^{103}Pd, ^{125}I, and ^{131}Cs. *Brachytherapy* 7 (4):327–35.

Rosengren, B. H., and B. Tengroth. 1963. A modified Cobalt 60 applicator for the treatment of retinoblastoma. *Acta Radiologica: Therapy, Physics, Biology* 1:305–13.

Rosengren, B., and B. Tengroth. 1977. Retinoblastoma treated with a 60CO applicator. *Acta Radiologica: Therapy, Physics, Biology* 16 (2):110–6.

Sahgal, A., B. A. Millar, H. Michaels, et al. 2006. Focal stereotactic external beam radiotherapy as a vision-sparing method for the treatment of peripapillary and perimacular retinoblastoma: Preliminary results. *Clinical Oncology* 18 (8):628–34.

Salmonsen, P. C., R. M. Ellsworth, and F. D. Kitchin. 1979. The occurrence of new retinoblastomas after treatment. *Ophthalmology* 86 (5):837–43.

Schipper, J. 1983. An accurate and simple method for megavoltage radiation therapy of retinoblastoma. *Radiotherapy and Oncology* 1 (1):31–41.

Schipper, J., S. M. Imhoff, and K. E. Tan. 1997. Precision megavoltage external beam radiation therapy for retinoblastoma. *Frontiers of Radiation Therapy and Oncology* 30:65–80.

Schipper, J., K. E. W. P. Tan, and H. A. van Peperzeel. 1985. Treatment of retinoblastoma by precision megavoltage radiation therapy. *Radiotherapy and Oncology* 3 (2):117–32.

Schueler, A. O., D. Flühs, G. Anastassiou, et al. 2006. β-Ray brachytherapy with 106Ru plaques for retinoblastoma. *International Journal of Radiation Oncology • Biology • Physics* 65 (4):1212–21.

Scott, I. U., T. G. Murray, W. J. Feuer, et al. 1999. External beam radiotherapy in retinoblastoma: Tumor control and comparison of 2 techniques. *Archives of Ophthalmology* 117 (6):766–70.

Sealy, R., E. Buret, H. Cleminshaw, et al. 1980. Progress in the use of iodine therapy for tumours of the eye. *The British Journal of Radiology* 53 (635):1052–60.

Shields, C. L., S. G. Honavar, J. A. Shields, H. Demirci, A. T. Meadows, and T. J. Naduvilath. 2002. Factors predictive of recurrence of retinal tumors, vitreous seeds, and subretinal seeds following chemoreduction for retinoblastoma. *Archives of Ophthalmology* 120 (4):460–4.

Shields, C. L., A. Mashayekhi, H. Sun, et al. 2006. Iodine 125 plaque radiotherapy as salvage treatment for retinoblastoma recurrence after chemoreduction in 84 tumors. *Ophthalmology* 113 (11):2087–92.

Shields, C. L., J. A. Shields, J. Cater, I. Othmane, A. D. Singh, and B. Micaily. 2001. Plaque radiotherapy for retinoblastoma: Long-term tumor control and treatment complications in 208 tumors. *Ophthalmology* 108 (11):2116–21.

Smith, L. M., S. S. Donaldson, P. R. Egbert, M. P. Link, and M. A. Bagshaw. 1989. Aggressive management of second primary tumors in survivors of hereditary retinoblastoma. *International Journal of Radiation Oncology • Biology • Physics* 17 (3):499–505.

Soares, C. G., S. Vynckier, H. Jarvinen, et al. 2001. Dosimetry of beta-ray ophthalmic applicators: Comparison of different measurement methods. *Medical Physics* 28 (7):1373–84.

Stannard, C., R. Sealy, E. Hering, et al. 2001. Localized whole eye radiotherapy for retinoblastoma using a 125I applicator, "claws." *International Journal of Radiation Oncology • Biology • Physics* 51 (2):399–409.

Stannard, C., R. Sealy, E. Hering, et al. 2002. Postenucleation orbits in retinoblastoma: Treatment with 125I brachytherapy. *International Journal of Radiation Oncology • Biology • Physics* 54 (5):1446–54.

Steenbakkers, R. J. H. M., M. D. Altschuler, G. J. D'Angio, J. W. Goldwein, and A. Kassaee. 1997. Optimized lens-sparing treatment of retinoblastoma with electron beams. *International Journal of Radiation Oncology • Biology • Physics* 39 (3):589–94.

Takeda, A., N. Shigematsu, S. Suzuki, et al. 1999. Late retinal complications of radiation therapy for nasal and paranasal malignancies: Relationship between irradiated-dose area and severity. *International Journal of Radiation Oncology • Biology • Physics* 44 (3):599–605.

Toma, N. M., J. L. Hungerford, P. N. Plowman, J. E. Kingston, and D. Doughty. 1995. External beam radiotherapy for retinoblastoma: II. Lens sparing technique. *British Journal of Ophthalmology* 79 (2):112–7.

Weaver, K. A. 1986. The dosimetry of 125I seed eye plaques. *Medical Physics* 13 (1):78–83.

Weiss, D. R., J. R. Cassady, and R. Petersen. 1975. Retinoblastoma: A modification in radiation therapy technique. *Radiology* 114 (3):705–8.

Wong, F. L., J. D. Boice, Jr., D. H. Abramson, et al. 1997. Cancer incidence after retinoblastoma. Radiation dose and sarcoma risk. *JAMA* 278 (15):1262–7.

Yu, C. L., M. A. Tucker, D. H. Abramson, et al. 2009. Cause-specific mortality in long-term survivors of retinoblastoma. *Journal of the National Cancer Institute* 101 (8):581–91.

Yue, N. C., and M. L. Benson. 1996. The hourglass facial deformity as a consequence of orbital irradiation for bilateral retinoblastoma. *Pediatric Radiology* 26 (6):421–3.

Index

H

I

J

K

L

S

T

U

V

W

X

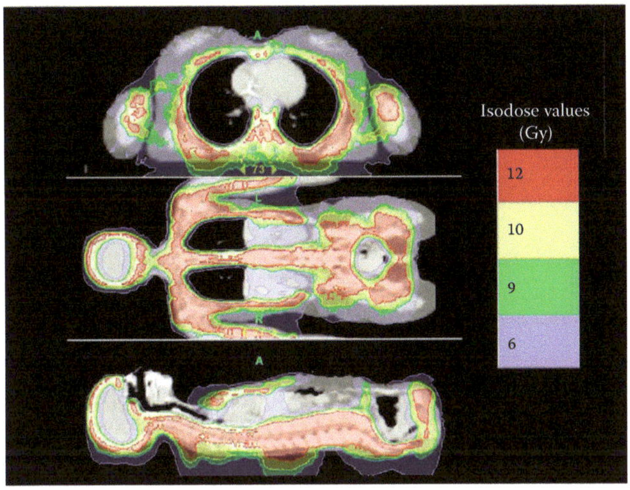

FIGURE 3.3 Color wash of a TMI TomoTherapy plan to a prescribed dose of 12 Gy in a 20-year-old adult female. Relative sparing of dose to brain, oral cavity, thyroid, lungs, heart, soft tissue, and gastrointestinal tract is seen. (From Wong, J. Y. C., A. Liu, et al., *Biology of Blood and Marrow Transplantation* 12 (3):306–15, 2006. With permission.)

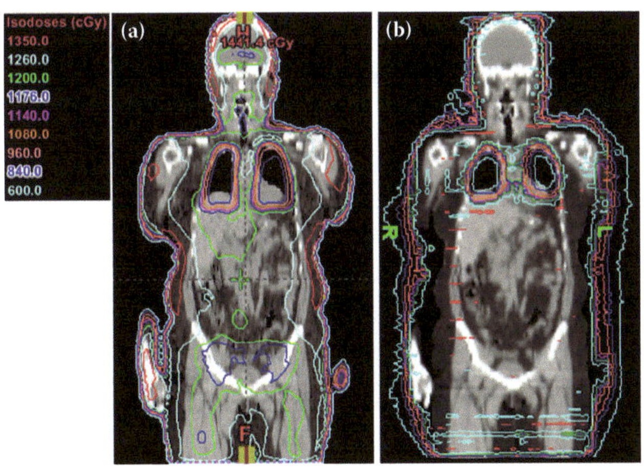

FIGURE 3.4 TBI dose distribution for (a) a conventional extended distance plan and (b) an HT plan for the same patient. (From Zhuang, A. H., et al., *Medical Dosimetry* 35 (4):243–9, 2010. With permission.)

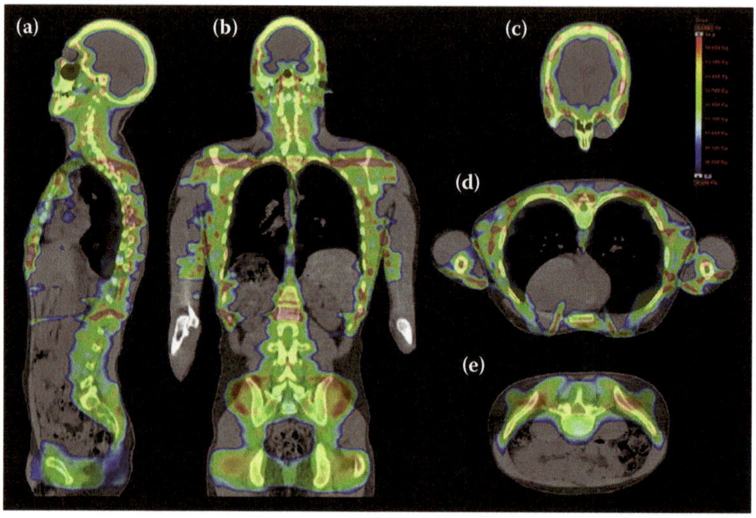

FIGURE 3.5 Intensity modulated total marrow isodose distribution using a conventional linear accelerator, (a) sagittal, (b) coronal, (c–e) transverse planes. Doses range from 9.6 Gy (blue) to 14.4 Gy (red); the prescribed dose is 12 Gy. (From Yeginer, M., et al., *International Journal of Radiation Oncology • Biology • Physics* 79 (4):1256–65, 2011. With permission.)

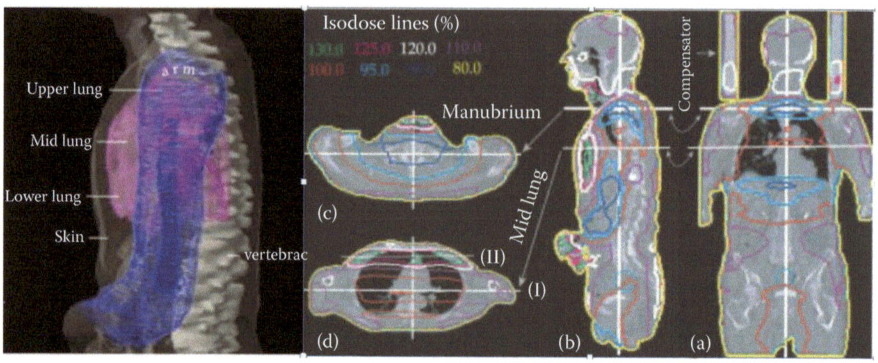

FIGURE 3.6 Beam's eye view of one lateral field. (Left) Lung is in pink and arm is in blue. (Right) Isodoses at shoulder and chest level. (After Hui, S. K., et al., *Journal of Applied Clinical Medical Physics* 5 (4):71–79, 2004.)

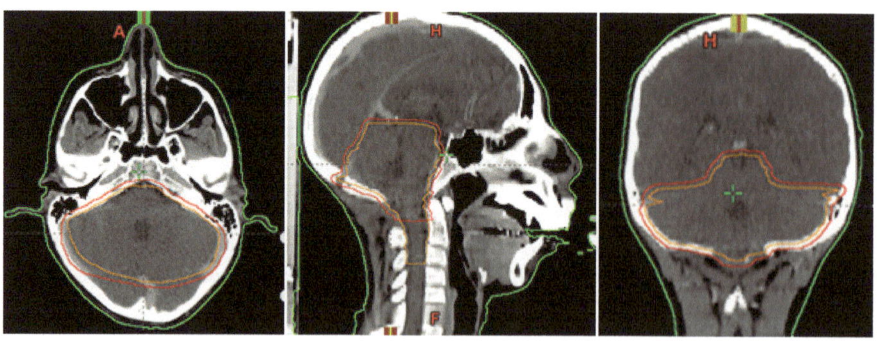

FIGURE 4.7 Medulloblastoma (whole posterior fossa) CTV and PTV on axial, sagittal, and coronal images.

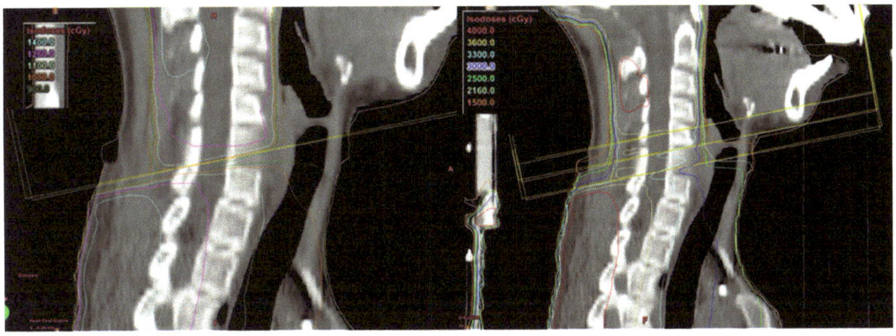

FIGURE 4.10 Two-millimeter gap between PA spine and whole brain fields, without (left panel) and with (right panel) two junction shifts.

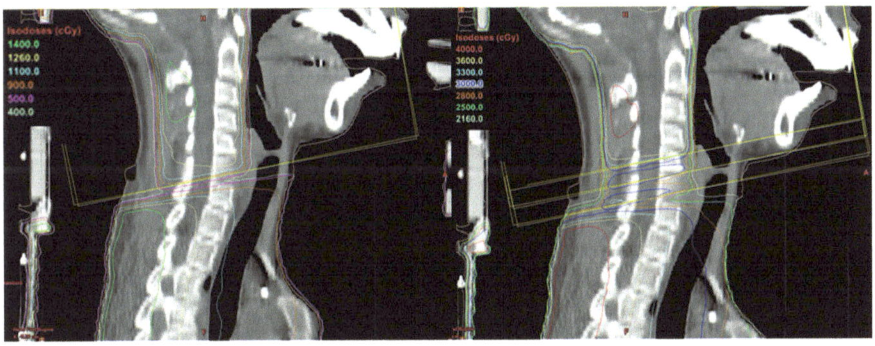

FIGURE 4.11 Five-millimeter gap between PA spine and whole brain fields, without (left panel) and with (right panel) two junction shifts.

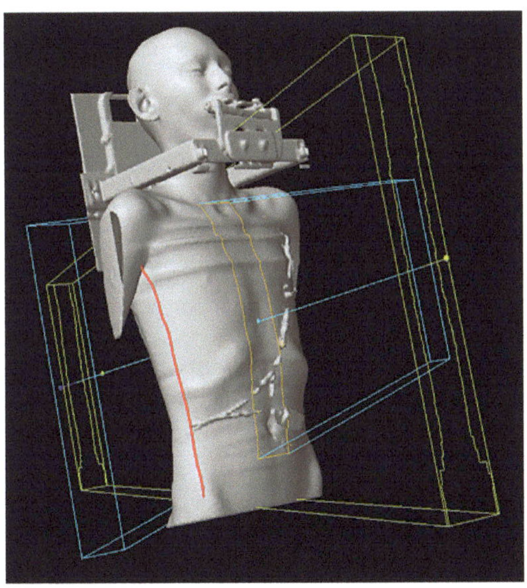

FIGURE 4.15 PA spine treatment field with AP setup field and horizontal laser leveling line.

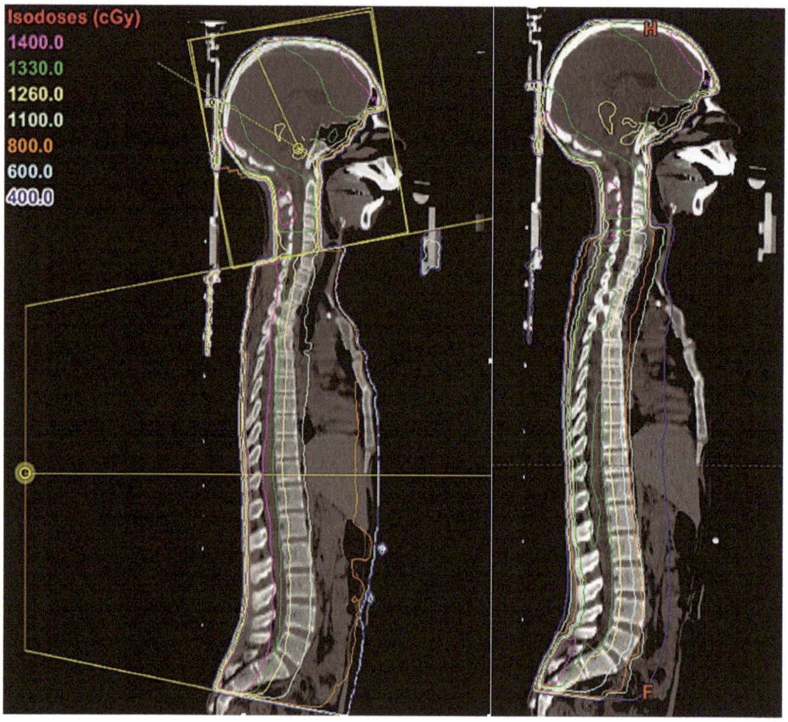

FIGURE 4.25 Three-field spine (right panel) versus single PA spine field isodoses. Note that the 86 cGy/fx (for 180 cGy/fx target dose) isodose line is at the anterior skin surface for the single PA field but at the anterior edge of the vertebral body for the three-field spine plan.

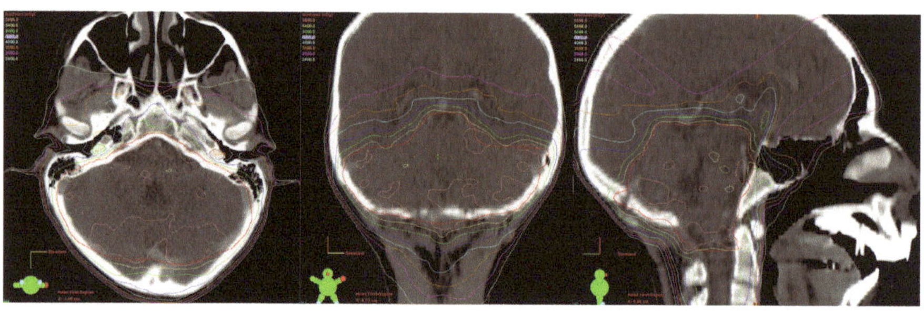

FIGURE 4.31 Average-risk medulloblastoma whole posterior fossa composite doses, 23.4 Gy CSI plus 30.6 Gy boost. Isodose lines range from 24–56 Gy.

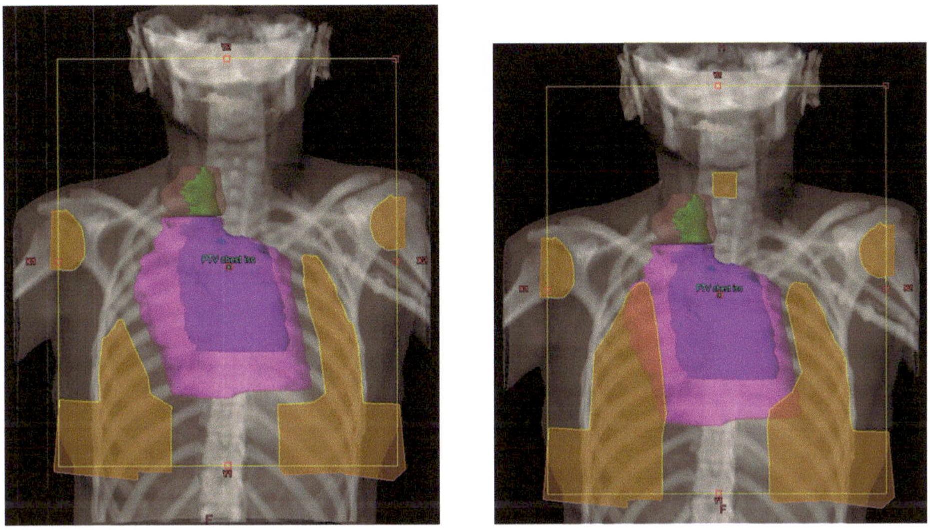

(a) (b)

FIGURE 5.2 (a) EFRT treating necessary lung. (b) EFRT with lung sparing.

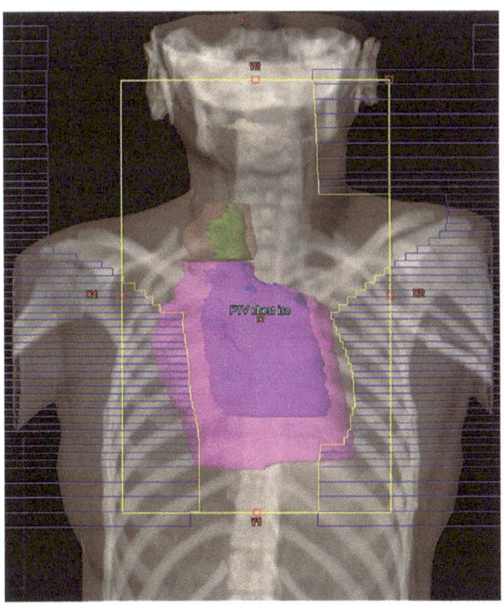

FIGURE 5.3 IFRT field design.

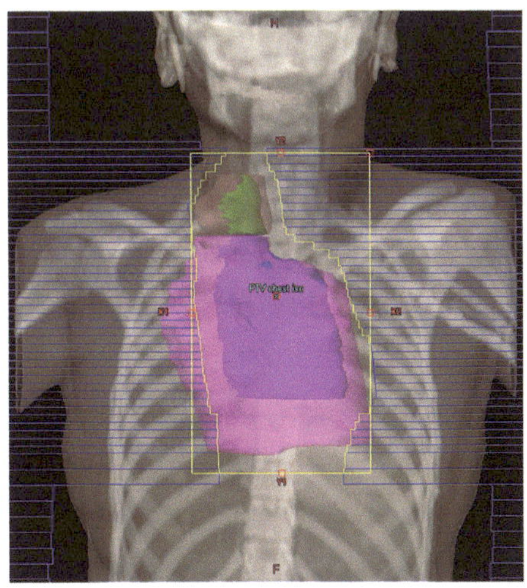

FIGURE 5.4 INRT field design.

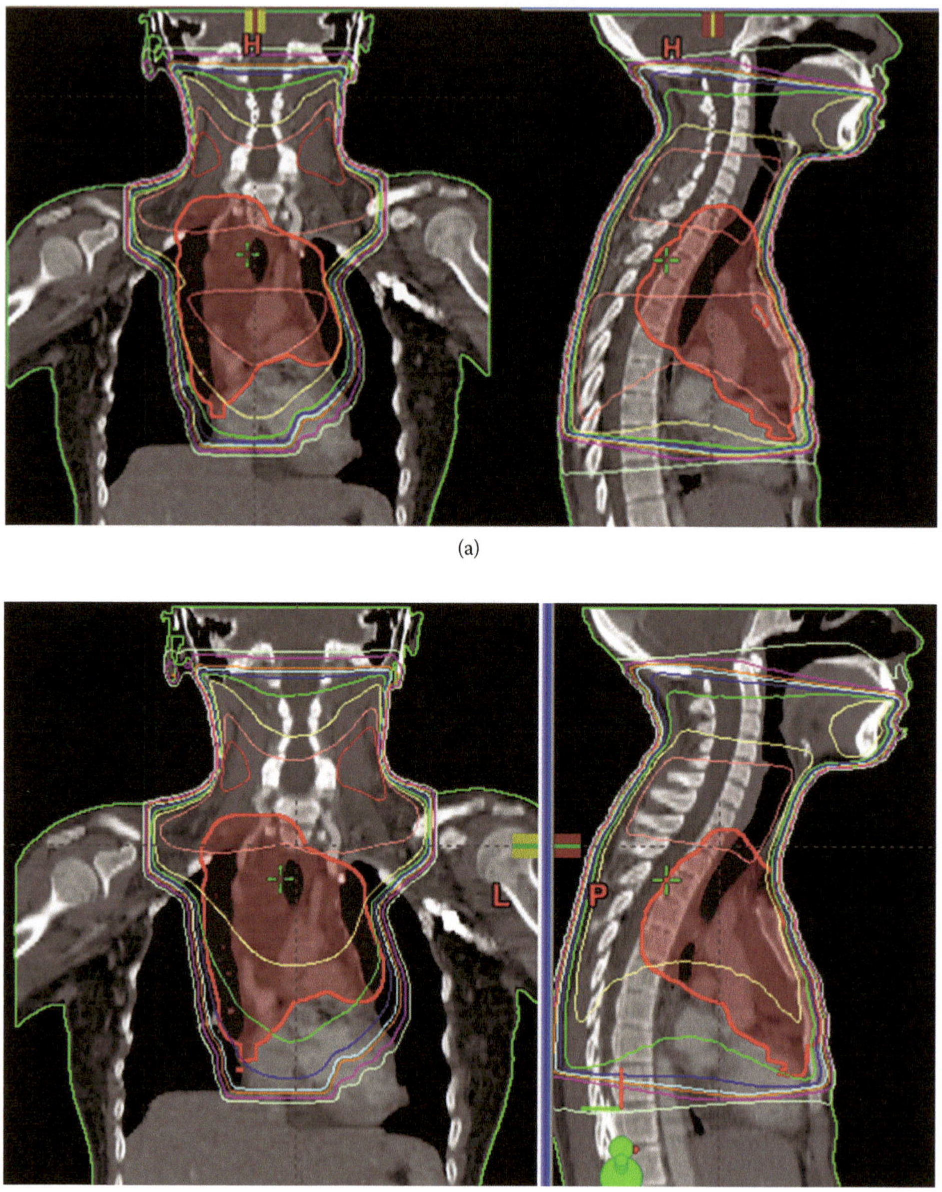

(a)

(b)

FIGURE 5.7 (a) IFRT with one mediastinal subfield (field-in-field). (b) IFRT without mediastinal subfield.

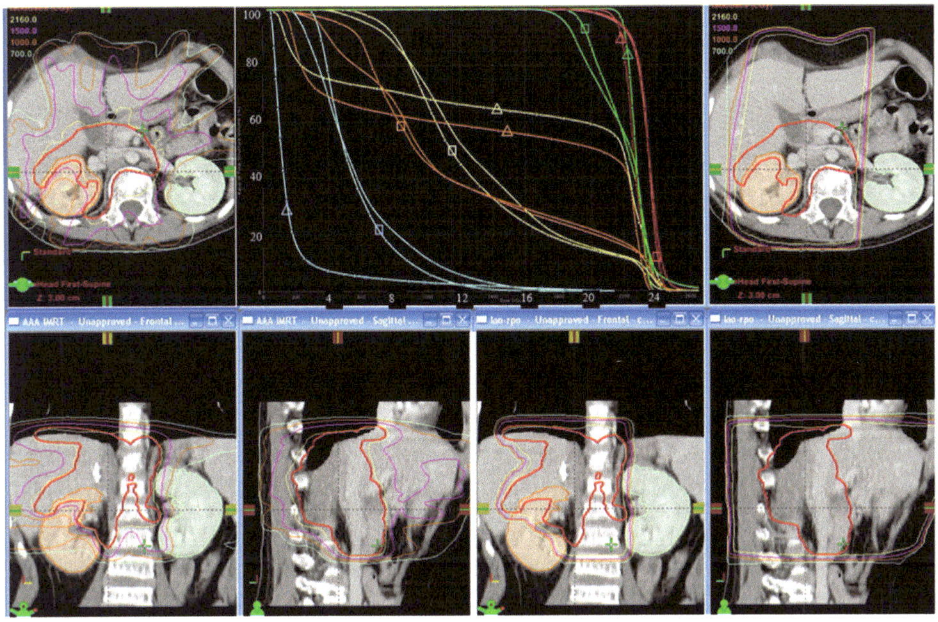

FIGURE 6.4 Case 1: Eight-beam IMRT (left-side panels, DVH = line) versus opposed obliques (right-side panels, DVH = triangles), VMAT isodoses not shown (DVH = boxes), 21.6 Gy target dose (red). Structures: green is vertebral body OAR, yellow is liver, orange is right kidney, cyan is left kidney. Isodoses: yellow, 21.6 Gy; magenta, 15 Gy; orange, 10 Gy; light blue, 7 Gy.

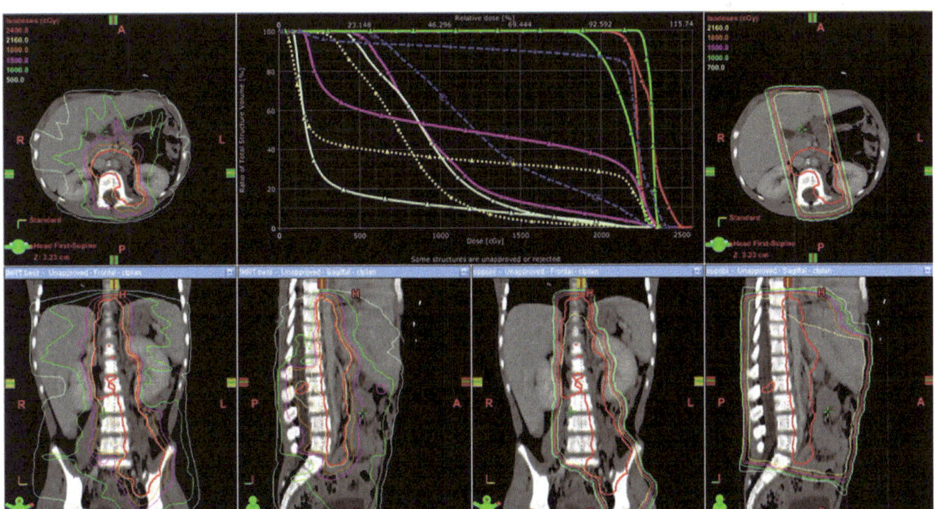

FIGURE 6.5 Case 2: IMRT (boxes) versus opposed obliques (triangles). Demonstrates longer volume including lymph nodes, VB sparing. Light green, right kidney; magenta, left kidney; yellow dotted, liver; green, vertebral body target; blue dash, vertebral body OAR; red, PTV. Isodoses: yellow, 21.6 Gy; orange, 18 Gy; magenta, 15 Gy; light green, 5 Gy.

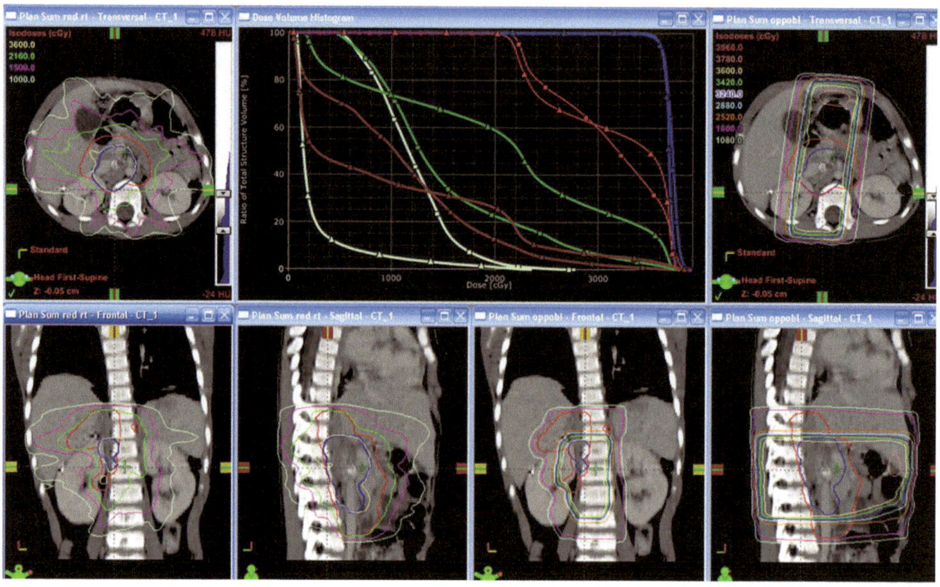

FIGURE 6.6 Case 3: 21.6 Gy + 14.4 Gy boost, IMRT (boxes) versus opposed obliques (triangles). Isodoses: light green, left kidney; dark green, right kidney; brown, liver; red, 21.6 Gy PTV; blue, 36 Gy PTV. Isodoses: yellow, 36 Gy; green, 21.6 Gy; magenta, 15 Gy; light blue, 10 Gy.

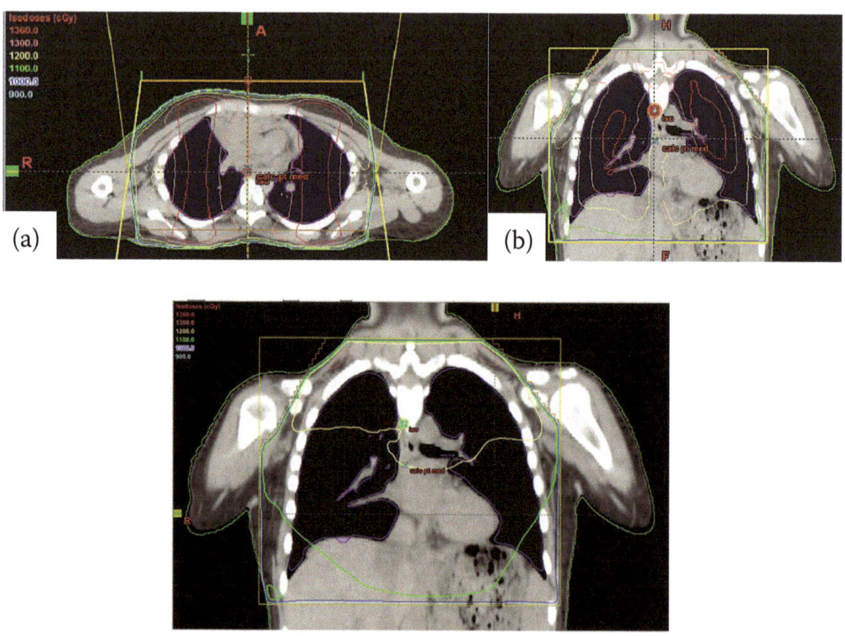

(c)

FIGURE 7.7 (a) Axial and (b) coronal AP–PA whole lung with prescription point in the mediastinum, inhomogeneity correction is turned on. (c) Same as in panel b without inhomogeneity correction.

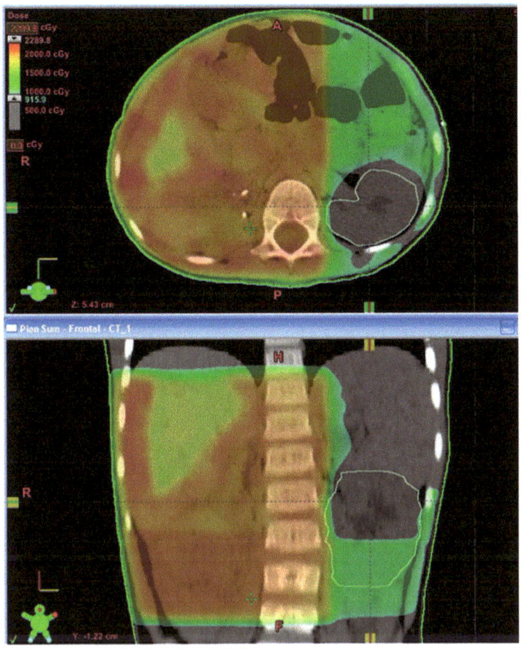

FIGURE 7.8 Bilateral Wilms', IMRT to right flank and partial left flank for 10.8 Gy used to spare partial left kidney. AP–PA fields boosted the right flank to 19.8 Gy. The central part of liver is spared such that ≥25% of the whole liver receives 20 Gy, green color wash = 10.8 Gy, brown = 19.8 Gy.

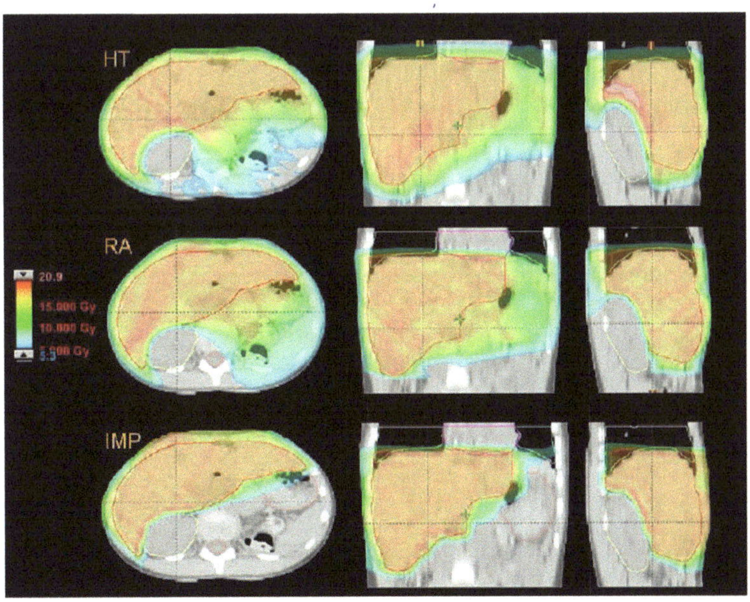

FIGURE 7.10 Protons versus RA versus HT for Wilms' case where the whole liver is treated with sparing of the ipsilateral kidney. (After Fogliata, A., et al., *Radiation Oncology* 4:2, 2009.)

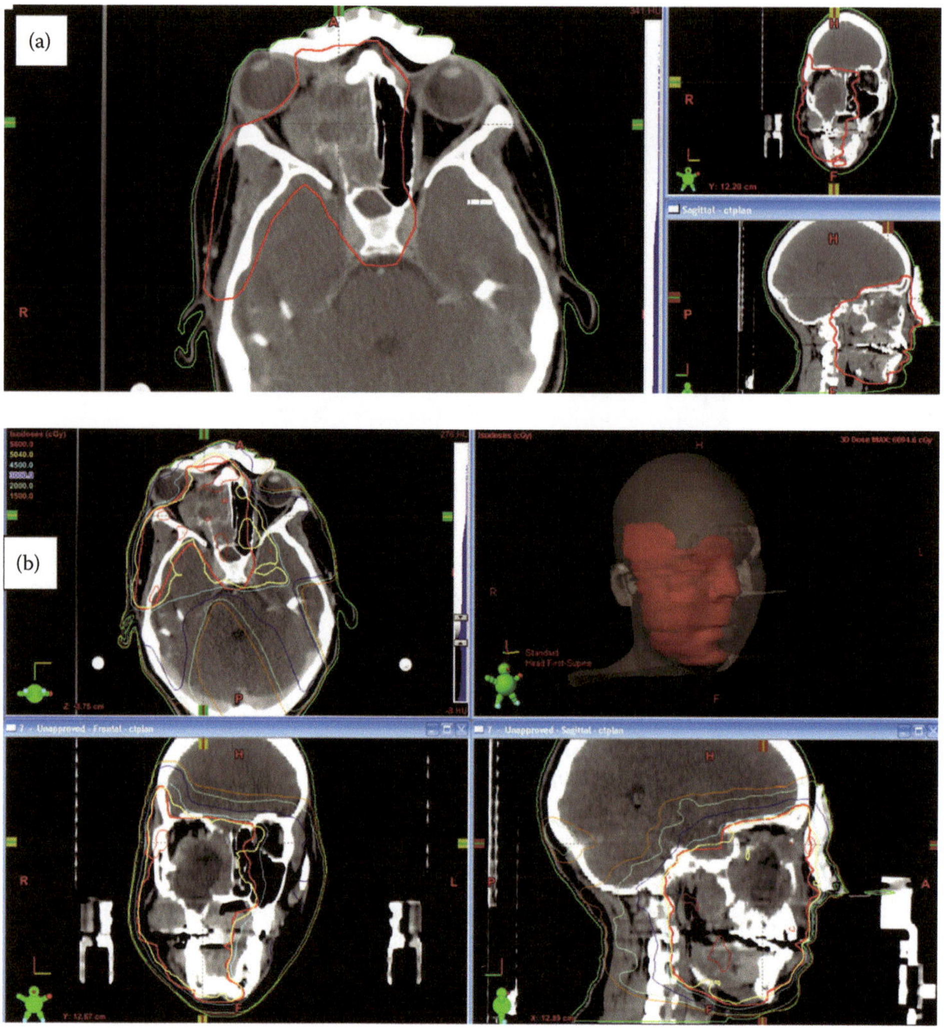

FIGURE 8.3 **Periorbital RMS. (a) PTV in three plane views and 3D view. (b) Isodoses in three views.**

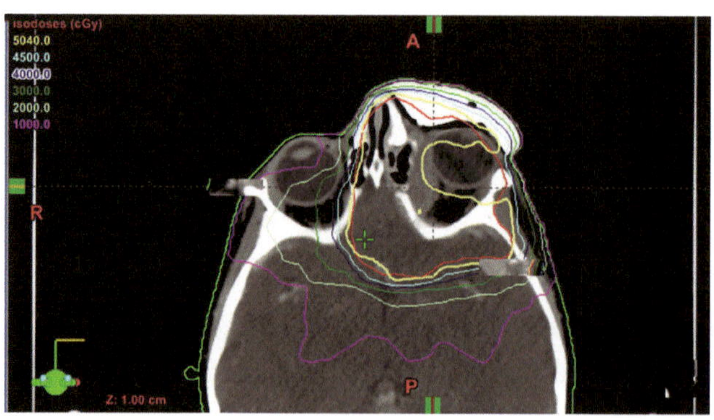

FIGURE 8.4 **Reducing the retinal dose to below 50 Gy using a nine-beam non-coplanar beam IMRT arrangement with lower dose constraint for the retina.**

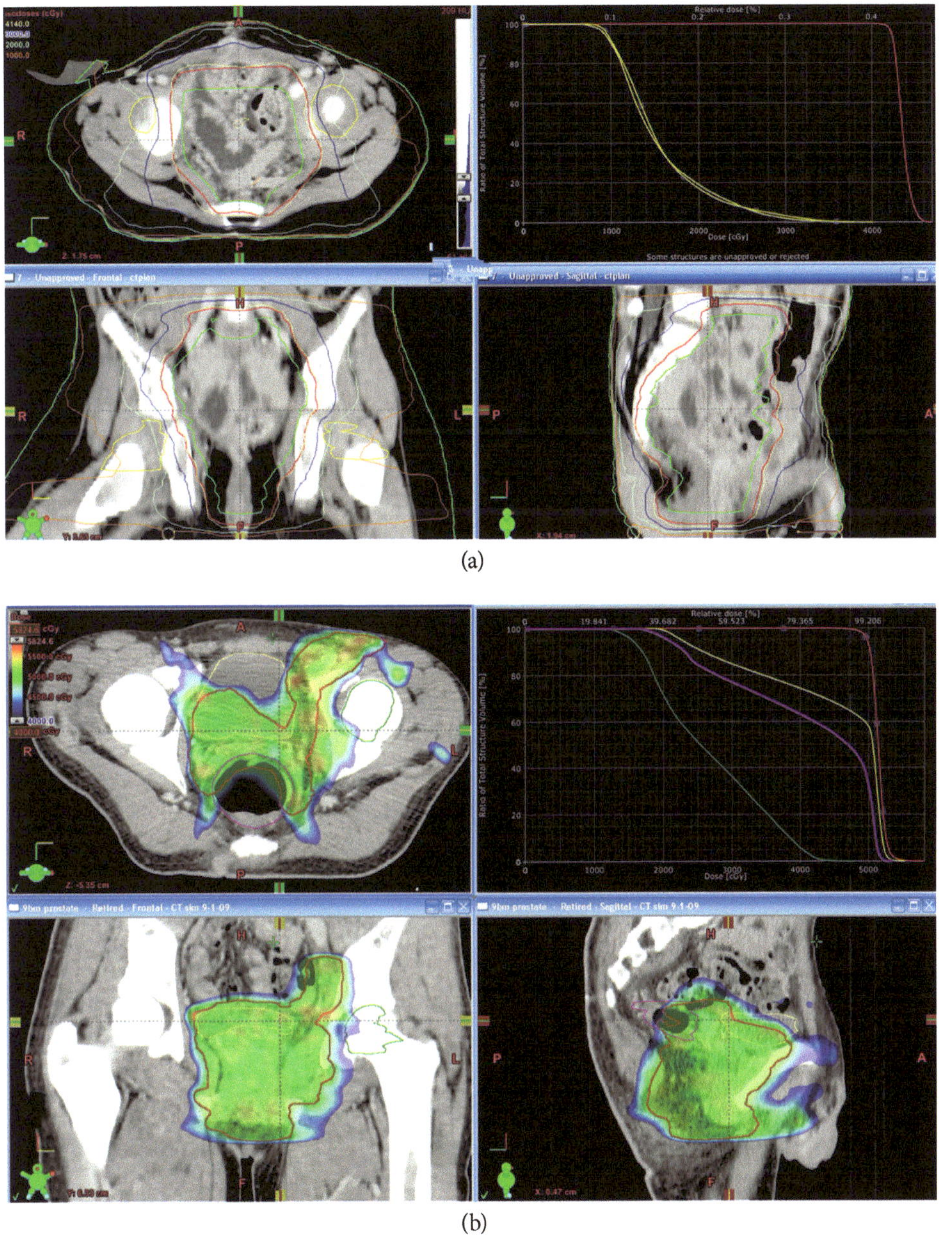

FIGURE 8.7 **(a) Prostate RMS dosimetry. 45 Gy prescribed dose. DVH: red, PTV; yellow, femoral heads. (b) Prostate RMS adjacent to bladder. 50.4 Gy prescribed dose. DVH: green, L femur; purple, rectum; yellow, bladder; red, PTV.**

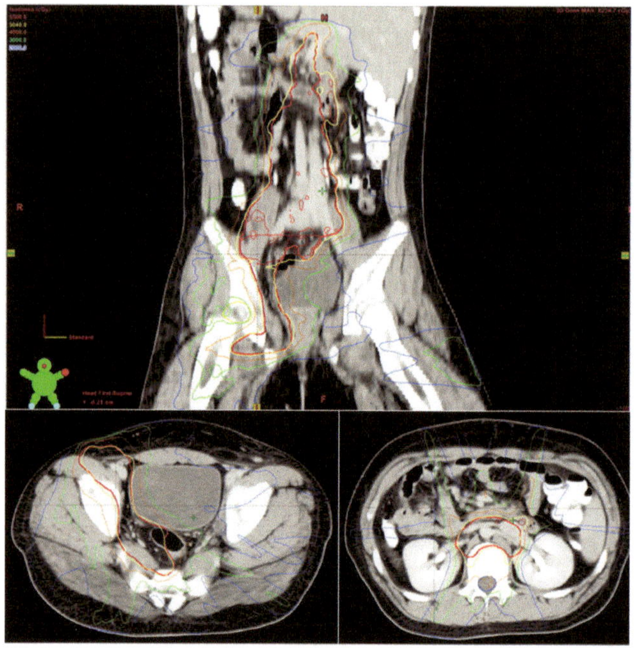

FIGURE 8.8 Extensive abdominal and pelvic RMS, coronal CT shows extent while axials show target (red) wrapping around bladder and adjacent to both kidneys. Isodoses: red, 55 Gy; yellow, 50.4 Gy (prescribed); orange, 40 Gy; green, 30 Gy; blue, 20 Gy.

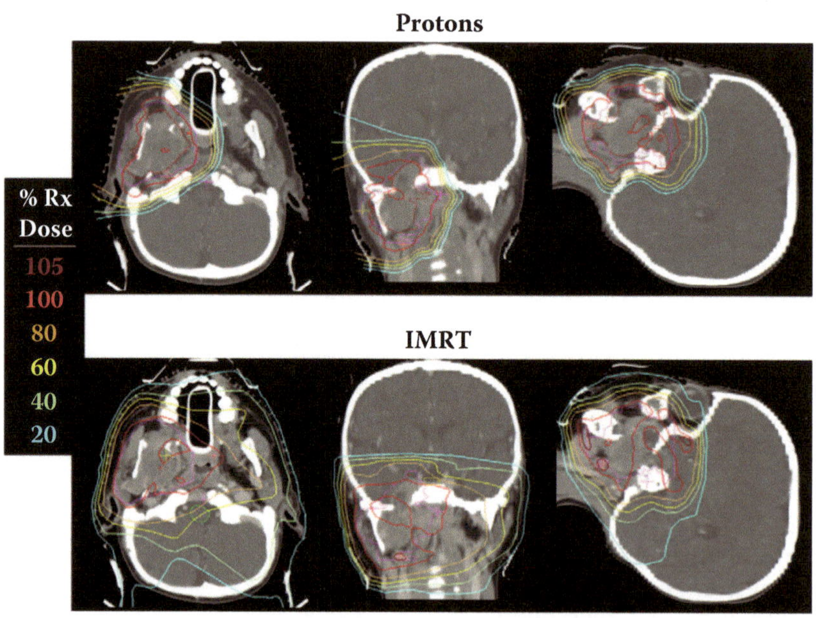

FIGURE 8.10 Protons versus IMRT for parameningeal RMS. Dose distributions in the axial, coronal, and sagittal planes. Upper panel is the three-field proton plan, lower panel is the five-field IRMT plan. (After Kozak, K. R., et al., *International Journal of Radiation Oncology • Biology • Physics* 74 (1):179–86, 2009.)

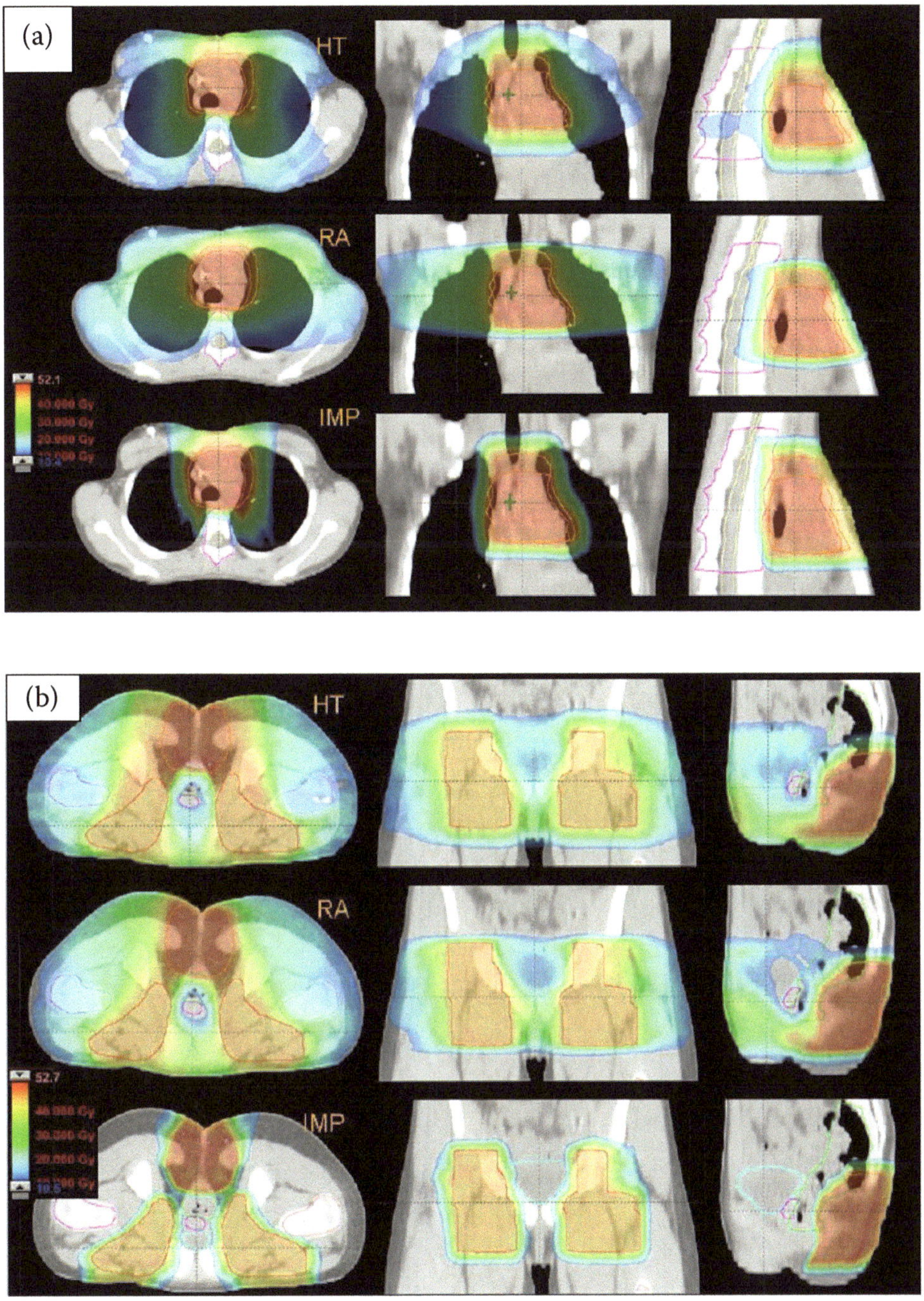

FIGURE 8.11 Helical TomoTherapy (HT) versus rapid arc (RA) versus intensity modulated protons (IMP) for (a) mediastinal RMS, (b) RMS of anus with metastases to lymph nodes. (After Fogliata, A., et al., *Radiation Oncology* 4:2, 2009.)

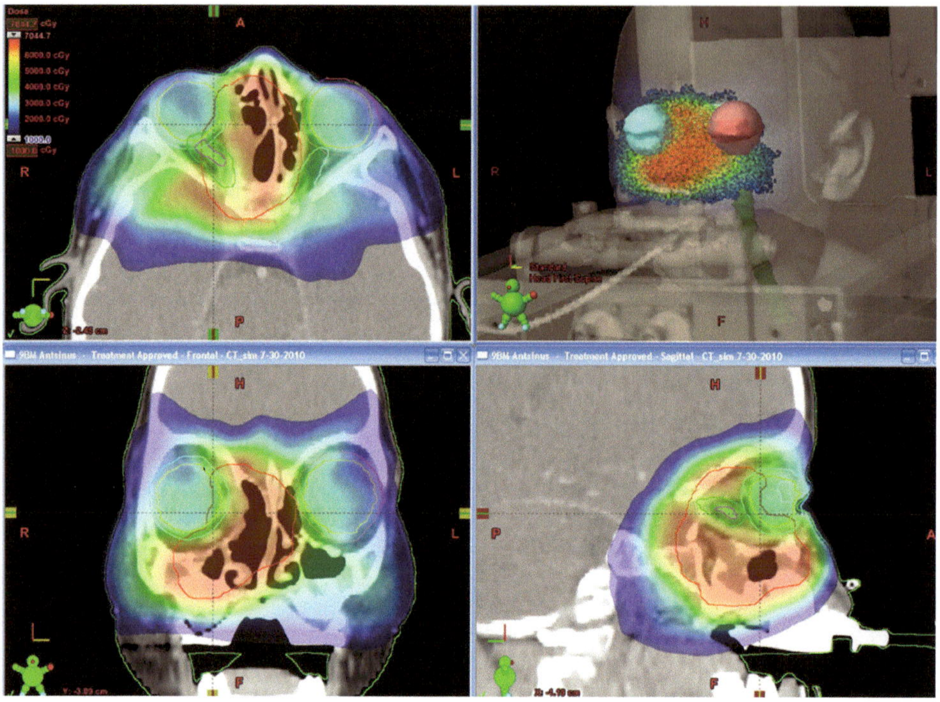

FIGURE 9.6 IMRT for osteosarcoma (66 Gy) of the sinus with color wash doses.

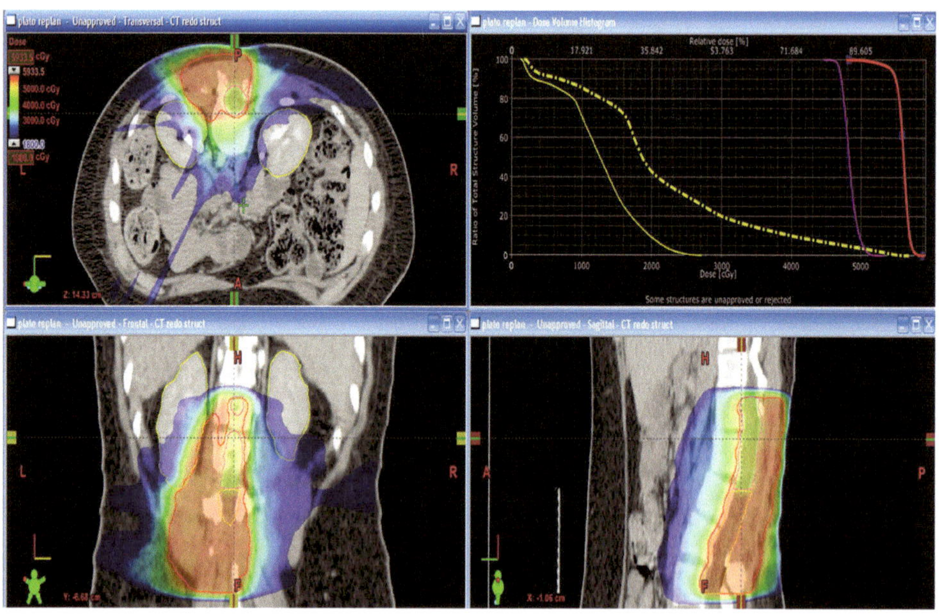

FIGURE 9.8 IMRT for paraspinal Ewing's sarcoma. DVHs are shown: red, cauda equina PTV; purple, spinal cord target/OAR; dotted yellow, left kidney; solid yellow, spinal cord target.

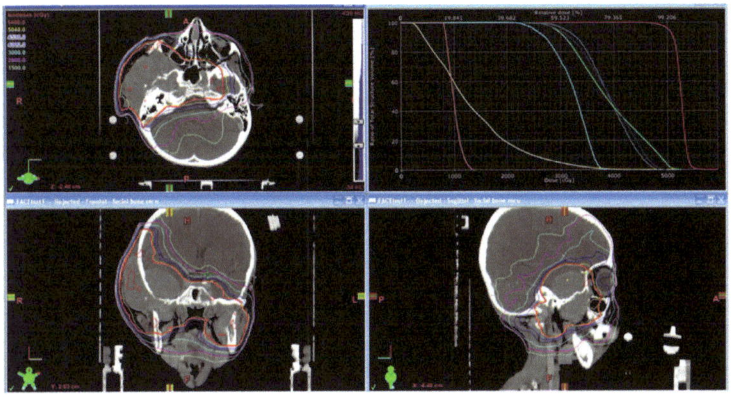

Gantry [deg]	Couch [deg]
266	19
345	94
303	0
310	80
36	337
80	0
39	304
271	39
329	27

FIGURE 9.11 Parameningeal Ewing's sarcoma involving facial bones and soft tissues and brain. Nine-beam noncoplanar IMRT plan. DVHs: red, PTV; yellow, brain; green, right eye; dark blue, optic chiasm; turquoise blue, right lens; pink, left lens; light blue, left cochlea.

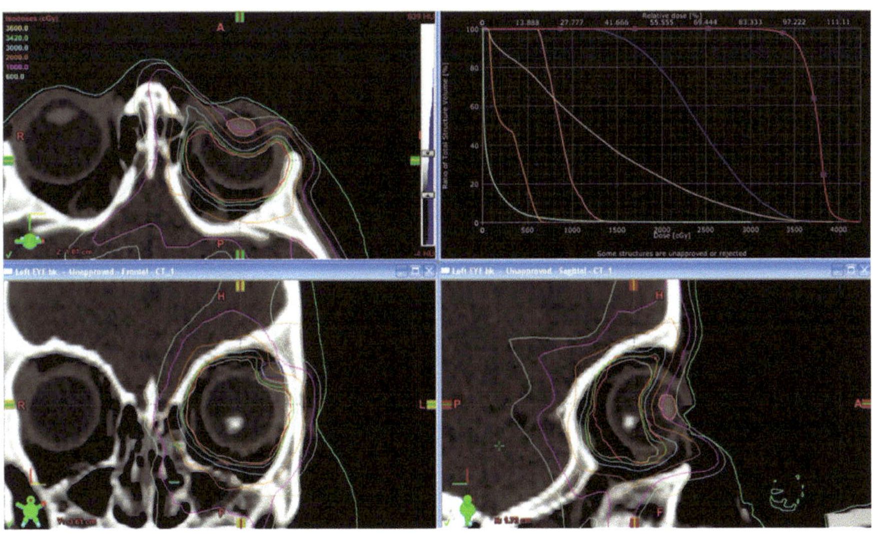

FIGURE 10.13 Whole retina IMRT with lens and lacrimal gland sparing. 36 Gy PTV, mean lacrimal gland dose (blue) = 24 Gy, mean lens dose (pink) = 9 Gy, V20 orbital bone (tan) = 20%, V15 orbital bone (tan) = 35%, mean pituitary dose (light green) = 3 Gy.

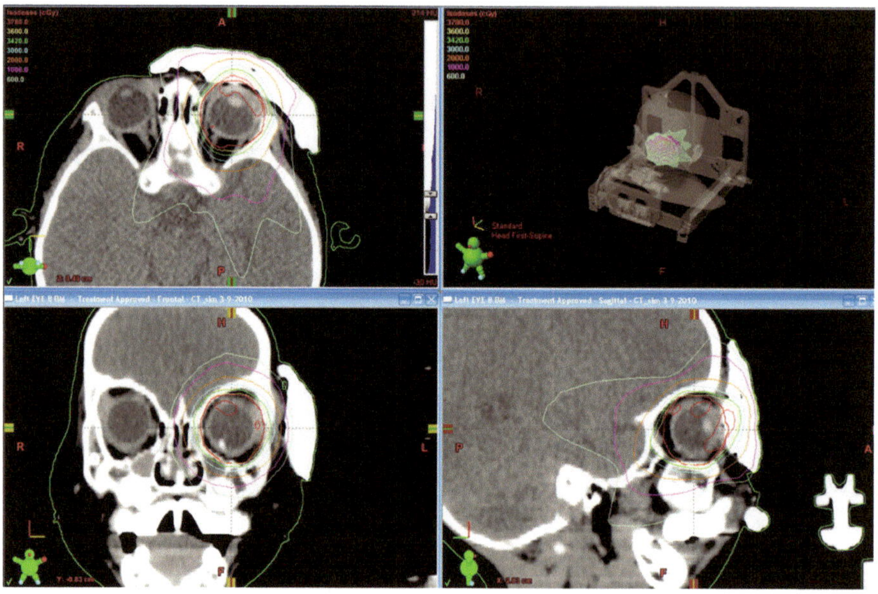

FIGURE 10.14 Left whole eye IMRT to 36 Gy. Dental impression material was used as bolus.

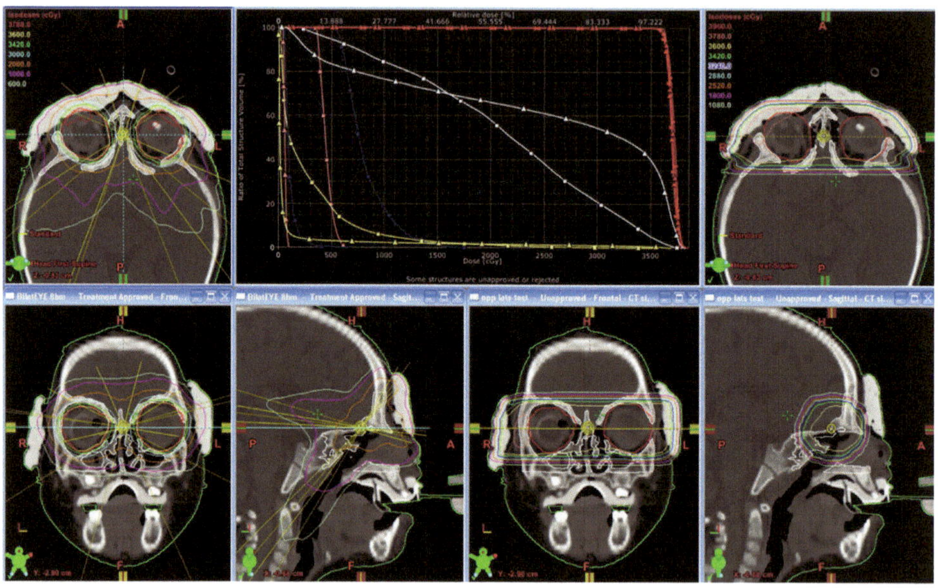

FIGURE 10.15 Bilateral whole eye RB treatment with 36 Gy PTV dose. Opposed laterals (right panels) versus IMRT (left panels). DVH triangles are opposed laterals; red, PTV; white, orbital bone; yellow, brain tissue; pink, pituitary; blue, optic chiasm.